Studies in Space Policy

Volume 14

Edited by
The European Space Policy Institute
Director: Jean-Jacques Tortora

T0414209

More information about this series at http://www.springer.com/series/8167

Marco Aliberti

India in Space: Between Utility and Geopolitics

 Springer

Marco Aliberti
European Space Policy Institute
Vienna, Austria

ISSN 1868-5307   ISSN 1868-5315   (electronic)
Studies in Space Policy
ISBN 978-3-319-89092-0   ISBN 978-3-319-71652-7   (eBook)
https://doi.org/10.1007/978-3-319-71652-7

Printed on acid-free paper

This Springer imprint is published by Springer Nature
The registered company is Springer International Publishing AG
The registered company address is: Gewerbestrasse 11, 6330 Cham, Switzerland

Foreword

Recounting the history of the Indian space programme is a good opportunity to take the necessary distance from the current and quick evolution of the global space sector.

As a matter of fact, it is actually quite a unique case of a developing country having identified decades ago the full potential of space for supporting its development in many areas. India has then decided to reap all the benefits it could get from space through the establishment of a complete mastery of the whole range of technologies needed for the design and manufacturing of launchers and satellites, the setting up of ground infrastructures and the deployment of space-based services addressing the full spectrum of space applications.

From the start, stakes were pretty high since India was confronted to major export control limitations from the most advanced space powers. However, it took up right away the challenge of building up an indigenous scientific, technological, and industrial base. Few shared the relevance of such strategy in the beginning. One of them was Professor Blamont, founder of the French "Centre National d'Etudes Spatiales (CNES)", who welcomed Indian ambitions and paved the way to fruitful collaborations and international recognition.

In the end, the persistency and the consistency demonstrated by Indian policymakers along the years, whatever the unavoidable failures, disappointments and difficulties on the way, proved to be the recipe for success. The visionary strategy they have defined and implemented must be saluted, and in many respects, India has invented a new approach to space long before the "New Space" era that is taking place at the moment worldwide, initiated from the United States.

This is certainly an exemplary source of inspiration for all the nations that are now considering entering the space sector, although the global context is pretty different nowadays and in many respects much less challenging.

But the Indian space programme is also a beautiful story since it is the quintessence of all the best that space can offer in terms of support to development. Since its inception, it has been driven by civil applications and peaceful purposes, and the efforts are now being paid back through improved connectivity throughout the vast Indian territory supporting a dynamic economy and innovative approaches to education, effective space imagery for meteorology, agriculture, resource management, etc. Now that many of the initial objectives have been achieved in terms of capacity building and deployment of infrastructures, India has become one of the members of the small club of fully fledged space powers.

This enviable statute is the outcome of decades of relentless efforts and investments. It comes along with increased ambitions to further take advantage of space capabilities for national security purposes, in particular to strengthen the position of India regionally. This move towards military applications of space is a major evolution of the Indian space programme, conducted by Indian policy-makers with a great sense of responsibility.

India is now a potential partner for balanced cooperation in space, having capacities and ambitions in all space application domains, including science and exploration. It is also asserting great ambitions towards business-oriented developments. Europe could take advantage of its long-established privileged relationship with India to build up a stronger collaborative framework for the benefit of both parties.

It is my hope that the book *India in Space Between Utility and Geopolitics* will prove a useful guide to navigate the exceedingly relevant transformations of the Indian space programme and identify key ways in which mutually beneficial Europe–India cooperation can substantiate.

Director of the European Space Policy Institute (ESPI)   Jean Jacques Tortora
Vienna, Austria

Acknowledgements

I would like to express my deep gratitude to the many people who have supported and contributed to the completion of this book. Much appreciation goes to Jean Jacques Tortora, director of the European Space Policy Institute (ESPI), for his patient guidance, enthusiastic encouragement and useful critiques of this research work. I would also like to extend my heartfelt thanks to David Nader, ESPI research intern, for his proactive and persistent contributions during the research and writing process, and to Sebastien Moranta, coordinator of ESPI studies, for his helpful inputs during the finalisation of the study and assistance in keeping my progress on schedule. More broadly, I wish to express great appreciation to the entire ESPI team for their useful feedback, assistance and cooperation.

The book has immensely benefited from the openness and wisdom of many befriended interlocutors outside ESPI. My gratitude goes to the incredibly talented Indian colleagues I had the privilege to interact with during my research activities. In particular, I would like to acknowledge Imtiaz Ali Khan, space counsellor and ISRO liaison officer in Europe; B.S. Bhatia, former director of the Development and Telecommunications Unit at ISRO; Rajeswari Pillai Rajagopalan, senior research fellow at the Observer Research Foundation (ORF); Ashok G.V, legal expert based out of Bangalore; Prateep Basu, senior research analyst at NSR; Rachana Reddy, ISRO engineer; and Susmita Muhanty, CEO of Earth2Orbit, for sharing their ideas and easing the understanding of the multifaceted dynamics surrounding the Indian space programme. A special mention goes to the co-founder of Dhruva Space, Narayan Prasad, who gracefully introduced me to a large and dynamic community of Indian space enthusiasts and professionals, in addition to illuminating me with his forward-looking wisdom. His kind assistance simply proved invaluable.

By the same token, I am also grateful to the many European stakeholders who accepted to be interviewed and provided substantial information for this book. Genuine thanks are in particular offered to Mathieu Weiss, CNES representative in India; Jean-Charles Bigot, ESA administrator of the International Relations Department; Paul Counet, head of Strategy and International Relations at EUMETSAT; Celine Bouhey-Klapisz, international relations advisor at CNES; Gabriella Arrigo, head of the International Relations Unit of ASI; Sveva Iacovoni, head of Multilateral Relations Office at ASI; and Marc Jochemich, regional manager at the International Cooperation Unit of DLR.

I would also like to highlight the precious assistance and suggestions offered by Jaque Grienberg during the review process. Of course, it remains that responsibility for errors and infelicities now rests with me. Finally, my appreciation is extended to Peter Hulsroj, former director of ESPI, for having envisioned this exciting project and to ESPI's editorial advisory board and Springer for having made this book possible. My greatest debt is, however, owned to my wife-to-be, Caterina, for the rare combination of patience and encouragement she offered throughout this effort. This book is for her.

Methodological Note

The book is the outcome of a year-long research conducted at the European Space Policy Institute (ESPI) between May 2016 and July 2017. Research activities have been based on an extensive review of publicly available documents, conference proceedings and bibliographic and electronic sources, spanning both sectorial and broader topics directly or indirectly related to India's space programme.

The research has been complemented and strengthened by more than 30 targeted interviews, generally off-the-record, with various stakeholders and space policy experts, from both Europe and India, including representatives from the respective space agencies, industries and academic community. Modern-day social media platforms, like in particular *New Space India*, have likewise been invaluable tools for receiving insight and exchanging ideas on the Indian space programme with a large group of Indian space professionals, astropreneurs, policy analysts, students and space enthusiasts.

The participation to conferences and events on the topics of the Indian space programme has also proved particularly useful for gathering relevant information and views from key personalities. A specific mention should be made to the well-attended event on the topic "India in Space: The Forward Look to International Cooperation" organised by the European Space Policy Institute (ESPI) at the margins of the 54th session of the Scientific and Technical Subcommittee of UNCOPUOS in February 2017. The event featured the participation of distinguished speakers representing the Indian, European, French, Italian and German space agencies, who contributed with valuable analyses and insights on the potential for stronger future cooperation between India and Europe in the space sector. From 16 to 18 February 2017, I also participated, in my capacity as resident fellow of ESPI, in the third ORF Kalpana Chawla Annual Space Policy Dialogue in New Delhi. The Observer Research Foundation (ORF) – a leading Indian think tank providing independent and in-depth analyses to decision-makers in governments, business communities and academia – organises this annual policy dialogue in honour of Ms. Kalpana Chawla, the first Indian–American woman in space, with the aim of creating a global platform for exchanging views on the growing socio-economic and political relevance that space activities have for India. The third edition of this annual space policy dialogue, held over two and a half days and attended by more than 50 speakers from around the world, included nine broad panels covering the topics of space finance, India's private players, satellite Internet for Digital India, transponder capacity for broadcasting and broadband over India, derivatives from space, emerging space actors, collective governance of global commons, current and evolving challenges to space sustainability, role of satellites in contingencies and the case for India's national space policy. In addition, special lectures were given on the topics of "Space Sustainability and Global Governance", "Outer Space and Strategic Stability", "A New Frontier: Boosting India's Military Presence in Outer Space" and "Elucidating the Components of India's National Space Policy". Notably, participation to the dialogue was also leveraged as an opportunity to further strengthen the study with some field research in New Delhi and Bangalore.

The book provides a variety of quantitative data and analyses, spanning both sectorial and general data concerning India. These were, however, mainly collected to support – in line with the general criteria proposed by Eisenhardt and Yin – the construction of a qualitative research. The basic research questions guiding this research were set during the planning phase of this project and were intentionally kept broad in order to better grasp the complexity and the many interrelated issues revolving around the Indian space programme. What are the political and socio-economic drivers informing India's policy-makers in their uses of outer space? Where is the country's space programme heading towards? How is its international behaviour changing? And, equally importantly, what is at stake for the various space actors in Europe, be they at national or at pan-European level? In this respect, it should be also highlighted that whereas the approach of the study is markedly empirical and its scope wide-ranging, the purpose has an inherent normative dimension and the perspective deliberately European: the study in concrete intends to describe and explain the dynamics surrounding the Indian space programme in order to elaborate on possible implications and ensuing policy actions and possible options for European decision-makers. This coincides with the mission of ESPI to provide European stakeholders with informed analyses in the field of space policy and to facilitate the decision-making process in Europe.

Policy studies are inherently fraught with normative biases. While at the beginning of the study I had no clear view on what the final policy implications and propositions would be, what certainly informed the research was both the recognition of the multifaceted and irking complexities surrounding India and the need to better capture the growing significance that this country is inescapably bound to play, here on Earth and in space. It is in my hope that this book will contribute to making the often-downcast European stakeholders more intrigued to learn more about – and from – this approaching behemoth.

Contents

List of Figures

List of Tables

Chapter 1
Introduction

अच्छे दिन आने वाले हैं।
Good days are coming

Narendra Modi

In his book, *The Argumentative Indian*, Nobel laureate Amartya Sen recounts that one of his teachers, the great Cambridge economist Joan Robinson, used to tell him: "the frustrating thing about India is that whatever you can rightly say about it, the opposite is also true" (Sen 2006). Whereas this poignant expression of the plurality and apparent contradictions that pervade the subcontinent proves to apply to almost any aspect of India's social, cultural and political life – from its philosophical and historical traditions to its democratic practice and economic development – there appears to be at least one compelling argument around which a widespread consensus can be found: the mounting importance of India.

1.1   Why India Matters

The most telling indicator of India's growing significance is perhaps encapsulated in the slogan used by the Bharatiya Janata Party leader Narendra Modi for the 2014 Indian general election: *Achhe din aane waale hain* ("Good days are coming"). Though the slogan was coined with the intention of conveying that a bountiful future was in store for India if the Bharatiya Janata Party won the election, India's future is indeed brimming with potential to fundamentally reshape the global landscape and generate transformative political, economic, social and technological revolutions across the board. For once, figures speak for themselves.

There are currently more than 1325 million people living in India, and this population continues to increase by 2403 new inhabitants every hour (MedIndia 2017). In fact, one-sixth of the world's population lives in the subcontinent – which is greater than the combined population of North America and Europe put together and more than in the whole of Africa (United Nations 2016). In a post-industrial digital age, where empowered individuals are becoming the most relevant natural

© Springer International Publishing AG 2018
M. Aliberti, *India in Space: Between Utility and Geopolitics*, Studies in Space Policy 14, https://doi.org/10.1007/978-3-319-71652-7_1

resource, India is well on track of turning into the most populous country on the planet. According to the United Nations (UN) World Population Prospects, India's population will surpass that of China as early as 2022, and, in stark contrast to the ageing populations of the United States, Europe, Japan and even China, the number of young people in India will continue to increase. By 2020, India will become the youngest country in the world, with two-thirds of its population in working age. With such a young population, India will be the country with the strongest demographic dividend in the twenty-first century and, therefore, the country with the greatest potential to seize its future. According to some estimates, the number of Indian graduates of working age – i.e. the most skilled and economically productive segment of society – will reach 117 million by 2030 and an impressive 208 million by 2050. In comparison, the United States will have 118 million graduates of working age and China 138 million, meaning that India is bound to have nearly twice the number of graduates of the United States and roughly 70 million more than the Central Kingdom (IndoGenius 2016b; University of Central Florida 2013; Mehrotra 2015). Young Indians, therefore, clearly represent the most significant human capital on the planet, a capital that will increasingly and more strongly drive India's path to prosperity (Ibid).

In parallel to this, the Indian economy has now emerged as the fastest-growing major economy in the world, and many projections by leading financial institutions, such as the OECD, Goldman Sachs, Credit Suisse and the International Monetary Fund, expect this trend to continue for the next few decades, turning India into the rising economic superpower that China is today (OECD 2017; International Monetary Fund 2017; Credit Suisse 2015). More importantly, these institutions also project that, by the middle of the century, India will join the United States and China as one of the three largest economies in the world in nominal as well as Purchasing Power Parity (PPP) terms. In the meantime, the economies of Europe, Japan and Russia will likely witness a relative decline, primarily owing to their shrinking and ageing populations.

This shift in the geo-economic barycentre of international affairs will most likely be also accompanied by a shift in political power. There is broad consensus that, over the next decades, India will reach the level of the United States and Europe in terms of national power based upon GDP, population size, military spending and technological investment (National Intelligence Council 2012). This transformation will turn India into an indispensable superpower exercising major influence and perhaps even vetoes over many international decisions (Meredith 2008). It is thus not possible to imagine solutions to the current and future global challenges without the engagement and support of India. Issues such as international security and terrorism, climate change, environmental sustainability, development and international trade will all require a strong partnership with New Delhi.

Moreover, in addressing global grand challenges, understanding of the Indian perspective can prove essential to deepen our own knowledge and provide a window into possible solutions. This is in part because India is that rare combination of a country that is simultaneously developed, emerging and underdeveloped and in part because "India today, with its highly dynamic and adaptive approach, has become a hub of creativity and all types of innovation. In areas as diverse as education, financial inclusion, healthcare, transportation, manufacturing and energy production, India's innovations are going a long way in solving important development problems and, in some cases, even positioning India to leapfrog entire developmental steps thanks to the adoption of new technologies" (IndoGenius 2016c). A cogent example of this "leapfrogging development" is the *Aadhaar* biometric identity card, which is being used by more than a billion people for presence-less, paperless and cashless e-governance services like the eSign and the Unified Payment Interface. Because India is simultaneously a developed, an emerging and an underdeveloped country, as it innovates for itself, its home-grown solutions might actually be more applicable in the rest of the developing world than Western innovations (Ibid). This makes India's experience of global relevance.

There are also other factors that urge the reconsideration of the importance of India to global development and affairs. Historian Ramachandra Guha has made a compelling case for why India is the most interesting country in the world (Guha 2010, 2014). He suggests that India's sheer scale and the unparalleled diversity arising from its population size, its cultural diversity and its traditions of pluralism combine with five simultaneous revolutions still underway to make India the most recklessly ambitious and busiest laboratory of socio-economic and political modernity in the world. The five revolutions identified by Guha are *national*, whereby princely states and colonies became an independent nation-state (or more properly state-nations); *democratic*, with the passage from a feudal society to a democratic polity with a multiparty system and universal suffrage; *urban*, with the progressive transformation from a rural society to a mostly urban one; *industrial*, whereby the country has been moving away from an agricultural society to an industrial one; and *social*, with deep-seated transformations of socially constitutive bonds such as caste, familiy, religion and ethnicity. The other key point Guha highlights is that, unlike Western countries, these revolutions began simultaneously in India, when the country gained independence from the British Raj in 1947. These revolutions are, in fact, still underway, further fuelled by India's young and growing population, the rapid economic growth and the transformative digital context, which are not only making change happen more quickly, but they are also impacting the overall trajectory of India's journey (IndoGenius 2016a).

This research project is, in a sense, the story of this journey's direction in one of the most relevant and promising endeavours India has embarked upon: space.

1.2   India's Achievements in Space

India's journey in space is unique in its inception in the sense of having been envisioned *from* and *for* a developing nation's perspective and purpose. Since the establishment of a space programme in the early 1960s, space activities have been understood as a very useful tool to address many of the societal issues faced by a huge and poor country like India. This vision was enunciated by the father of the Indian space programme, Dr. Vikram Sarabhai, in his well-known statement:

> There are some who question the relevance of space activities in a developing nation. To us there is no ambiguity of purpose.... We are convinced that if we are to play a meaningful role nationally and in the comity of nations, we must be second to none in the applications of advanced technologies to the real problems of man and society.

Founded in 1969, the Indian Space Research Organisation (ISRO), a government institution under the Department of Space (DOS), has piloted all Indian space activities guided by the two-pronged aim of demonstrating the applications of space assets for the development and betterment of society and achieving self-reliance and indigenous capability development in what was deemed a technologically strategic field.

Another equally impressive element of India's journey in space has been the consistent way in which the programme has instilled user involvement and consultation among various stakeholders as integral to the process. From the early 1970s – at a time when technological development in the space sector was mostly pushed rather than pulled – ISRO started to implement a unique approach based on the concrete needs of space technologies' users. Indeed, the utilisation of space technology began with pioneering experiments in the areas of remote sensing and telecommunications, such as detection of coconut root-wilt disease and the use of satellite-based television broadcasting to reach out to rural areas under the Satellite Instructional Television Experiment (SITE). Since then, Indian space activities have made tremendous progress with the successful development of indigenous capabilities and one of the largest space, launch and ground infrastructures in the world. This comprises the Indian Remote Sensing (IRS) constellation, the Indian National Satellite (INSAT) System for telecommunications and meteorology and the Indian Regional Navigation Satellite System (IRNSS) for PNT services, as part of ISRO's operational space infrastructure, and the Polar Satellite Launch Vehicle (PSLV) and later the Geosynchronous Satellite Launch Vehicle (GSLV) and its variants, as part of its launch infrastructure. As of 2016, ISRO had conducted 124 launch and satellite missions with a total of 34 active satellites in orbit. In parallel, a ground infrastructure of several centres for data reception and processing was set up to track and control these space assets and provide relative applications to Indian citizens. With all these elements in place, the Indian space programme has witnessed significant scaling of activities in space and has matured high-class, end-to-end systemic capability in the design, development and operations of Indian space assets and applications (Prasad 2016a, b; Rao et al. 2015).

The provision of space-based services has over the years created an impressive and ever-growing user base, which has in turn contributed to making the Indian space programme an internationally outstanding model for the successful development of technology and its utilisation for achieving societal goals in a variety of fields such as access to education and healthcare, crop monitoring and natural resource management and rural and urban development, among others. ISRO, in addition, has been setting "national benchmarks for skills in project management, openness to the public, transparency in its knowledge transactions and, in general, for the development of a unique corporate culture among its employees" (Comptroller and Auditor General of India 2012). The country's achievements in space have been widely supported and admired by political leaders of all kinds as well as ordinary citizens – and highly acclaimed by the international space community, which has frequently hailed ISRO's prolific space programme and success despite its shoe-string budget.

International attention to India's space programme increased further after the 2008 launch of the spacecraft Chandrayaan-1 to the Moon, which became a source of great national pride but also set off intense discussions on India's use of space, primarily about whether the purely utilitarian path should be abandoned at a time when huge poverty and terrestrial infrastructure issues had still not been overcome. Lately India scored an even bigger success with the Mangalyaan Mars probe, launched in 2013 and entering a Mars orbit in 2014 – the first country to reach a Mars orbit on the first try. Moreover, plans for more daring exploration missions are in the works.

The Indian space programme is currently gathering a swelling traction with a comprehensive roadmap ahead of over 70 satellites to be built within the next five years and the target of tripling the launch frequency of Indian rockets from the current 4–6 to 12–18 annually. These developments have been strongly encouraged by legislators within the country, with several Parliamentary Standing Committees (e.g. Science and Technology, as well as Environment and Forests) recurrently recommending to the federal government a 50% rise in ISRO's annual budget to realise these missions (Prasad 2016a, b).

Propelled by this ambitious roadmap, the Indian space programme is now emerging as one of the fastest-growing programmes in the world and is in parallel moving towards using space technology products and services not only for societal applications but also to support commercial, diplomatic, defence and security objectives. This unprecedented acceleration and progressive diversification is bound to generate important consequences, both internally and internationally. It is hence essential for the international space community to better appreciate the unfolding trajectory and impact of this space programme in order to seize the "India opportunity" in the best possible manner.

1.3 Objectives and Structure of the Book

This book narrates the story of India's renewing strategic vision and progressive diversification of its space programme at the nexus socio-economic development, commerce and geopolitics. It disentangles India's evolving rationales for engaging in space, by embedding New Delhi's space agenda in the context of the changing geopolitical situation and in the context of domestic Indian politics, and provides novel and in-depth assessment of the factors influencing the pace and directions of the country's space activities. The study hence includes an extensive analysis of

India's path forward, the toolbox India has at its disposal and the broader implications this diversification will generate for the international space community. In doing so, the book offers a holistic reflection on the outlook of Indo-European relations and elaborates on what the opportunities for Europe are to nurture closer and mutually beneficial space cooperation with India, both at the pan-European and individual nation level, with the goal of sorting out possible elements for a comprehensive European long-term strategy towards India.

The study is comprised of eight chapters. After providing an introductory overview of India's growing importance in the chapter at hand, the next chapter will examine the historical path of India in space.

The objective is to provide a reflection on the broader evolution of the Indian space programme to enable understanding of India's unique place in the world space arena, of its rationales for engaging in space with a narrow utilitarian focus and of how support for that focus has been structured and adjusted in India's in the context of India's domestic needs as well as regional and international drivers.

Chapter 3 will subsequently offer a succinct overview of the current landscape for space activities in the country. Particular attention will be paid to the organisational setup, decision-making processes and budgetary allocations, to the space policies and regulatory framework for space activities and to India's technological space capabilities and infrastructure.

The study will then examine India's rationales for engaging in space by embedding the country's space activities within the wider national, social, economic, political and technological context. This will be the objective of Chaps. 4 and of 5. Whereas the former will more specifically elaborate on India's uses of space as a tool of its developmental and economic emersion strategy, the latter will put the spotlight on the emerging uses related to India's evolving diplomatic and security objectives.

Chapter 6 will assess the historical evolution and current status of India's relations with the major spacefaring nations. The focus will be in particular put on India's relations with the United States, Russia, China, Japan and Europe, at both pan-European and at individual nation level. Specific attention will be also given to France, since it has been a key partner in the development of the Indian Space Programme and one of the most active European nations internationally.

Chapter 7 will then provide an extensive analysis of India's long-term pathway in space. The objective will be to identify and examine long-term factors that will likely drive the evolution of the Indian space programme and to describe a most probable scenario for the future of space activities in this country. Consistently, this chapter will focus first on the country's future outlook and subsequently provide a reflection on the evolution of its civil, military and commercial space efforts. Considerations on the toolbox India will have at its disposal, on how India can be assumed to choose its partners and in which fields and on how the space ecosystem will be adapted to address these evolutions will also be provided.

Finally, in Chap. 8, the book will offer a comprehensive reflection of the implications and prospects for Europe. More specifically, the first section will gauge current status of Europe–India cooperative ties, or lack thereof, in the political and space arena. The second part will then elaborate on the stakes, drivers and opportunities underpinning the case for closer space cooperation with India, both at the pan-European and individual nation level, and probe whether such cooperation can be a catalyst for broader political cooperation. The chapter concludes with the identification of key domains for future cooperation and with an analysis of the requirements that must be fulfilled to achieve this cooperation.

References

Comptroller and Auditor General of India. (2012). *Report of the comptroller and auditor general of India on hybrid satellite digital multimedia broadcasting service agreement with devas*. New Delhi: Department of Space.

Credit Suisse. (2015). *Global wealth databook 2015*. Zurich.

Guha, R. (2010). *Makers of modern India*. New Delhi: Penguin Books.

Guha, R. (2014, February 24). *Why India is the most interesting country in the world?* Retrieved from Youtube: https://www.youtube.com/watch?v=6PteQsyUAbg

IndoGenius. (2016a, August). The most interesting country on earth. *The importance of India* (Online Lecture).

IndoGenius. (2016b, August). The youngest population on earth. *The importance of India*. (Online Lecture).

IndoGenius. (2016c, August). Creative solutions to grand challenges. *The importance of India*. IndoGenius.

International Monetary Fund. (2017, April). *World economic outlook*. Retrieved May 10, 2017, from International Monetary Fund: https://www.imf.org/external/pubs/ft/weo/2017/01/weo-data/index.aspx

MedIndia. (2017). *Indian population clock*. Retrieved December 21, 2017, from MedIndia: http://www.medindia.net/patients/calculators/pop_clock.asp

Mehrotra, S. (2015). *Realising the demographic dividend policies to achieve inclusive growth in India*. Delhi: Harvard University Press.

Meredith, R. (2008). *The elephant and the dragon: The rise of India and China and what it means for all of us*. New York: Norton.

National Intelligence Council. (2012). *Global trends 2030: Alternative worlds*. Washington, DC: National Intelligence Council.

OECD. (2017). *GDP Long-Term Forecast*. Retrieved April 29, 2017, from OECD data: https://data.oecd.org/gdp/gdp-long-term-forecast.htm

Prasad, N. (2016a). Diversification of the Indian space programme in the past decade: Perspectives on implications and challenges. *Space Policy, 36*, 38–45.

Prasad, N. (2016b). Industry participation in India's space programme: Current trends & perspectives for the future. *Astropolitics: The International Journal of Space Politics and Policy, 14*(2–3), 234–255.

Rao, M. K., Murthi, S., & Raj, B. (2015). *Indian space – Towards a "National Eco-System" for future space activities*. Jerusalem: 66th International Astronautical Federation.

Sen, A. (2006). *The argumentative Indian. Writings on Indian history, culture and identity*. London: Penguin Books.

United Nations. (2016). *World population prospects. The 2016 revision*. New York: United Nations Department of Economic and Social Affairs.

University of Central Florida. (2013). *Global perspectives: India; population, technology and education*.

Chapter 2
India's Path in Space: A Brief History of an Evolving Endeavour

Often regarded as a recent entrant into the space arena, India is, in fact, an "early adopter" of space technology. Its space programme has deep roots, and it is nearly as old as those of the two superpowers. Yet, India's place in the realm of the major space powers is unique in the ways in which this country has sought to leverage its space programme. In sharp contrast to the United States, the Soviet Union and China, India's initial space priorities did not revolve around the pursuit of national security and defence objectives, but were rather expressions of a leading-edge vision about exploiting space for lifting India's people out of poverty and escaping dependency positions with advanced nations through technological leapfrogging. For India's space leaders, space was primarily intended as a very useful tool for addressing the many socio-economic issues facing a huge and developing country.

Significantly, this view was engrained in India's perception of its space programme even before the launch of the first communication and remote sensing satellites and before any international recognition of the great potential that space technology had for accelerating national development and "getting away from percolation and trickle down models of development for which developing countries do not have the time" (United Nations 1982).

Another equally impressive element has been the consistent way in which the programme has instilled user involvement and consultation among various stakeholders as integral to the process. From the early 1970s – at a time when the development of space technologies was predominantly pushed rather than pulled – ISRO started to implement a unique approach based on the concrete needs of space technologies' users. This process has been structured along multiple phases that include need assessment, joint experimentation and technology development, institutionalisation and operationalisation. Interestingly, this feature is not specific to single projects but is more broadly engrained in the evolution of the entire Indian space programme.

This chapter provides an introductory account of India's historical path in space in the context of its domestic, regional and international drivers. The objective is to disentangle the broader evolution of the Indian space programme to enable better

© Springer International Publishing AG 2018
M. Aliberti, *India in Space: Between Utility and Geopolitics*, Studies in Space Policy 14, https://doi.org/10.1007/978-3-319-71652-7_2

understanding of India's unique place in the world space arena, of its rationales for engaging in space with a utilitarian focus and of how support for that focus has been structured and adjusted in India's domestic and international political life.

From an overall perspective, India's journey into space can be seen as having undergone four distinct phases of development over the past five decades. In the first phase, which began in 1962 with the creation of INCOSPAR and ended in 1972, with the institutionalisation of the Space Commission and Department of Space, the rationales for establishing a national programme were identified and its directions spelled out. This period also saw the sowing of the seeds of future space application programmes as well as the harnessing of resources and key manpower for subsequent efforts. Understandably, India's status as a postcolonial, non-aligned and developing country strongly affected the very visualisation of its space profile and the pace of its progress too. While the country's poverty levels and lack of technological infrastructure hampered a steady growth of space activities, its position as a leading member of the non-aligned movement enabled the programme to benefit from international cooperation with all the major spacefaring nations.

Following this first decade of formulation of the country's space profile and directions, India's space programme entered into a phase of consolidation and experimentation during which demonstration of end-to-end capability in the operationalisation of space systems and associated applications for the users was realised. During this phase, which lasted until the mid-1980s, India's first satellites were successfully manufactured, the societal applications of space technology fully tested and the foundations of an operational launch vehicle programme laid.

At the beginning of the 1990s, when all three major components of the programme had been successfully tested, the programme entered its operational stage. During this phase, major space infrastructure was created under two broad categories: one for communication, broadcasting and meteorology through the multipurpose Indian National Satellite (INSAT) System and the other for Earth observation through the Indian Remote Sensing (IRS) satellite system. The development and operationalisation of India's Polar Satellite Launch Vehicle (PSLV) and the development of the Geosynchronous Satellite Launch Vehicle (GSLV) – components necessary for the deployment of the INSAT and IRS systems – were also realised in this period. In stark contrast to the experimental phase of the 1970s and early 1980s, this phase was characterised by the emergence of major international complications, particularly from the United States and other Western countries, which backed away from previous cooperative stances (in space and elsewhere) for fear of promoting the development of a long-range ballistic missile for nuclear weapons.

Eventually, with the political, economic and military "rise of India" at the beginning of the twenty-first century and the broader evolution of the regional and international context, the space programme entered a new phase of maturity characterised by an unprecedented scaling up of activities that has also impacted its overall direction. Indeed, the programme's focus started to move beyond the idea of harnessing space for socio-economic development and to explore uncharted territories, such as space exploration and the military uses of space.

2.1   The 1960s: Envisioning the Programme

India's space programme was initiated within a few years of the launch of Sputnik in 1957. Although at that time India could already boast a long historical legacy in terms of interests in astronomy and rocketry, there appeared to be no ostensible reason to initiate a space programme. After all, India was a recently independent state with a huge largely rural population facing tremendous development challenges. Given the novelty and complexity of space technology and the harsh realities of India's economic situation, the Nehru government's decision to establish a national programme was "an act of extraordinary foresight and courage" (Sheehan 2007). But in fact, that decision was the result of a well-thought-out vision that would in time succeed in reversing perceptions from "why should poor India should have a space programme" to "India should have a space programme exactly because it is poor" (Lele 2013a, b). The inceptor of this forward-looking vision was Vikram Sarabhai, later known as the founder of the Indian space programme.

2.1.1   The Mark of a Genius

Whereas India's quest for nationhood in the post-independence era has in many instances brought a rejection of the positivism-inspired concept of modernity, India's science remained strongly conditioned by the West. As documented by historian David Arnold, "the authority that Western science had come to enjoy in India by the late nineteenth century was too great to be ignored in Indians own programmes of reform and revitalisation" [Consistently, Indian scientists of the period were inclined] "to believe that science and technology, rightly applied, could rapidly improve India's fortunes and remove many of its long-standing problems" (Arnold 2000). Vikram Sarabhai, a Cambridge-educated physicist, was no exception. As poignantly recalled by Jacques Blamont following his first meeting with Sarabhai:

> I had just become the scientific and technical director of the fledging French Space Agency, CNES, born on 1 March 1962 with great ambitions and small means. Vikram, a cosmic rays physicist, had already conceived an original doctrine which he explained to me with convincing enthusiasm. For him, the only method by which India could catch up with developed countries was to bypass the usual stages by exploiting the most modern technology, a process he called leapfrogging. He considered nuclear energy, electronics and space as domains where governmental investments would quickly provide major improvements to the life conditions of the poor people who formed the majority of the Indian nation.
>
> If the relevance of the first two was obvious, space did not appear at that time to provide applications to leapfrog, at least in my opinion. But Vikram convinced me by an exposition of his views on education through satellite. For him, illiteracy represented the major obstacle to progress in India. Imagine, he said, a central station transmitting to a satellite education programmes prepared by excellent professors somewhere in a specialized organisation, and the satellite relaying these programmes to tens of thousands of villages, where television receivers would display them to the local farmers. [...] You have to realize that when I heard Vikram explain his dream, the first telecommunication satellite Telstar had just been

placed in orbit, and the first geostationary satellite Syncom had not yet been launched. Sarabhai's vision appeared to me as a mark of a genius" and I decided to help him as much as I could (Blamont 2015).

With his pioneering vision, Vikram Sarabhai was well aware of the different path that India – as a huge and developing country – had to follow as compared to the two superpowers. Consistently, he rejected any notion of competing with the economically advanced nations through prestigious undertakings and rather urged India to use the nascent space technologies as a means for addressing concrete societal needs and as a means to escape dependence by bypassing the intermediate technological state and moving directly into the high technology era.

In 1961, Sarabhai presented a formal proposal to the government, articulating the priorities for a national space programme oriented towards functional benefits for Indian national development, educational system and healthcare provision. Vikram Sarabhai's ideas immediately encountered a positive reception in the government led by PM Jawaharlal Nehru, who had always heralded that "science alone can solve the problem of hunger, insanitation and illiteracy". The following year, the Indian government formalised its participation in space research by creating – under the supervision of the development-oriented Department of Atomic Energy (DAE) – the Indian National Committee for Space Research (INCOSPAR), with a mandate to advise the government on space policy.

As a first step, it was decided to build a launch site for conducting scientific studies of the high atmosphere dynamics through the creation of sodium vapour clouds by sounding rockets. Homi J. Bhabha, generally known as the father of India's nuclear science programme, assisted Vikram Sarabhai in setting up India's first launching station. This centre was established at Thumba, near Thiruvananthapuram, primarily because of its contiguity to the magnetic equator (Indian Space Research Organisation 2016a, b, c).

However, as recounted by Vikram Sarabhai in the famous *Atomic Energy and Space Research: A Profile for the Decade 1970–1980*, "it was clear at the outset that space research could not progress without the simultaneous development of advanced space technology" (Atomic Energy Commission 1970). Since nobody in India could build sounding rockets – despite the country's recorded prowess in rocketry during the eighteenth century – that meant that "space technology was [to be] acquired from abroad under licence" (ibid, p. 27). As a firm advocate of indigenisation, Sarabhai knew that at least in the early stage of the programme, the country would be necessarily reliant on assistance from other countries "to acquire initial technological and engineering skills in an unfamiliar realm" (Sheehan 2007).

The choice of foreign partners was representative of India's resolve to be even-handed and not be part of one superpower bloc or the other. As a leading member of the non-aligned movement, Nehru's India indeed sought technical assistance from both the United States and the USSR in order to avoid a dependency position (Moltz 2012). This would prove a great advantage to its emerging space programme, which greatly benefited from cooperation with a large number of countries from both blocs.

Between 1962 and 1963, cooperative links were established with the National Aeronautics and Space Administration (NASA), the Soviet Union's Hydro-Meteorological Service (HMS) and also the newborn French space agency – the Centre National d'Etudes Spatiales (CNES) – for launching their scientific sounding rockets from Thumba. The three agencies also helped INCOSPAR to complete the setting up of the Thumba Equatorial Rocket Launching Station (TERLS), and on 21 November 1963, a US-provided Nike Apache rocket successfully launched a CNES-built sodium ejector payload at a height of 200 km, marking India's entry into the space age.[1] Overshadowed by the co-temporal assassination of President Kennedy, the launch of the first sounding rocket from Indian territory did not attract much attention, neither at a national nor an international level. The launch was, however, followed by three other sodium vapour payload experiments fired on-board French Centaure rockets in January 1964.

Given the importance Vikram Sarabhai attached to the development of indigenous capacities, in May 1964, DAE also reached an agreement with CNES for licencing the manufacturing of its Bélier and Centaure sounding rockets to India, with the accompanying transfer of solid propulsion technologies. The subsequent setting up of a propellant plant for the Centaure and the indigenisation of the related technology proved a major boost for the development of the first Indian sounding rocket, the *Rohini*, which, in fact, was very similar to the French Dragon (a more powerful version of Centaure).

Overall, more than 65 French Centaures, American Nike, Soviet M100 and British Skua Petrel rockets were fired form Thumba between 1963 and 1968. In 1968, the site was turned into an international launch centre dedicated to the United Nations (UN) and renamed International Equatorial Sounding Rocket Facility. The following year, ISRO started launching the indigenously made *Rohini* sounding rockets, and the experience gained proved to be of immense value to the process of mastering solid propellant technology.[2] Thus, the Rohini sounding rockets programme not only provided Indian space scientists with the means to study the upper regions of the atmosphere but more importantly laid the foundation for future launch development in ISRO (Blamont 2017).

2.1.2 Sowing the Seeds for the Application of Space Technology

Despite this seminal hands-on experience in the field of rocket technology, India's major efforts during the inception phase focused on sowing the seeds of India's future applications programme. As aptly argued by James Clay Moltz, one of the

---

[1] Interestingly, the person in charge of the payload integration at TERLS was Abdul Kalam, who would become the president of India in 2002.

[2] In the course of its development, the altitude reached by the Rohini rose from 90 to 360 km, and the payload capacity grew from 16 to 90 kg.

reasons for India's willingness to put off development of its own long-range missiles and to focus on space applications from rockets was its lack of nuclear weapons and its rejection of the superpower arms race. By choosing its policy of non-alignment, Nehru also sought to avoid being drawn into a possible World War III. Thus, unlike the Soviet Union, the United States and China, national space priorities did not initially hinge around the development of an intercontinental delivery system (Moltz 2012). Also, Indian scientists were highly sensitive to the possible accusation that India should not be devoting precious resources to space technology, when there were so many problems on the ground demanding government attention (Sheehan 2007). Hence, they simply rejected the idea that a poor country could exploit space for prestige and military purposes, focusing instead on the development of space technologies that could help address many of the societal issues facing post-independence India. Due to their enormous potential for enhancing the quality of Indian citizens' life, three sectors were specifically identified, namely, space-based communications, meteorology and remote sensing.

India's application-oriented focus of space efforts became fully manifest at the dedication of TERLS to the United Nations on 2 February 1968. It was, in fact, on that occasion that Vikram Sarabhai provided the most famous and enduring articulation of the country's space objectives:

> There are some who question the relevance of space activities in a developing nation. To us, there is no ambiguity of purpose. We do not have the fantasy of competing with the economically advanced nations in the exploration of the moon or the planets or manned spaceflight. But we are convinced that if we are to play a meaningful role nationally, and in the community of nations, we must be second to none in the application of advanced technologies to the real problems of man and society, which we find in our country. The application of sophisticated technologies and methods of analysis to our problems is not to be confused with embarking on grandiose schemes whose primary impact is for show rather than for progress measures in hard economic and social terms. (Indian Space Research Organisation 2016a, b, c)

Even though Vikram Sarabhai originally wanted to develop an indigenous satellite programme, he knew that it would have taken a long time before India could design, build and launch its own satellites. To prove the worth of space-based applications and kick-start their utilisation, it was therefore clear that India could not wait for its own satellites, but that foreign satellites should be used instead in the initial stages. Accordingly, Sarabhai's idea was to begin with an experimental programme for direct broadcasting to Indian villages by tapping into existing satellites, which in time would allow first operating a foreign-built communication satellite, then a domestic-built satellite launched by one of India's partners at TERLS (the United States, France, the USSR or Britain) and finally a domestically built and launched satellite.

However, before starting a satellite-based television experiment, Indian scientists deemed it essential to carry out some controlled experiments using conventional television in order to demonstrate the effectiveness of television as a tool for supporting national development. Consequently, an instructional TV programme called *Krishi Darshan* was set up to provide Indian farmers with a range of agricultural information on a weekly basis. The results of the experiments proved encouraging:

through comparison with a control group of farmers not exposed to the programme, the experiment showed that participating farmers improved their use of fertilisers and had higher wheat harvests (Harvey et al. 2010). In parallel to this, Indian scientists also undertook – with the support of NASA – a cost–benefit assessment to prove the efficacy of space-based solutions as compared to terrestrial ones. The study demonstrated that a space-based system would, in fact, be cheaper than ground masts in relaying educational television to India's villages (Atomic Energy Commission 1970).

In order to tap into the signals of Soviet and American communication satellites and enable India to gain competence in global satellite communications, a team of Indian engineers led by Vikram Sarabhai began the construction of a receiving station in Ahmedabad. Thanks to the equipment provided by Japan's Nippon Electric and the support of the United Nations Development Programme (UNDP) and the International Telecommunications Union (ITU), the first Experimental Satellite Communication Earth Station (ESCES) was opened as early as August 1967 and immediately started to receive the signals from the American Technology Satellite-2 (ATS-2). A second ground station would be built shortly thereafter in Arvi, in the state of Maharashtra, and a third one in New Delhi.

Meanwhile, Vikram Sarabhai's discussions with NASA had led to the signature of the *Television for Development* agreement between NASA and the Atomic Energy Commission. Under the agreement, NASA would move its upcoming ATS-6 to 20°E to enable an experimental programme of direct broadcasting to some of India's remotest areas (Atomic Energy Commission 1970).

While awaiting the launch of the communication satellite experiment (ATS-6), Vikram Sarabhai also started outlining the possibilities offered by remote sensing technology to monitor and manage the country's natural resources in a more efficient manner. Because of the damage done to agriculture by floods, droughts, pests, famine and associated matters, remote sensing technology was not perceived as a luxury but as a necessity that could address these problems and benefit India more than would have a developed country.

Interestingly, the genesis of the remote sensing applications can be traced to a request made to Vikram Sarabhai by M.S. Swaminathan (generally acknowledged as the father of India's Green Revolution) to help in studying the spread of coconut wilt-root disease in the Kollam area of the state of Kerala. The request was made in 1968, long before any satellite was built or the remote sensing programme was even conceived.[3] Given the lack of equipment, Indian scientists used the helicopter supplied by the USSR for the TERLS and an aircraft equipped with a set of cameras and false-colour films that was supplied by NASA to carry out the study. The experiment successfully detected the affected areas and hence provided a basis for the "acceptance" by the Indian political community of remote sensing technology as a developmental tool (Pisharoty 2015).

---

[3] For Swaminathan's personal account of the genesis and growth of remote sensing aplications in Indian agriculture, see Swaminathan (2015).

As a clear indication of the stronger attention attached to the development of space technology and applications, in August 1969 the Indian government established the Indian Space Research Organisation (ISRO), with Vikram Sarabhai becoming its first chairman. A few years later, in June 1972, Indira Gandhi's government passed a resolution to set up a supervisory body called the Space Commission and an administrative body called the Department of Space (DOS) while bringing the ISRO under the Department of Space in September 1972, thus linking the programme directly to the prime minister's office. The DOS was now responsible for policy formulation and implementation in all areas of space activities.

Consistent with Sarabhai's vision, the 1972 resolution that established the Space Commission declared that India attached the highest priority to the development of space applications and justified the new organisational structure and governmental attention to space activities on the basis of "the sophistication of this technology, the newness of the field, the strategic nature of its development and the many areas in which it has applications" (Sheehan 2007).

Unfortunately, Sarabhai would not live enough to see the eventual institutional consolidation of the programme, having suddenly passed away on 30 December 1971. His death symbolically marked the end of the inception phase of the Indian space programme. His vision, however, would loom large over the country's space efforts for at least the three subsequent decades.

2.2 Building the Programme

Following the first decade of visualisation of the country's space profile and directions – what R. Aravamudan called "Vikram Sarabhai's decade" – India's space programme entered into a phase of consolidation, competence building and experimentation during which the demonstration of end-to-end capability in the realisation of space systems and the associated ground systems for users was realised (Aravamudan 2015). During this phase, which lasted until the early 1980s, India's first satellites were built, the societal application of space technology was fully tested, and autonomous access to space was eventually achieved.

Under the guidance of Satish Dhawan, who was appointed chairman of ISRO by Prime Minister Indira Gandhi after the death of Vikram Sarabhai, the formative years of the newly founded ISRO witnessed considerable growth in capacity, as evidenced by the steady increase in the programme's staffing, which rose from 110 people in 1965 to 4500 in 1975 and 7000 in 1980 (Atomic Energy Commission 1970).

During these formative years, ISRO's efforts were concentrated around three major lines of activity, namely, experimentation of space-based application programmes, the development of India's first satellites for communication and remote sensing and the space transportation system. For each of these areas of activity, assistance provided by the major spacefaring nations – including France, West Germany and Japan, in addition to the United States and the USSR – continued to be of utmost importance for India. Yet, it was in this second phase that the Indian space programme started to witness the progressive development of indigenous capabilities and to move towards self-reliance.

2.2.1 Experimenting Space Applications

The application of space technology for development started in the early 1970s with various experiments in the field of space communications with television broadcasting to reach out to rural areas and the field of remote sensing with small experiments using aerial photography to study crops. In both cases, these experiments provided valuable experience for the subsequent development, testing and management of a satellite-based system for supporting development, particularly in rural areas.

Satellite Communications
In the field of satellite applications, ISRO scientists primarily worked towards bringing into fruition the efforts initiated under Vikram Sarabhai, the most relevant of which was the Satellite Instructional Television Experiment (SITE).

Hailed as "the largest sociological experiment in the world" and as the "cradle of the direct-to-home", SITE was designed to broadcast instructional TV programmes to some 2400 villages and hence demonstrate the potential of satellite technology as an effective mass communication tool for education and development in rural areas.

The experiment was made possible thanks to the 1969 agreement with NASA to use its ATS-6 satellite for 1 year[4] and was conducted between August 1975 and July 1976 in partnership with the Ministry of Information and Broadcasting and with the Ministry of Education. The core tasks, however, were all executed by ISRO. Through the adoption of an end-to-end approach, ISRO indeed took on the responsibility not only of providing the space-related infrastructure but also of designing, developing and manufacturing the required ground hardware. It had to further take on responsibility for deploying the ground receiver terminals in the remote villages and arranging for their operations, upkeep and maintenance. Further, as broadcasting studios needed to be set up, ISRO also became involved in the design of low-cost educational studios and helped the Ministry of Education and the Ministry of Information and Broadcasting in setting up. As highlighted by Bhatia, "this was probably much beyond the normal scope of a space agency, but ISRO took on the responsibility because the experiment would just not have happened without this and the adaption of satellite broadcasting in India would not have been that rapid" (Bhatia 2016).

Recognising that "the effectiveness of Satellite Instructional Television was greatly dependent on the quality of the programmes besides the functioning of the hardware, ISRO took upon itself the responsibility of producing science programmes for school children and agriculture-oriented programmes for the adult audience" participating in SITE (Ibid). Overall, between August 1975 and July 1976, 1200 h

---

[4]The satellite was launched in May 1974 and subsequently moved to 20°E to enable the conduct of the experiment.

of developmental programmes were beamed to an estimated 2400 villages and watched by 5 million people (Harvey et al. 2010). The programme concentrated on general education, teacher training, occupational skills, agriculture, family planning and health, with a focus on maternal and child healthcare, nutrition, hygiene and birth control. As noted by Harvey et al., "contrary to the expectations that people would only watch entertainment programmes, the highest uptake was for 'hard' instructional programmes" (Harvey et al. 2010).

Following the conclusion of SITE, "ISRO also took upon itself the evaluation of all aspects of the experiment, including the performance of the hardware systems and the impact of the programmes on school children and adult audiences" (Bhatia 2015). The social impact of the experiment was assessed also thanks to the work of 100 social scientists, who went to 27 selected villages before, during and after the experiment, and resident anthropologists in 9 villages for 19 months (Harvey et al. 2010).

The results of SITE "clearly indicated the success and effectiveness of the programmes[5] and gave a strong push to the operational acceptance of satellite based broadcasting systems in the country".[6] Equally important, SITE provided "valuable experience in the development, testing and management of a satellite-based instructional television system" (Indian Space Research Organisation 2015a, b, c, d) and would later result "in the evolution of wider application programmes", including a "unique satellite concept, Gramsat (gram meaning village), tailored to disseminating culture-specific knowledge to the vast and diverse rural India" (Sheehan 2007).

SITE was followed by the Satellite Telecommunication Experiments Project (STEP), a joint project of ISRO and the Post and Telegraphs Department (P&T) carried out between 1977 and 1979, using the Franco-German Symphonie satellite. Conceived as a sequel to SITE, the aim of STEP was "to provide a system test of using geosynchronous satellites for domestic communications, enhance capabilities and experience in the design, manufacture, installation, operation and maintenance of various ground segment facilities and build up requisite indigenous competence for the proposed operational domestic satellite system, INSAT, for the country" (Indian Space Research Organisation 2015a, b, c, d). Hence, although STEP was

---

[5] According to ISRO, SITE directly benefited around 200,000 people and takes credit for training 50,000 science teachers of primary schools in 1 year (Indian Space Research Organisation 2015a, b, c, d).

[6] As articulated by Bhatia, "the agencies responsible for education adopted satellite broadcasting in a big way. For school education, the CIET (Central Institute of Educational Technology) was set up and several SIETs (State Institutes of Educational Technology) were created to provide the necessary infrastructure for ongoing operational use of satellite technology. For higher education, the UGC (University Grants Commission) approached ISRO for setting up the CEC (Consortium for Educational Communication) and several EMMRCs (Educational Multi-Media Research Centres). For Open and Distance Learning, the Indira Gandhi National Open University approached ISRO for the setting up of its broadcasting networks. All these networks were made operational on INSAT where bandwidth was provided to all educational agencies without any cost. In the meanwhile, ISRO experimented with the use of one-way video and two-way audio networks for training. These were found to be most effective for teacher training and all the above educational agencies were oriented and trained to utilize these features" (Bhatia 2016).

only a temporary experiment, it greatly contributed – just like SITE – to the laying of the foundations on which satellite-based communications applications in India would mature.

Remote Sensing Applications

In parallel with these seminal efforts in the area of satellite communications, ISRO began to work on major experiments for also validating the effectiveness of remote sensing applications.

Interestingly, the first experiments in the field were initiated upon a request of the director general of the Indian Council of Agricultural Research, M.S. Swaminathan, who – building on the success of the coconut wilt-root studies of 1968 – asked ISRO to study the area under paddy cultivation in a district in Andhra Pradesh and a district in Punjab (Swaminathan 2015). Accordingly, two survey experiments using infrared aerial photography were undertaken by ISRO together with the Indian Agriculture Research Institute (IARI) between 1974 and 1976: the Agricultural Resources Inventory and Survey Experiment (ARISE) at Anantapur in Andhra Pradesh and the Crop Acreage and Production Estimation (CAPE) at Patiala in Punjab. The two surveys gave very useful information and more importantly demonstrated that remote sensing technology could be used to assess yield, measure the acreage under forest, determine soil types and moisture levels and quantify different crops and the extent of irrigated area (Sheehan 2007; Navalgund 2015).[7]

During the same period, efforts were made for the indigenous development of an airborne thermal scanner to measure sea surface temperature (SST) and to monitor the spread of disease in crops. Thanks to an agreement with the United States, ISRO also began to use the data from Landsat-1 and Landsat-2 Earth resource satellites (launched in 1972 and 1975) in a number of demonstration studies carried out in cooperation with other governmental organisations (Harvey et al. 2010). In 1978, ISRO also began the construction of a data reception centre at Shadnagar, near Hyderabad, to receive the data directly from Landsat.[8] The following year the centre became the base of the Indian National Remote Sensing Agency (NRSA), the establishment of which acted as a catalyst for the use of satellite remote sensing in a significant way by various central and state government departments.

It was around this time that ISRO realised the need to involve stakeholders, right from the beginning at the planning level, to develop an interface with user agencies in the government and make the programme user-driven (Navalgund 2015). Eventually, in order to make remote sensing data regularly available to the different ministries/departments at various levels, it was deemed necessary to create a national system of resource management in which remote sensing technology would be integrated with traditional techniques, integrating space technologies with conventional

---

[7] The procedures used for CAPE were subsequently revised and upgraded to improve the accuracy and timeliness of crop estimates. The wide swath coverage and the ability to quantify different crops eventually led to the national-level crop forecasting programme for a number of crops including sugar cane, potato, cotton, jute, mustard, sorghum, etc. (Harvey et al. 2010).

[8] Although the data of Landsat proved very useful, it was limited by the sporadic coverage of India, given the nature of the satellite's orbital trajectory (Moltz 2012, p. 116).

techniques. To this end, a preparatory committee under the chairmanship of the Planning Commission was constituted in 1982 and "59 well-defined experiments were conducted to demonstrate the end utilisation of remote sensing in various application areas", including groundwater targeting, mineral exploration and fisheries (Navalgund 2015). Building on the successful conclusion of these experiments, the Natural Resource Management System (NRMS) was eventually established, with DOS acting as its nodal agency and nine standing committees under the chairmanship of secretaries of major user ministries/departments addressing the application needs and the use of remote sensing data in specific areas (Bhatia 2016).[9]

In addition, a widely extended infrastructure was created to promote the use of remote sensing data for specific regional and local issues. Starting in 1985, DOS established Regional Remote Sensing Service Centres (RRSSCs) at Bangalore, Dehradun, Kharagpur, Jodhpur and Nagpur to provide digital image processing facilities to a larger segment of users in various regions of the country (Navalgund 2015). Thus, under the NRMS "a national organizational structure with the user ministries/departments taking lead in the applications of remote sensing imagery was created and a national infrastructure of Regional Service Centres was created to meet the local needs of the states" (Bhatia 2016).

The 59 experiments initiated under the NRMS grew to some 160 projects in a vast area of applications, including wetland inventory, geology and mineral exploration, coastal zone management, urban mapping, atmospheric and ocean studies, cyclone monitoring, disaster management support, etc. Over time, this list would only continue to grow (Bhatia 2016).

2.2.2 India's First Satellites

Parallel with the testing of satellite-based applications, the experimental phase during the 1970–1980s also saw end-to-end capability demonstration "in the design, development and in-orbit management of space systems together with the associated ground systems for the users".

Remote Sensing Satellites
Planning for the first satellite began in the early 1970s. Although it was clear that it would take some time before India could mature indigenous launch capacities for its satellites, ISRO held that it would be possible to develop an Indian satellite earlier and have it launched by one of India's partners at Thumba (Emerging Space Powers 2010). Indeed, in May 1972, an agreement was reached with the Soviet Union. The agreement provided "for the launch by the USSR of an Indian satellite and for Soviet use of Indian ports by tracking ships and vessels launching sounding

---

[9] The nine standing committees cover the areas of agriculture and soils, bio-resources, geology and mineral resources, water resources, ocean resources and meteorology, cartography and mapping, urban management, rural development and training and technology.

rockets" (Harvey et al. 2010). As noted by Sheehan, "while this clearly demonstrated a reliance on a major external power, the cooperation with the Soviet Union was also a way of balancing the earlier cooperation [agreement for SITE] with the United States, an important consideration given that India was one of the leading states in the Non-Aligned Movement" (Sheehan 2007, pp. 146–147).

Named after the fifth-century Indian astronomer and mathematician Aryabhata, India's first satellite was conceived as a test-bed for the development of indigenous satellite technology. It was designed and fabricated by a team of 200 people led by Prof. U.R. Rao.[10] Although ISRO initially thought that the payload would carry television, telephone and remote sensing equipment, the weight of such equipment eventually pushed ISRO to opt for lighter scientific instruments developed by the Physical Research Laboratory (PRL).

Aryabhata was dedicated to the study of stellar X-rays, solar physics and radiation in the Earth's ionosphere. It was launched aboard a Cosmos-3M rocket from Volgograd Launch Station on 19 April 1975, but its solar electric system ceased operating after only 41 orbits. This failure, however, was helpful for ISRO scientists in redesigning their second satellite, *Bhaskara*, for which they developed "a more sophisticated instrumentation, including an electrical system that benefited from 3500 Soviet solar cells" (Moltz 2012).

The satellite was named after a seventh-century Indian astronomer. Although it was conceived as a back-up model to Aryabhata and was hence very similar in its general configuration, Bhaskara-1 carried remote sensing experiments and is considered the first Indian Earth observation (EO) satellite. It had two types of sensor systems, namely, two television cameras with a resolution of 1 km aimed at examining changes in water and vegetation and three microwave radiometers to study the oceans around India and obtain information on sea surface temperature, ocean and wind velocity moisture content (Rao 2015). The satellite also carried a data collection and relay package designed to receive data from eight meteorological stations and retransmit them to the central receiving station in Ahmedabad.

Bhaskara I was launched on 7 June 1979 by a Soviet Intercosmos rocket. Although this satellite also faced some technical problems that limited its functionality, it provided valuable information on agriculture, weather and vegetation, including the "snow melting in the Hymalays, river flooding in Northern India, desertification in Rajasthan, rainfall off the coast of India and mineral resources in Gujarat" (Harvey et al. 2010).

The follow-up satellite, *Bhaskara II* (launched on 20 November 1981), provided similar information but had a new channel for the improved estimation of atmospheric and oceanographic parameters. It was the most successful of the three satellites launched by the Soviet Union and returned the most data (Navalgund 2015). In addition, it was a major contribution towards the establishment of indigenous capacities to build operational remote sensing satellites.[11]

---

[10] For an insightful account of the story of Aryabhata by U.R. Rao himself, see Rao (2015).

[11] As Navalgund recounts "with the successful completion of the Bhaskara programme, the capability to build operational satellites for remote sensing was well established and this in conjunction

An equally important contribution was provided, during the same years, by the development of the first series of domestically launched satellites: the Rohini. These 35–40-kg multisided capsules with a power handling capability of just 16 W were primarily designed as technology demonstrators to test Indian-made solar cells and to monitor the flight performance of SLV-3, the first Indian launch vehicle (see Sect. 2.3). However, they also carried instruments for Earth observation.

The first Rohini Technology Payload (RTP) was launched on-board SLV-3 on its maiden flight from SHAR Centre on 10 August 1979. However, the satellite could not be placed into its intended orbit, due to a launch failure. The follow-up Rohini Satellite (RS-1) was instead successfully launched aboard the SLV-3 on 18 July 1980 into a low Earth orbit (LEO) of 305×919 km with an inclination of 44.7°. As reported by ISRO, "all the fourth stage parameters of SLV-3 were successfully tele-metered to the ground stations by RS-1 during the launch phase" (Indian Space Research Organisation 2015a, b, c, d). The satellite had an orbital life of 9 months, and, even if the SLV-3 payload capacity limited the instrumentations available to it, the television camera and temperature sensors of RS-1 were operated to provide three complete sets of India (Harvey et al. 2010).

The second Rohini satellite (RS-D1), also a 38-kg experimental spin-stabilised satellite designed with a power handling capability of 16W, was launched on-board SLV-3 from SHAR Centre on 31 May 1981. The satellite carried a solid-state cam-era using a linear array of detectors for remote sensing applications, but it was only a partial success, as it was unable to reach the intended height and stayed in orbit for only 9 days (Indian Space Research Organisation 2015a, b, c, d).

The launch of the third Rohini (RS-D2), on 17 April 1983, proved the most rewarding. Its smart sensor camera, which had on-board processing capability for classifying ground features such as water, vegetation, bare land, clouds and snow, "sent more than 5000 pictures frames in both visible and infra-red bands for identi-fication of features and demonstrated the technique of determining altitude and orbit using images" (Indian Space Research Organisation 2015a, b, c, d).

Overall, both the Rohini and Bhaskara satellite programmes – together with the experience gained through the joint experiments carried out by the NNRMS – were pivotal steps towards the establishment of the future Indian Remote Sensing (IRS) satel-lite programme, which was activated in 1988 with the launch of the first IRS satellite.

Telecommunication Satellites

During the same years, however, another major breakthrough for the operationalisation of indigenous satellite systems was achieved: the development of the Ariane Passenger Payload Experiment (APPLE), India's first experimental communication satellite.

The programme was initiated pursuant to a proposal made in 1977 by the European Space Agency (ESA) for the provision of an experimental satellite to be flown aboard the third qualification flight of Ariane-1 in June 1981. As highlighted by Harvey et al., this was seen as "a major opportunity to develop the experience

---

with the experience gained through JEP laid the foundation for the Indian Remote Sensing Satellite Programme" (Navalgund 2015).

necessary for controlling satellites in 24-h [geostationary] orbit, and later in constructing a domestic communications satellite. Already, Indian space planners were thinking ahead to a domestic-built comsat" (Harvey et al. 2010, p. 186).

Designed and built in industrial plants with reduced infrastructure over a period of just 2 years (1978–1980), APPLE integrated a C-band communication payload compatible with the infrastructure developed for the Satellite Telecommunication Experiments (STEP) and was intended to carry out advanced satellite communication experiments beyond those conducted using ATS-6 and Symphonie satellites (Vasagam 2015).

On 19 June 1981, the Ariane-1 rocket successfully launched APPLE into a geosynchronous transfer orbit (GTO), and thanks to an apogee motor derived from the fourth stage motor of SLV-3, it was then boosted into geosynchronous orbit (GSO). On that date, India became the fifth country with the technology for operating a geostationary satellite, after the superpowers, France and Canada (Sheehan 2007). As an indication of the importance of this feat, then Prime Minister Indira Gandhi dedicated the satellite to the nation on 13 August 1981 (Vasagam 2015).

However, beyond its symbolic relevance, APPLE also accrued a plethora of important tangible benefits. For one, it gave ISRO valuable hands-on experience in designing and developing a three-axis stabilised geostationary communication satellite as well as knowledge in in-orbit raising manoeuvres, in-orbit deployment of appendages, station keeping, etc. (Indian Space Research Organisation 2016a, b, c). In addition, it enabled experimentation with advanced communications technology for a number of purposes, including the testing of emergency communications for cyclones, the tracking of Indian railway wagons and the first telemedicine projects. APPLE was extensively used for over 2 years "to carry out extensive experiments on time, frequency and code division multiple access systems, radio networking computer interconnect, random access and pockets switching experiments" (Indian Space Research Organisation 2016a, b, c). As such, it became the forerunner for the eventual establishment of the planned Indian National Satellite System (INSAT).

The development of INSAT – a joint project between the Department of Space, the Department of Telecommunications, All India Radio and the Meteorological Department – had been envisaged since 1976 with the aim of combining communications and weather services for the country in a single system.

The programme was approved by the government in 1977. However, given that India did not have any know-how to build such a system by itself,[12] it was clear that not only the launch but also the construction of the two INSAT-1 satellites had to be procured from abroad, although with the ultimate intention of building the following series in India itself. This was consistent with the roadmap delineated 10 years earlier by Vikram Sarabhai. The Space Commission contracted the two satellites of the planned INSAT-1 series to the US company Ford Aerospace and Communications Corporation and then signed an agreement with NASA for their launch, despite the free assistance offered by ESA and CNES for the development of APPLE and its free ride aboard Ariane-1.

---

[12] This know-how would only be acquired after the APPLE missions a few years later.

As reported by Harvey, Smid and Pirard, "INSAT 1 was built with 12 television transponders, two TV direct broadcasting antennae and, for weather forecasting, a Very High Resolution Radiometer to image the Earth every 30 minutes. The radiometer had a resolution of 2.5 m in the visible band and 10 km in the infrared. The TV transponders could reach up to 100,000 small Earth terminals, or, instead of transmitting television, the system could handle up to 8,000 telephone calls at a time or be used for radio. [...Hence, it] was intended to make a big difference, for, in 1982, India had only 12 television transmitters" (Harvey et al. 2010).

INSAT-1A was launched on 10 April 1982 on a Delta rocket instead of the Space Shuttle, because of the various delays in its entry into service. While the satellite was successfully placed into orbit and began to provide weather and communications services in May 1982, it subsequently encountered thermal management problems due to non-deployment of the solar sail, which curtailed the S-band transponder operation time. The non-deployment of the solar sail also "resulted in yaw build-up that eventually led to a catastrophic failure of the satellite in September 1982" (Kale 2015).

The second satellite, INSAT-1B, was launched from the US Space Shuttle *Challenger* on 31 August 1983. Unlike INSAT-1A, the mission was a total success, with the satellite remaining operational beyond the designed 7-year life and extending television coverage to 70% of the Indian population (Harvey et al. 2010; Kale 2015).

Although the INSAT-1 series had originally been conceived as a two-satellite system, two more satellites were eventually commissioned, because of the "much higher than expected demand on satellite television" (Harvey et al. 2010). The first, INSAT-1C, was initially planned to be launched by an Indian astronaut on a Space Shuttle mission, as part of a cooperative undertaking initiated by the Reagan administration to counterbalance India's participation in the USSR human spaceflight programme. Following the Challenger disaster, however, India decided to relocate the satellite onto Europe's Ariane rocket instead of a Delta rocket, hence contributing to the deterioration of the progressively lukewarm United States–India relations (Moltz 2012). Ariane successfully launched INSAT-1C on 21 July 1988, but within a few days after the launch, the satellite lost half of its power because of a power bus failure. As a result, only a part of the telecommunications systems could be used (Kale 2015). In addition, in November 1989, the satellite command receiver encountered a temperature drift problem, and the satellite had to be abandoned. Consequently, India had to fill the gap by leasing transponder capacity from the Soviet Intersputnik system. Fortunately, the last satellite of the series, INSAT-1D, turned out to be the most successful of the 4: it remained operational for 5 years and sent back 24,500 weather images and relayed television to some 2000 community receiving sets (Harvey et al. 2010) (Table 2.1).

Table 2.1 India's first satellites

Name	Date	Weight (kg)	Power (W)	Launcher	Orbit	Type
Aryabhata	19 Apr 1975	360g	46	C-1 Intercosmos	LEO	Experimental
Bhaskara I	07 June 1979	442g	47	C-1 Intercosmos	LEO	EO, experimental
RTP	10 Aug 1979	35g		SLV-3E1		EO, experimental
RS-1	18 July 1980	35	16	SLV-3E2	LEO	EO
RS-D1	31 May 1981	38	16	SLV-3D1	LEO	EO
APPLE	19 June 1981	670	210	Ariane-1 (V-3)	GSO	Comm, experimental
Bhaskara II	20 Nov 1981	444	47	C-1 Intercosmos	LEO	EO, experimental
RS-D2	17 Apr 1983	41,5	16	SLV-3	LEO	Earth observation

Source: (ISRO)

2.2.3 Access to Space: Moving Towards Self-Reliance

Together with satellite systems and satellite applications, a third major line of activity during the experimental phase of the 1970s and 1980s aimed to lay the foundation for India's future operational launch vehicles. Even if the country's satellite programmes and application experiments could take place without independent access to space, India did not want to depend on other countries for too long.

Following the transfer of solid propulsion technology in the 1960s, in 1972, ISRO also signed an agreement with French company Société Européenne de Propulsion (SEP) for the technology transfer of liquid propulsion systems. Under the agreement, ISRO acquired the technology of the Viking liquid engine that was being developed for the Ariane launch vehicle programme, and in return it provided SEP with the services of 100 man-years of ISRO scientists and engineers for its work on the Ariane development (Prasad 2015).[13] The Viking engine, subsequently

---

[13] As detailed by Blamont "to acquire the Viking engine technology, ISRO engineers worked in all areas of development activities of the Ariane programme. They participated in design reviews, progress reviews and even had interaction with European industries. They received all detailed design drawings and documents, and participated in inspection and quality assurance of systems, subsystems and components. They were also part of assembly and integration, checkout and testing operations in SEP facilities. They had discussions with SEP specialists and received clarifications to understand the technology fully. Some 40 engineers, working under a five-year contract, participated in the technology acquisition program at Vernon and Brétigny in France". In 1980 the ISRO Chairman created a Liquid Propulsion Project (LPP) which organised three teams under the leadership of three SEP trained experts – the first team developed the system, the second was tasked to realise all hardware in India in association with Indian industries, and the third team was to establish all development facilities at Mahendragiri (Blamont 2017).

indigenised under the name "Vikas", had a great impact on the development and very configuration of India's future launchers.

However, throughout the 1970s, ISRO preferred to leave liquid propulsion on the slow burner and to focus instead on the development of a launch vehicle powered by solid propulsion in all its stages: the Satellite Launch Vehicle (SLV). As recounted by Gupta, this was not only because it was easier to develop solid propellant motors than to develop liquid engines but also because "the whole ambience at that time was suffused with solid propellants!" (Gupta 2015). For one thing, all the sounding rockets being launched from TERLS were using solid propellants. In addition, Hideo Itokawa, the father of the Japanese space programme and a close friend of Vikram Sarabhai, had invited Indian engineers to the Institute of Space and Aeronautical Science (ISAS) of Tokyo to gain first-hand knowledge of the Japanese launcher programme, all of which used solid propellants. Most importantly, Sarabhai had already arranged for the indigenous manufacture of the French Centaure rockets, production of which had started as early as 1969.[14] There are, however, also allegations that the real reason for India developing launch vehicles using solid propellants was to develop intermediate-range ballistic missiles (IRBMs), because such missiles could be stored and deployed at short notice (Gupta 2015).

The development of India's first indigenous launcher, named Satellite Launch Vehicle (SLV), began in 1973 under the lead of Abdul Kalam, later president of India. As he recently recalled (Kalam 2015):

> Realisation of SLV-3 from concept to development through design was a challenge in both technology and management. Equally important was building the infrastructure and human resources in the areas of propellant, propulsion, avionics, materials, motor testing, vehicle assembly, vehicle checkout and ground telemetry/tracking. With the industry in the country in a very nascent stage to take up such sophisticated technology, harnessing them was a unique effort by itself. ISRO establishments in the country were geared to these mammoth efforts through time-bound projects. Even as centres of excellence in rocketry were getting evolved in ISRO centres, specifically in VSSC and SHAR (Sriharikota Range), partnering with national laboratories and academic institutions grew rapidly.

Largely inspired by the American Scout missile, the SLV was designed as a four-stage rocket with a take-off weight of 17 tonnes capable of launching a 40-kg-class satellite into an orbit of 300–900 km. The SLV had approximately 10,000 components, 85% of which were domestically manufactured. Several academic institutions and industries provided important support to the work of the ISRO centres, as did foreign partners such as the United States, Japan, France and West Germany.[15]

---

[14] The propellant plant of the Centaure rocket was set up in Thumba in 1968 and the first indigenous Centaure rocket launched on February 1969. The propellants were successively indigenised and used for the development of the two-stage *Rohini* sounding rocket, which was, in fact, very similar to the French Dragon (a more powerful version of Centaure). The first all-Indian Rohini rocket was launched on 27 January 1973, reaching an altitude of 350 km with a payload of 150 kg (Blamont 2017).

[15] Indian scientists visited the launch base of the Scout at Wallops Island, Virginia and used the Porz Wahn wind tunnel in Cologne to simulate various altitudes and pressures (Harvey et al. 2010).

The development of the SLV also required the construction of an appropriate launch base. A new site was chosen on the southeast coast of India: the remote island of Sriharikota. The centre, named SHAR (acronym for Sriharikota Range, but also meaning arrow in Sanskrit), was equipped with the entire infrastructure required for flight testing the SLV-3 rocket, test facilities for the evaluation and ground qualification of solid propellant rocket motors and other subsystems, a plant for manufacturing solid propellant, a mission control centre and range instrumentation for telemetry, tracking, data reception and commanding (Narayana 2015).[16]

Originally, ISRO planned the first launch for 1978, the 30th anniversary of India's independence, but the first attempt was eventually made on 10 August 1979 from the new launch site at Sriharikota, with the satellite being the small Rohini Technology Payload (RTP). The rocket, however, crashed into the Bay of Bengal a few minutes after the launch.

Thanks to the incorporation of recommendations made by the Failure Analysis Committee, the SLV-3's second attempt on 18 July 1980 was successful: India became the sixth country to successfully launch a satellite using its own launch vehicle.

Two other launches of the SLV-3 were conducted (on 31 May 1981 and on 17 April 1983, respectively) before ISRO embarked on the development of a new launcher capable of putting a 150-kg-class payload into orbit. While ISRO initially wanted to proceed directly to a much larger launcher capable of placing a tonne into orbit, the construction of an intermediate rocket was seen as a necessary step for demonstrating and validating critical technologies for future launch vehicles (Harvey et al. 2010). This intermediate rocket was the Augmented Satellite Launch Vehicle (ASLV).

In the meantime, the SLV was adapted for the development of the Agni intermediate-range ballistic missile (IRBM). In this respect, it is of interest to note that "unlike in most countries, India's technology transfer moved from the civil side to the military side, rather than the reverse" (Moltz 2012, p. 119). Indeed, it has been widely reported that the Defence Research and Development Organisation (DRDO) largely borrowed human resources and technology from ISRO from the development for its missile programme, including "the wholesale transfer of the SLV-3's project leader, Abdul Kalam, to DRDO, whose work succeeded in 1989 in bringing about the flight of the first Agni-1 ballistic missile" (Ibid, p. 119). As explained in the next section, this would, however, have major repercussions for the subsequent development of cryogenic engine technology by India.

Like the SLV, the ASLV was configured as a multistage, all solid-fuel vehicle, but with the addition of two strap-on boosters and equipped with control and guidance

---

[16] A Radar Development Project (RDP) was undertaken for "the development and realisation of two C-band medium-range radars for installation at SHAR. The project was successfully completed with Radar I commissioned in the year 1977 and Radar II in 1978. These radars formed the backbone of the tracking system for many years. Radar II continues to be in operation at SHAR" (Narayana 2015). A dedicated unit named ISRO Telemetry, Tracking and Command Network (ISTRAC) was subsequently created at SHAR for operating telemetry stations at SHAR, Trivandrum, Port Blair, Mauritius, Brunei and Biak. In 1986, ISTRAC shifted its headquarters to Bangalore (Ibid).

Table 2.2 India's first indigenous launches

Name	Date	Payload	Orbit	Outcome
SLV-3E1	10 Aug 1979	RTP	LEO	Failure
SLV-3E2	18 July 1980	RS-1	LEO	Success
SLV-3D1	31 May 1981	RS-D1	LEO	Success
SLV-3	17 Apr 1983	RS-D2	LEO	Success
ASLV-D1	24 Mar 1987	SROSS-1	LEO	Failure
ASLV-D2	13 July 1988	SROSS-2	LEO	Failure
ASLV-D3	20 May 1992	SROSS-C	LEO	Success
ASLV-D4	05 May 1994	SROSS-C2	LEO	Success

Source: (ISRO)

systems to orbit a satellite. With a lift-off weight of 40 tonnes, this 24-m tall rocket augmented the payload capacity to 150 kg – three times more than the SLV – for a low Earth orbit (Indian Space Research Organisation 2016a, b, c).

Since the ASLV was intended as an intermediate rocket, only four of them were commissioned. Just like the SLV, the first launch of the ASLV, on 24 March 1987, was a failure. The second launch, on 13 July 1988, also failed, necessitating the introduction of some modifications to make it less vulnerable to instability. Incorporating these modifications took almost 4 years, but the third flight of ASLV, on 20 May 1992, was successful in placing a 106-kg SROSS-C satellite into orbit. The fourth ASLV, launched on 4 May 1994, was also successful, perfectly placing into orbit the SROSS-C2 satellite (Indian Space Research Organisation 2016a, b, c).

Overall, while through the SLV-3 programme, "competence was built up for the overall vehicle design, mission design, material, hardware fabrication, solid propulsion technology, control power plants, avionics, vehicle integration checkout and launch operations", the more complex ASLV "demonstrated newer technologies such as the use of a strap-on, bulbous heat shield, closed loop guidance and digital autopilot". This paved the way for learning many nuances of launch vehicle design for complex missions, leading to realisation of operational launch vehicles such as PSLV and GSLV (Indian Space Research Organisation 2015a, b, c, d) (Table 2.2).

2.3 Operating the Programme (1990s)

Between the late 1980s and the early 1990s, all the three major components of the programme had been successfully tested. The SLV and ASLV rockets had enabled India to achieve independent access to space, while the first satellite programmes and related application development had ripened Indian skills in both the field of telecommunications and Earth observation. The programme could hence enter the operational stage and actively pursue the goal for which it had been originally conceived: leverage space technology as a tool for India's socio-economic development. Towards this, from the 1990s, ISRO deployed two major systems of space infrastructure: one for communication, broadcasting and meteorology through the multipurpose Indian National Satellite (INSAT) System and the other for Earth observation through the Indian Remote Sensing (IRS) satellite system. The development and operationalisation of India's Polar Satellite Launch Vehicle (PSLV) and the development of the Geosynchronous Satellite Launch Vehicle (GSLV)" – necessary components for the deployment of the INSAT and IRS systems – were also realised during this phase.

In stark contrast to the experimental phase of the 1970s and early 1980s, during which India greatly benefited from international cooperation with all the major spacefaring nations, this phase was characterised by the emergence of major international obstacles, particularly in the field of launch technology. Indeed, as India began to progress in its missile programme by drawing on the experience and technologies gained through its civilian rockets, the United States and other Western countries backed away from previous cooperative stances (in space and elsewhere) for fear of aiding the development of a long-range ballistic missile for nuclear weapons.

Also on the domestic front, India started to face an increasing number of socioeconomic problems, which in a sense were the inevitable by-product of India's state-led socialist model. In effect, the widespread Ghandian belief in self-reliance and Nehruvian socialism had left India's economy weak and ill-equipped to deal with the incipient globalisation tendencies, resulting instead in economic stagnation and isolationism. Against this backdrop, the new Congress-led government of Narasimha Rao, which had entered office in 1991, introduced a series of major economic reforms, including the cutting of state spending and the privatisation of several state corporations, which would also have an impact on the space programme. As James Clay Moltz observes, "the thrust of these reforms placed a new emphasis on privatising technologies developed by ISRO, leading the Department of Space to create the Antrix Corporation for this purpose in September 1992" (Moltz 2012).

2.3.1 Two Operational Satellite Systems: INSAT and IRS

Introducing the INSAT-2 System

In the field of satellite telecommunications, India worked towards meeting its goal of developing a successor satellite system for the first INSAT series, which had been procured by the Ford Aerospace Corporation and launched by either European or American rockets. For the INSAT-2 series, the plan was to build the satellites in India itself, with some limited assistance from abroad, although foreign launch vehicles would still be required. Consistent with the Sarabhai roadmap, ultimately the idea was that future series would be launched aboard Indian geosynchronous rockets.

The preliminary studies for the INSAT-2 system were completed in 1981, and the programme was approved in 1985, with an in-service date set for 1992. The four geostationary satellites forming the INSAT-2 series were designed as multifunctional satellites carrying both communications and meteorological payloads. The communications payload included a package for the provision of television programmes, telephone, fax, data and telegraph, among other services. What made the configuration and overall design unique, however, was the accommodation of a very high-resolution radiometer (VHRR) as a meteorological payload to derive pictures of cloud-cover imageries with an accuracy of 2 metres per second. The satellites were also designed to collect weather information from several hundred data collection points on the ground and activate warning sirens in selected villages under the threat of impending cyclones.

The faultless Ariane-4 was chosen as the reference launcher for the INSAT-2 series. INSAT-2A was launched on 10 July 1992 and began operations in August of that year, while INSAT-2B was launched just 1 year later, on 23 July 1993. As recalled by P.S. Goel, "INSAT-2A and 2B were initially called INSAT-2 TS (TS standing for Test Satellites). They were not meant to be operational satellites. They were more for technology demonstration and were not part of any planned capacity for communication transponders. However, there was such a sudden increase in the demand for transponder capacity in the early 1990s that not only these two satellites were declared operational soon after launch, but ISRO also was compelled to plan for INSAT-2C and 2D in quick succession. To increase the number and power of the transponders of these two satellites, the VHRR was removed" (Goel 2015). As a result, they became dedicated communication satellites. The same was the case with the just-started INSAT-3 series, in which 3A and 3B were multipurpose satellites, while 3C and 3D had communications payload only.[17]

However, thanks to the success of the PSLV launcher (see next section), ISRO would eventually decide to develop an exclusive meteorological satellite, initially called METSAT. Launched on 12 September 2002 and renamed KALPANA-1 in February 2003 after the death of the Indian-born American astronaut Dr. Kalpana Chawla in the US Space Shuttle Columbia disaster, this first all-meteorology satellite accommodated a VHRR and a data relay transponder (DRT) as main payloads and was integrated in a dedicated 1-tonne-class spacecraft bus built by ISRO for PSLV launches to GTO.[18] In 2004, ISRO also launched a geostationary satellite called EDUSAT, built exclusively for serving the educational sector.

Over the course of the previous 20 years, the multipurpose INSAT system became extensively used for a number of applications, enabling the pervasive and ubiquitous development of direct-to-home (DTH) broadcasting, very small aperture terminal (VSAT) network and mobile satellite services (MSS), in addition to societal-oriented services such as telemedicine, tele-education, Village Resource Centres (VRC) and Disaster Management Support (DMS) programmes. A more detailed overview of these applications is provided in Chap. 4.

---

[17] INSAT-3B was the first of the series to be launched – on 22 March 2000 – followed by INSAT-3 on 24 January 2002, INSAT-3A on 10 April 2003 and INSAT-3E on 28 September 2003. All these launches were performed by Ariane-5 from the Guiana Space Centre in Kourou.

[18] Four types of spacecraft bus would be built by ISRO for its communication satellites: I-1K (KALPANA), I-2K (INSAT-2), I-3K (INSAT-3) and more recently I-4K for GLSV Mark III rocket (see Chap. 3).

Initiating the IRS System

The INSAT system was one of the two operational classes of satellites established by ISRO between the late 1980s and the mid-1990s: the other was the Indian Remote Sensing (IRS) satellite system. From the onset, the IRS was conceived as an integral pillar of the national system for managing natural resources in India, the NNRMS, and hence for supporting the work of agriculturalists, geologists and hydrologists in estimating crop yields, combating droughts, mapping wasteland, managing water resources, etc.

Drawing on the experience gained with the Bhaskara and RS-1 satellites, the first series of IRS satellites was designed to "weigh up to a tonne, cross the same stretch of Earth every 21 days and carry out systematic survey of the Earth's surface" (Harvey et al. 2010). The bus could accommodate different kinds of remote sensing payloads and operate for a period of 3 years thanks to 16 small thrusters with 80 kg of hydrazine fuel.

The first series comprised two satellites, IRS-1A and IRS-1B, carrying multi-spectral cameras with a spatial resolution of 72.5 and 36 m, respectively, which were synchronised to provide 11-day revisits. Since the satellites were too heavy to be flown aboard the ASLV, they were launched by a USSR Vostok rocket, respectively, in March 1988 and August 1991. During its 6 years of operations – twice its design life – IRS-1A made more than 50 complete maps of India and took part in a number of remote sensing experiments. As for IRS-1B, it provided India with a devoted stream of data for natural resource management, particularly crop forecasting.

During the operational lifetime of IRS-1A and IRS-1B, the IRS second generation (IRS-1C and IRS-1D) was developed and then successfully launched by Russian rockets between 1995 and 1997. The two satellites offered improved spatial resolution, stereo viewing and frequent site revisits, making them "the most advanced remote sensing satellite in the world at that time". This second generation saw the introduction of the Linear Imaging Self-Scanning Sensor (LISS), which "operated in the visible and infrared wavebands using French-made charge-couple device scanners, with a panchromatic camera able to provide six metres resolution from 600 km, then thought to be the highest available in the world" (Harvey et al. 2010, p. 176).

A clear indication of the quality of IRS data is the fact that by the late 1990s, the American company EOSAT and the Japanese National Space Development Agency (NASDA – now Japan Aerospace Exploration Agency, JAXA) had applied to receive and process IRS data. Internally, the IRS data products were used for the definition and implementation of a national Geographical Information System (GIS), in addition to applications covering urban planning, rural development, mineral prospecting, environment, forestry, ocean resources and disaster management.

With the eventual introduction of the PSLV, India's Earth observation programme further witnessed a significant expansion, with almost yearly launches. It also became more specialised, with dedicated satellites for oceanography, cartography and resources. In outlining India's strategy through 2025, the then ISRO Chairman Krishnaswamy Kasturirangan (1994–2003) paid specific attention to the launch of remote sensing satellites, including spacecraft such as Oceansat (ocean colour mon-

itor and a multi-frequency scanning microwave radiometer for oceanographic studies), Cartosat (with improved high-resolution imaging and mapping capabilities),
Resourcesat (with an advanced multispectral camera), RISAT (for radar-based
assessments of crop and moisture levels) and SCATSAT (with a scatterometer for
wind readings), Megha-Tropiques (a joint Indo-French mission to map atmospheric
distribution of water vapour, clouds, rainfall and water evaporation in three dimensions in the tropical belt) and SARAL (a second Indo-French mission dedicated to
altimetry). A more detailed account of these missions is provided in subsequent
chapters, for now suffice it to highlight the steady growth of Earth observation
capacities in India, which can today boast one of the largest constellations of remote
sensing satellites in operation.

   All in all, during the operational phase of the 1990s, India's capacity to develop
and operate satellite system greatly matured, fully matching the level of the other
main actors. However, as P.S. Goel also points out, "[whereas…it is appropriate to
say that ISRO has mastered the technology of building and launching operational
satellite buses for remote sensing and communication [...it is undeniable that] the
expertise, the knowledge base and the competence have largely remained internal to
ISRO, without percolating to Indian industry [and hence without improving] the
general competence of the country" (Goel 2015).

2.3.2   India in the Big Rocket Club: The PSLV and GSLV

The 1990s also saw the introduction of India's first operational launch vehicle, the
Polar Satellite Launch Vehicle (PSLV), which substantially drew on the experience
gained in the development and launches of the SLV and ASLV. Even though the
ASLV could launch a payload of 150 kg into LEO, it had always been considered as
an intermediate functional step towards the development of more capable rockets
that could allow the country to place its much larger satellite systems first in LEO
and eventually also in GEO, thus removing India's dependency on foreign
launchers.

   In its configuration, the PSLV was roughly ten times heavier than ASLV, having
a weight of 275 tonnes and a height of 44 m. It was designed for placing 1000-kg
remote sensing satellites into 900-km sun-synchronous polar orbit (SSPO) and
more specifically for the IRS series, which had been so far launched by the Soviet
Union (Harvey et al. 2010).

   While its development was approved as early as 1982 – long before the first
qualification flight of the ASLV – it took more than a decade "to design, develop and
demonstrate its flight-worthiness. This was because of the large number of new
technologies involved which in turn called for the establishment of a large number
of state-of-the-art facilities, both within ISRO and in the Indian industries"
(Narayanamoorthy 2015).

   Among these new technologies, the PSLV marked the eventual introduction of
liquid propellants in Indian rockets, making an unprecedented combination of solid

and liquid-fuelled engines (solid-liquid-solid-liquid).[19] More specifically, while the first and the third stages employed solid propellants, the second and fourth stages had liquid-fuel engines. Notably, the second stage was powered by Vikas, the indigenised version of the engine obtained in 1972 through the technology transfer agreement with SEP, the Viking. Tested in 1985 in SEP laboratories, the Vikas greatly impacted on the very configuration of the PSLV and remains a crucial staple of India's launch technology to this day, thus giving Ariane-4 a "second lease of life" (Bouhey-Klapisz 2017).

The fourth stage was, by contrast, a completely indigenous effort, realised with no external help. It had two identical engines burning storable liquid fuels – oxide of nitrogen and monomethyl hydrazine – and propellant tanks built of titanium. It took more than 7 years to develop this stage, which was eventually test-fired in 1990.

Like its two predecessors (the SLV and ASLV), the PSLV failed its maiden flight on 20 September 1993. The second flight, on 15 October 1994, was, however, successful, eventually marking India's graduation into the big launcher league. With the third launch, in 1995, the PSLV's development phase was completed.

Operational launches began in 1997, and since then, the PSLV has been increasingly deployed for numerous missions in LEO, SSPO, GTO and lunar orbit, earning the title of "workhorse of ISRO". Systematic improvements in propulsion systems, avionics and structures have in addition enabled enhancement of its payload capacity from 1000 to 1600 kg in polar orbit (Narayanamoorthy 2015). A more detailed account of this launch vehicle and of its various configurations is provided in Chap. 3. For now it suffices to highlight that by the turn of the century, the PSLV also started to earn a certain reputation on the commercial launch market for its reliability and relatively low-cost launch services. Following the first commercial contracts for launching South Korea's Kitsat-3 and Germany's TubeSat, commercial flights have been carried out by the PSLV for an increasingly long list of clients, including Argentina, Belgium, Canada, Denmark, France, Germany, Indonesia, Japan, the Netherlands, Singapore, South Korea, Switzerland and Turkey.

The GSLV Programme

The introduction of the PSLV certainly marked an invaluable step forward for the future advancement of the Indian space programme. However, to achieve full-spectrum capabilities, India still needed to develop a launcher capable of placing its INSAT communication satellites in GEO, so as to end the country's dependence on foreign launch service providers. The first 10 INSAT satellites, in effect, would be all launched aboard foreign rockets – seven aboard Europe's Ariane, two aboard the Delta and one on the Space Shuttle.

The plans for the development of this new launcher – named Geosynchronous Satellite Launch Vehicle (GSLV) – were approved as early as 1987. According to the configuration studies, the GSLV would be a three-stage launcher comprising one solid rocket motor stage derived from the PSLV first stage, one storable liquid

---

[19] In effect, until that time, rockets had been either solid- or liquid-fuelled or alternatively liquid-fuelled with small strap-on boosters.

stage using the Vikas engine and one cryogenic stage, in addition to four liquid-fuel engine strap-ons using one Vikas engine each. While much of the technology was designed to employ heavier derivatives of the PSLV rocket, the planned introduction of a 12-tonne cryogenic, liquid-hydrogen powered engine for the third stage would prove a major technological – and indeed highly political – feat for India. According to Indian engineers, it would have taken up 15 years' work before the country could fully master cryogenic technology. Understandably, India initially sought to acquire such technology from abroad.

Negotiations began in 1987 with the French company Snecma (now Safran) – a historic and hence natural partner for ISRO – and with the US General Dynamics Corporation, but in both cases, the cost of the technology transfer proved prohibitive. In 1988, ISRO thus sided with the Soviet commercial agency Glavkosmos, which offered the transfer of cryogenic technology at a very competitive price (just ₹2.35 billion crores, then €188 million). ISRO and Glavkosmos formalised the technology transfer agreement in June 1991, under which the USSR would deliver two KVD-1 cryogenic engines and associated technology by 1995.[20]

The agreement, however, would never fully see the light, as it immediately prompted the wrath of the United States, hence becoming caught up in issues of *high politics*. Indeed, following India's use of the SLV-3's first stage for the development of the Agni ballistic missile – which was successfully tested in 1989 – the United States raised non-proliferation concerns to deny India the right to acquire such technology. In May 1992, the George Bush administration condemned the agreement as a violation of the Missile Technology Control Regime (MTCR) and imposed sanctions on both ISRO and Glavkosmos that barred US firms from cooperating with them. Initially, the Yeltsin government rejected the US request to cancel the deal, primarily because of the urgent need for hard currency imposed by the collapse of the Soviet Union. However, as detailed by Moltz, "after steady American lobbying and the U.S. delivery of a substantial aid package to Russia at the Vancouver Clinton-Yeltsin summit in April 1993, the Russian government eventually agreed to amend the Indian deal by withholding the cryogenic production technology and selling only the completed boosters themselves" (Moltz 2012).[21] Accordingly, in July 1993, Russia backed off the initial contract with ISRO, proposing a revised agreement for the transfer of two off-the-shelf engineering models and seven ready-to-fly KVD-1 engines (Harvey et al. 2010). While the new contract was signed, the failed technology transfer contributed to exacerbating India's irritation vis-à-vis both the United States and Russia and its resolve to proceed on its own for the development of cryogenic technology.

Between 1997 and 2000, the seven KVD-1 engines were delivered to India, enabling the first set of flights with the GSLV, starting in 2001. The first attempt was made on 28 May 2001 (though it failed), while the first operational flight was suc-

---

[20] The KVD-1 was an old engine developed in 1964 by the Isayev Design Bureau for the USSR lunar landing programme. It was test-fired in 1967, but it was never used as the Moon landing programme was cancelled.

[21] However, it is debated whether technology transfer had already occurred at that point.

cessfully carried out in September 2004, with the 1950-kg Edusat as payload. However, at the same time, decisions were taken favouring accelerated indigenisation, particularly in view of the GSLV's underperformance.[22]

On the one hand, emphasis was put on reaching operational readiness of the GSLV Mark II, a more powerful version of the GSLV (GTO capacity of approximately 2500 kg) on which the Russian upper stage would be replaced by an Indian manufactured stage, the cryogenic upper stage (CUS).[23] On the other hand, development of a new GSLV version with an even higher GTO performance of approximately 4400 kg (the GSLV Mark III) was approved by the Indian government in April 2002, with a budget allocation of roughly ₹2500 crores (€ 500 million) (Lele 2013a, b).

Contrary to what the name may suggest, GSLV Mk III is designed as an entirely different vehicle from GSLV Mark II and not just as an evolution. Following firing test series for the strap-on boosters and first stage that started in 2010, the GSLV Mk III maiden flight was carried out successfully on 18 December 2014. For this flight, a passive non-functional upper stage with a dummy engine was used. The vehicle carried a 3775-kg prototype of the Indian crew module, re-entry and landing of which were also successfully tested during the mission. Concerning the upper stage, development activities progressed at a slower pace due to funding being bound by the procurement of foreign launch services for missions that could not be launched on GSLV Mk II. In early 2014, however, tests were reported to be progressing on the cryogenic engine, which had undergone three successful firing tests at that time with tests continuing throughout 2014.

In parallel, evolutions of the PSLV launch vehicle enabled its first launch into GTO in 2011. Having a light GTO launch vehicle at hand with the enhanced PSLV-XL version and eventually a 4 t–class GTO launcher with GSLV Mk III, India now plans to abandon the GSLV Mk II once GSLV Mk III becomes fully operational.

2.4   Expanding (and Reorienting) the Programme

With the turn of the century, some major developments on both the domestic and international front would mark a new major shift for the overall direction of the Indian space programme.

On the international front, the terrorist attacks of September 2001 generated a strategic reorientation of the US posture vis-à-vis India. The common threat posed by Islamist-inspired terrorism and the shared democratic values led the Bush admin-

---

[22] Setbacks in the GSLV programme as a consequence of four subsequent launch failures between 2006 and 2010 triggered a fundamental review of the entire GTO launch service programme. While theoretically one final GSLV Mk I launch could have been carried out using the remaining Russian upper stage in stock, decisions had been made to not use it any longer, de facto ending the GSLV Mk I programme.

[23] Also the GSLV Mk II, however, experienced a launch failure on its maiden flight, requiring modifications over a period of 3 years following which a GSLV Mk II flight was successfully carried out in January 2014.

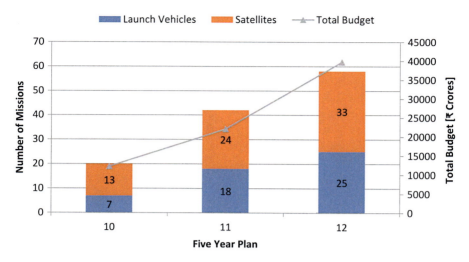

Fig. 2.1 ISRO's budget and missions (*Source*: Prasad 2016)

istration to give up non-proliferation sanctions and form a new strategic alliance with India, which had a positive spillover effect also on space activities. As recorded by Moltz, "after a visit by Prime Minister Vajpayee in November 2001, the two administrations released a list of priority areas for cooperation, which included space. These plans resulted in the formation of the Indo-U.S. High Technology Cooperation Group in 2002, which in turn led to the Next Steps in Strategic Partnership (NSSP) agreement in January 2004" (Moltz 2012). The NSSP spelled out nuclear and space cooperation as two key areas for advancement, providing the groundwork for US loosening of export controls for NASA cooperation in space science with ISRO, which has since then flourished.

In addition to this rapprochement with the United States, another major international development exercising structural pressures on India's strategic posture, and eventually also on the overall profile of the its space programme, was the concomitant rise of China – in space and elsewhere. However, in sharp contrast to India's growing ties with the United States and Japan, Sino-Indian relations seemed to head towards a completely different direction, that of increasing confrontation (see Sect. 6.2).

On the domestic front, the post-liberalisation period entailed a huge expansion of India's economy (see Sect. 4.1), which would positively impact national space activities as well. In effect, since the early 2000s, India's space programme has seen a remarkable acceleration, with an almost vertical take-off in both the number of missions and the budgetary allocation of ISRO (see Fig. 2.1).[24]

Indeed, whereas in previous Five-Year Plans – and specifically during the seventh (1985–1990), the eighth (1992–1997) and the ninth (1997–2002) Five-Year Plan – ISRO had typically conducted a total of 10–12 missions, this figure rose to 20 missions during the tenth Five-Year Plan (2002–2007), to 42 in the eleventh

---

[24] It is important to note that ISRO considers launch vehicle and satellite building as separate missions.

Five-Year Plan (2007–2017) and to an impressive total of 58 missions during the twelfth Five-Year Plan (2012–2017). Similarly, the budget allocation grew from ₹12,165 crores during the tenth Five-Year Plan to ₹20,268 crores during the eleventh and ₹39,750 crores during the twelfth (in annual terms, it grew from ₹2263 crores to ₹7509 crores).[25]

Remarkably, this unprecedented acceleration of Indian space activities not only entailed a change in velocity but also in trajectory, with an equally relevant transition from an application-based and societal-oriented space programme to a fully fledged endeavour covering the full spectrum of activities. The transition point can be symbolically fixed in 2007, when the then chairman of ISRO stated: "the vision of Sarabhai has been fulfilled and today we are at a turning point" (Foust 2008). There were at least two major developments that accounted for this declared policy shift, namely, the opening towards daring space exploration and human spaceflight endeavours and India's increasing interest in the military uses of space, which are briefly introduced hereafter.

2.4.1 Beyond Sarabhai Roadmap: Space Exploration and Human Spaceflight

India's looming departure from its narrow utilitarian focus on Earth applications became apparent to the international community with the launch of the Indian spacecraft Chandrayaan-1 to the Moon in 2008–40 years after Vikram Sarabhai presented its roadmap to the world at TERLS.

The mission was India's first effort to move beyond Earth orbit and reach a celestial body and, interestingly, was used by ISRO as a beacon of international cooperation with European and American partners.[26] Understandably, Chandrayaan-1 became a source of great national pride but also set off intense discussions on India's use of space and primarily about whether the purely utilitarian path should be abandoned at a time when huge poverty and terrestrial infrastructure issues had still not been overcome. One interpretation of India's increased involvement in space exploration was, however, that India did not feel that it could allow China to seize all initiative in this geopolitical significant field, given India's long history in space and its interest in protecting its international and regional stature (Moltz 2012; Sheehan 2007; Lele 2013a).

The interpretation that India was driven towards a rethink of its strategy in the glaring light of China's surging space ambitions was corroborated by the almost simultaneous launch of Chandrayaan-1 with similar Chinese and Japanese missions,

---

[25] See Chap. 3 for a detailed analysis of the budget profile of ISRO and of its recent evolution.

[26] For the Chandrayaan-1, ISRO worked in cooperation with the United States, Germany, the United Kingdom and Sweden through the European Space Agency (ESA) and with Bulgaria, offering the possibility to fly their experiments. The mission, however, also experienced some technical difficulties and underpinned ISRO's lack of full transparency with foreign partners, with India refusing to immediately report difficulties to partners.

respectively, the Chang'e-1 and Kaguya-1 lunar probes in 2007 and 2008. The proximity of the launching dates and the nature of the missions immediately fed the impression that the three countries were competing against each other, a competition that would likely extend also to manned space capabilities and, ultimately, end up in an intra-Asian space race (both civil and military). In effect, there was no shortage of indications about this possible path.

Soon after the historic flight of Jiang Liwei in 2003, ISRO set up several study groups on space exploration and later produced a formal proposal for the development of a human spaceflight programme. The proposal was presented to Indian Prime Minister Manmohan Singh on 17 October 2006, and, on the latter's advice, it was submitted by ISRO Chairman Gopalan Madhavan Nair to a cross-section of the scientific community that met in a brainstorming session in Bangalore on 7 November 2006 (Peter 2008). The conference agreed to immediately initiate a human spaceflight programme and to autonomously launch its first manned flight by 2014 and land an Indian astronaut on the Moon by 2020 (Harvey et al. 2010).

The decision represented a major change in Indian space policy which, since its inception, had always explicitly declined any potential involvement in human spaceflight. Manned spaceflight and planetary exploration were exactly the goals that the space programme's founder, Vikram Sarabhai, had rejected in favour of a *developmental rationale*. As he insisted, the application of space technology to addressing development goals "is not to be confused with embarking on grandiose schemes" and specifically that "we do not have the fantasy of competing with the economically advanced nations in the exploration of the Moon or the planets, or manned spaceflight".

As many analysts have argued, although India would insist that its "manned programme was a logical next step for the Indian space programme, and not a reaction to the emergence of the Chinese manned programme, the shift in policy was so dramatic in comparison to the almost ideological opposition to manned flights, that the timing seems hardly coincidental" (Sheehan 2007).

India's manned spaceflight programme initially received strong political backing, as A.P Abdul Kalam, then the country's president, had for 20 years served as a director of ISRO and was a strong supporter of the human spaceflight programme. Initial funding began in April 2007 (beginning of the fiscal year), with the objective of sending the first Indian astronaut to orbit by 2014 (Peter 2008). To support this ambitious exploration programme, ISRO took the decision to human-rate its Geosynchronous Satellite Launch Vehicle (GSLV III) and to develop a two-person manned capsule. Actions were initiated accordingly. By the middle of 2007, ISRO had already validated its re-entry technology with the successful recovery of a space capsule and had started to work on pre-projects, including long-lead items for human missions such as spacesuits and simulation facilities (Peter 2009).

Following a working group on space cooperation with Russia, an agreement was reached between the two countries in December 2008 for the Indian manned spacecraft to be built following the trusted Russian Soyuz design, thus echoing the

Chinese path with its *Shenzhou* programme.[27] In that context, India also considered sending one of its citizens into space on-board a Russian spacecraft to acquire the skills necessary for future manned space missions, but that plan never materialised.

In spite of an overall increase in the space budget, the financial commitment to human spaceflight has remained rather limited. ISRO had estimated that the project leading to a first manned flight would cost from US$2.5 to $3 billion a year (more than three times the agency's annual budget of that period) (Jayaraman 2006). However, the funds so far allocated to the programme have not exceeded a few hundred million dollars.[28] In addition, cooperation with Russia in the development of a Soyuz-based manned capsule has not proceeded as planned, while the development of the enhanced GSLV Mark III launch vehicle, whose first flight was scheduled for 2013, has encountered considerable delays and is still under development.[29] Probably as a result of these adverse developments, in August 2013, ISRO Chairman Koppillil Radhakrishnan announced that a human spaceflight mission was not an ISRO priority (NDTV 2013),[30] a position that had already been clear since the release of the 12th Five-Year Plan, in which human spaceflight did not figure among the list of projects to be implemented over the period (Planning Commission – Government of India 2014). Although in his announcement Mr. Radhakrishnan was keen to emphasise that the country would not be starting from scratch should a decision for human spaceflight eventually be taken, he avoided stating a definite time frame for the potential launch, limiting himself to specifying: "We are not going to see the human spaceflight as a programme in the 12th Five-Year Plan. *We will see maybe later*" (NDTV 2013).[31]

At the same time, India has been pursuing alternative paths to gain international prestige and capitalise on the techno-nationalist benefits stemming from space, including a much-heralded reorientation of its space exploration programme towards Mars. In November 2013, it launched the Mars Orbiter Mission (MOM), which made India the first and only Asian state to successfully reach the Red Planet on the first attempt. Unsurprisingly, the undertaking proved a landmark achievement for India, demonstrating its technological prowess and providing the country with a significant source of national pride, as confirmed by the speech made by Prime Minister Modi in September 2014, immediately after the Mangalyaan spacecraft entered Mars orbit: "History has been created. We have dared to reach out into the unknown

---

[27] Also echoing the Chinese approach, it was decided that the first mission would be for a day, while the second for a week (Harvey et al. 2010, p. 238).

[28] See Chap. 3 for ISRO's budget breakdown during the past years.

[29] The GSLV experienced problems in April 2010 when the main cryogenic engine on India's domestically produced third stage failed to ignite. See the previous section in this chapter.

[30] The position was also clarified in a personal interview with Ajey Lele, where he stressed, "As of today, a human mission is not in our space agenda. We are in a very early phase of developing a few critical technologies required for realising a human mission" (Lele 2013a).

[31] Emphasis added. Quoted from NDTV (2013). In ISRO Space Vision 2025, the objective is limited to the development of a space vehicle capable of putting two humans into LEO and returning safely to Earth, something that China achieved as early as 2003.

and have achieved the near impossible […] the success of our space programme is a shining symbol of what we are capable of as a nation" (Mars Daily 2014).

2.4.2 Opening to the Military Uses of Space

The second major development during the 2000s that appeared to evidence India's progressive departure from its developmental approach to space activities is offered by its increasing opening towards the military uses of space.

Before 2001, the involvement of India's military in the country's space activities was essentially non-existent. India's armed forces did not even have access to dedicated imagery, satellite-based communications or data services but had to rely foreign assets (e.g. by purchasing imagery from US companies and the Israeli government, among other sources).

The management of the space system was exclusively entrusted to the civilian ISRO and DOS. Behind the government's decision to firewall any military component from the management of the programme, there was the view – articulated in the early 1980s by Satish Dhawan – that the "Indian space programme would have been unnecessarily slowed if it had been a combined military and civil space programme, and that the cooperation with other states which had benefited ISRO would have been impossible if India had possessed an obvious (military space) programme" (Sheehan 2007). India got a clear demonstration of this when SLV technologies were transferred to the development of Agni ballistic missile.

Things started to change – once again – because of regional dynamics, more specifically the 1999 Kargil, which demonstrated the ineffectiveness of Indian space assets to warn the military about Pakistani incursions in Kashmir, and, more importantly, China's ASAT test of 2007, which served as a wake-up call vis-à-vis the "rising regional and international technology demonstration in military space technology" (Rajagopalan 2013).

The following year, some major changes were introduced at both institutional and operational level. On the former front, an Integrated Space Cell within the Headquarters of the Integrated Defence Staff was formed to coordinate India's military space activities. The move was more specifically intended to create synergies between the three branches of the Indian Armed Forces (Air Force, Army and Navy), the Ministry of Defence, the civilian Department of Space and ISRO, so as to facilitate dialogue between these different institutions about requirements, capabilities and relevant policy and pave the way for the establishment of operational space command capabilities. The Headquarters of the Integrated Defence Staff also issued a new doctrine called "Defence Space Vision 2020", outlining the roadmap for the armed forces in the space sector (Sharma 2011). The plan called for a phased growth of dual-use assets, beginning with space-based surveillance and communication capabilities and the establishment of military-run operational capabilities under its new Integrated Space Cell. It also highlighted that the three branches of India's armed forces should each receive a dedicated satellite from ISRO and that "additional training activities, ground- and space-based hardware, and command/control capabilities should be developed in subsequent years" (Moltz 2012).

On the operational front, ISRO accordingly put a new priority on the development of dual-use assets for space-based intelligence gathering, as well as regional navigation and communication capacities. In April 2008, it began to launch a series of dual-use remote sensing missions with high-resolution imaging and SAR capability, which would be then used by the military to address some of the required intelligence gathering requirements (Rajagopalan 2013).

The first dual-used satellite was Cartosat-2A, a 0.8-metre high-resolution optical satellite launched in April 2008. It was complemented in 2009 by RISAT-2, an all-weather X-band SAR reconnaissance satellite built for ISRO by Israel Aerospace Industries (IAI) for addressing surveillance and defence purposes, including tracking of hostile ships at sea. A second optical imaging satellite, Cartosat-2B, was launched in July 2010 for providing scene-specific spot imagery in high resolution, while in April 2012, ISRO orbited RISAT-1, the first indigenous satellite mission to use an active C-band radar sensor system providing both civil and military users in the country with an all-weather as well as day-and-night SAR observation capability.

In order to further expand these still-limited capabilities, in April 2013, the Technology Perspective and Capability Roadmap of the Headquarters Integrated Defence Staff, Ministry of Defence, defined the requirements of the armed forces, including their space-based requirements. The roadmap specifically called for the development of space assets for tackling meteorological, reconnaissance, surveillance, navigation and communication requirements and the need for strategic forces to be supported by real-time information in the area of interest, including sub-metric resolution imagery, aid in communications, all-day–all-weather capability and signal intelligence, among others (Headquarters Integrated Defence Staff 2013). Remarkably, the document also called for the development of ASAT capabilities "for electronic or physical destruction of satellites", not only in LEO but also GEO.

Due to the perceived proliferation of offensive military space capabilities at both regional and international level, some organisations indeed started to review their position and consider developing deterrence capabilities to protect their space-based assets. In January 2010, important confirmation was given by the DRDO Director General V.K. Saraswat, who openly recognised Indian efforts to develop ASAT capabilities, claiming "we are also working on how to deny the enemy access to its space assets" (Brown 2010). And by the following year, he also added "we have all the technology elements which are required to integrate a system (and) defend our satellites" (Bagchi 2011). Parallel statements by several military commentators and analysts also highlighted the need for India to kinetically test such weapons in order to establish a precedent in case of future space arms control talks and hence avoid a Non-Proliferation Treaty (NPT)-like situation (Harsh 2016).

While it remained very unclear whether the Indian government was supportive about the development (and eventual testing) of offensive space capabilities – also considering the fundamental political, legal and technological complexities in operationalising a space security deterrent – its commitment towards ensuring military support functions from space became more apparent and robust.

This new posture was inter alia confirmed by the launch of India's first dedicated military satellite in August 2013. Built and launched by ISRO for supporting the maritime communications' purposes of the Indian Navy, GSAT-7 carries communication capabilities in UHF, S-band, C-band and Ku-band (Lele 2013a, b). Its launch marked the beginning of a new chapter of the Indian space programme: the use of outer space for strategic and national security-related operations. In 2015, with the induction of India's foreign secretary to the Space Commission for the first time, India's focus on space from a foreign policy and national security perspective received its eventual confirmation (Rajagopalan 2016).

2.5   Summarising Remarks

India has been an "early adopter" of space technology, and, since the official inception of INCOSPAR in the early 1960s, its space programme has made impressive strides forward, which are even more impressive when recalling that, at the end of the Cold War, India was one of the poorest countries in the world. Equally important is "the consistent way in which India has sought to use space as a crucial mechanism for lifting India's people out of poverty through education and social and economic programmes" (Sheehan 2007).

The vision enunciated in the early 1970s by Vikram Sarabhai about harnessing space for meeting national development goals and escaping dependency through technological leapfrogging has loomed large over Indian space policy over the past four decades, enabling the country to achieve tremendous progress and self-reliance in this strategic area but also making it reluctant to embark upon grandiose schemes. Moreover, throughout most of its history, the programme has had a remarkably peaceful orientation, having placed emphasis only on civilian applications to address many of the societal issues facing a huge and developing country. This was in stark contrast to the space superpowers Russia and the United States, both of which have been heavily investing in defence, space exploration and human spaceflight since the inception of their respective space agencies more than five decades ago. Over time, India's "developmental ideology has become firmly embedded in the self-perception of the space programme" (Sheehan 2007); a programme that, thanks to an accountable and transparent user-driven approach, has in addition become part of India's everyday life and has opened the door to big public support for ISRO (Kasturirangan 2011).

In its execution, India's application-driven space programme has followed what Perumal sees as a "strategy of backward integration". Starting with "the demonstration of the efficacy of space application with borrowed satellites, it progressed through bought-out and indigenous satellites with procured launch services and finally building the indigenous launch capability" (Perumal 2015). While this path has been marked by a slow – yet steady – pace of development that has lasted more than three decades, in the process, India has grown to become an influential space-faring nation with many remarkable achievements to reflect on: an autonomous access to space with the PSLV and GSLV launch vehicles, the INSAT communication satellite system, the IRS remote sensing constellation and more recently the IRNSS positioning and navigation satellite system as well as significant science and exploration missions such as the Chandrayaan-1 mission to the Moon and Mangalyaan probe to Mars, among others.

Indeed, as India's has expanded its technical capabilities, its space programme has seen opportunities "to step up to the next level and begin to compete (and cooperate) with other major spacefaring countries in previously neglected areas, such as space science and exploration" (Moltz 2012). While these areas tend to have less of a problem-solving effect than education satellites or disaster mitigation projects through Earth observation, for instance, for the many Indians still struggling with poverty, lack of education and overall life quality, they provide the opportunity to demonstrate a certain level of dominance and technological advancement which in turn provides geopolitical and diplomatic leverage to the nation. As also stressed by Bharath Gopalaswamy, the space programme's broadening to space exploration indicates that "India no longer views space as only enhancing the living conditions of its citizens but also as a measure of global prestige".[32]

All in all, although the inclusion of exploration in India's space portfolio suggests that techno-nationalist considerations have become an important factor, it cannot be argued that they have come to dominate the space programme at the expense of the development rationale. An examination of the priorities and goals listed in the 12th Five-Year Plan for the period 2012–2017 (see Chap. 4) shows a sharp contrast between the great attention devoted to the "societal use of space" and the scant attention paid to *grandeur* projects. This indicates that the "needs-based approach" is still *the* overwhelming principle of India's space policy. The same Chandrayaan-1 or Mars Orbiter Mission could be interpreted using different lenses, including that of technological innovation, which can, in fact, have positive spill-over effects on societal advancement through spin-offs.

From an overall perspective, the programme has, however, clearly began to expand its focus, as demonstrated by the increasing attention paid towards military uses of space and the commercial exploitation of commercial space assets and the improvement of industrial competencies. This ongoing diversification of the Indian space programme will be analysed in greater detail in the following chapters. Before diving into that, however, an overview of the current framework for space activities in India is provided in the next chapter.

---

[32] Quoted from Moltz (2012).

References

Aravamudan, R. (2015). Evolution of ISRO: A personal account. In P. M. Rao (Ed.), *From fishing hamlet to red planet: India's space journey.* New Delhi: Harper Collins.

Arnold, D. (2000). *Science, technology and medicine in colonial India.* Cambridge: Cambridge University Press.

Atomic Energy Commission. (1970). *A profile for the decade 1970–1980 for atomic energy and space research.* New Delhi: Government of India.

Bagchi, I. (2011, March 6). India working on tech to defend satellites. *Times of India.*

Bhatia, B. S. (2015). Satcom for development education. The Indian experience. In P. M. Rao (Ed.), *From fishing hamlet to red planet: India's space journey.* New Delhi: Harper Collins.

Bhatia, B. (2016). Space applications for development – the Indian approach. In *ESPI 10th autumn conference.* Vienna: European Space Policy Institute.

Blamont, J. (2015). Starting the Indian space programme. In P. M. Rao (Ed.), *From fishing hamlet to red planet: India's space journey.* New Delhi: Harper Collins.

Blamont, J. (2017). Cooperation in space between India and France. In R. P. Rajagopalan & N. Prasad (Eds.), *Space India 2.0 commerce, policy, security and governance perspectives* (pp. 215–233). New Delhi: Observer Research Foundation.

Bouhey-Klapisz, C. (2017, February 1). Space cooperation between India and France. In *India in space: The forward look to international cooperation.* Vienna: European Space Policy Institute.

Brown, P. J. (2010, January 22). *India's targets China's satellites.* Retrieved November 18, 2016, from Asia Times: Brown, Peter J "India's targets China's satellites. Asia Times. http://www.atimes.com/atimes/South_Asia/LA22Df01.html

Foust, J. (2008, February 11). *India and the US: Partners or rivals in space?* Retrieved September 11, 2016, from The Space Review. http://www.thespacereview.com/article/1056/1

Goel, P. (2015). Operational satellites of ISRO. In M. Rao (Ed.), *From fishing hamlet to red planet: India's space journey.* New Delhi: Harper Collins.

Gupta, S. (2015). Beginnings of launch vehicle technology in ISRO. In P. M. Rao (Ed.), *From fishing hamlet to red planet: India's space journey.* New Delhi: Harper Collins.

Harsh, V. (2016, June 14). *India's anti-satellite weapons.* Retrieved June 15, 2016, from The Diplomat. http://thediplomat.com/2016/06/indias-anti-satellite-weapons/

Harvey, B., Smid, H. H., & Pirard, T. (2010). *Emerging space powers.* Chichester: Springer – Praxis.

Headquarters Integrated Defence Staff. (2013). *Technology perspective and capability roadmap.* New Delhi: Ministry of Defence.

Indian Space Research Organisation. (2015a). *Genesis.* Retrieved July 3, 2016, from ISRO. http://www.isro.gov.in/about-isro/genesis

Indian Space Research Organisation. (2015b). *Rohini satellite RS-1.* Retrieved July 18, 2016, from ISRO. http://www.isro.gov.in/Spacecraft/rs-1-1

Indian Space Research Organisation. (2015c). *Rohini satellite RS-D2.* Retrieved July 18, 2016, from ISRO. http://www.isro.gov.in/Spacecraft/rohini-satellite-rs-d2

Indian Space Research Organisation. (2015d). *Space applications centre celebrates Ruby year of SITE.* Retrieved July 18, 2016, from ISRO. http://www.isro.gov.in/space-applications-centre-celebrates-ruby-year-of-site

Indian Space Research Organisation. (2016a). *APPLE.* Retrieved July 20, 2016, from IRSO. http://www.isro.gov.in/Spacecraft/apple

Indian Space Research Organisation. (2016b). *ASLV.* Retrieved July 20, 2016, from ISRO. http://www.isro.gov.in/launchers/aslv

Indian Space Research Organisation. (2016c). *Vikram Sarabhai (1963–1971).* Retrieved June 10, 2016, from ISRO. http://www.isro.gov.in/about-isro/dr-vikram-ambalal-sarabhai-1963-1971

Jayaraman, K. (2006, November 13). *ISRO seeks government approval for manned spaceflight program.* Retrieved September 21, 2016, from Space News. http://www.spacenews.com/article/isro-seeks-government-approval-manned-spaceflight-program/

Kalam, A. A. (2015). India's first launch vehicle. In P. M. Rao (Ed.), *From fishing hamlet to red planet: India's space journey.* New Delhi: Harper Collins.

Kale, P. (2015). Origins of INSAT-1. In P. M. Rao (Ed.), *From fishing hamlet to red planet: India's space journey*. New Delhi: Harper Collins.

Kasturirangan, K. (2011). *Relating science to society*. Mysore: University of Mysore.

Lele, A. (2013a). *Asian space race: Rhetoric or reality?* New Delhi: Springer.

Lele, A. (2013b, September 9). *Commentary on GSAT-7: India's strategic satellite*. Retrieved November 8, 2016, from Space News. http://spacenews.com/37142gsat-7-indias-strategic-satellite/

Mars Daily. (2014, September 24). *Quoted from: "India wins Asia's Mars race as spacecraft enters orbit.* Retrieved September 30, 2016, from Mars Daily. http://www.marsdaily.com/reports/India_wins_Asias_Mars_race_as_spacecraft_enters_orbit_999.html

Moltz, J. C. (2012). *Asia's space race. National motivations, regional rivalries, and international risks*. New York: Columbia University Press.

Narayana, K. (2015). The spaceport of ISRO. In P. M. Rao (Ed.), *From fishing hamlet to red planet: India's space journey*. New Delhi: Harper Collins.

Narayanamoorthy, N. (2015). PSLV: The workhorse of ISRO. In P. M. Rao (Ed.), *From fishing hamlet to red planet: India's space journey*. New Delhi: Harper Collins.

Navalgund, R. (2015). Remote sensing applications. In P. M. Rao (Ed.), *From fishing hamlet to red planet: India's space journey*. New Delhi: Harper Collins.

NDTV. (2013, August 16). *Human space flight mission off ISRO priority list*. Retrieved September 28, 2016, from NDTV. http://www.ndtv.com/article/india/human-space-flight-mission-off-isro-priority-list-406551

Perumal, R. (2015). Evolution of the geosynchronous satellite launch vehicle. In P. M. Rao (Ed.), *From fishing hamlet to red planet: India's space journey*. New Delhi: Harper Collins.

Peter, N. (2008). Developments in space policies programmes and technologies throughout the world and Europe. In K.-U. Schrogl, C. Mathieu, & N. Peter (Eds.), *ESPI yearbook 2006/2007: A new impetus for Europe* (p. 96). Vienna: Springer.

Peter, N. (2009). Developments in space policies programmes and technologies throughout the world and Europe. In K.-U. Schrogl, C. Mathieu, & N. Peter (Eds.), *ESPI yearbook on space policy 2007/2008: From policies to programmes* (p. 81). Vienna: Springer.

Pisharoty, P. (2015). Historical perspective of remote sensing. Some reminiscences. In P. M. Rao (Ed.), *From fishing hamlet to red planet: India's space journey*. New Delhi: Harper Collins.

Planning Commission – Government of India. (2014, May 31). *GDP at factor cost at 2004–05 prices, share to total GDP and % rate of growth in GDP*. Retrieved September 20, 2016, from Planning Commission Database: Planning Commission. http://planningcommission.nic.in/data/datatable/0814/table_4.pdf

Prasad, M. (2015). ISRO and international cooperation. In M. Rao (Ed.), *From fishing hamlet to red planet: India's space journey*. New Delhi: Harper Collins.

Prasad, N. (2016). Diversification of the Indian space programme in the past decade: Perspectives on implications and challenges. *Space Policy, 36*, 38–45.

Rajagopalan, R. P. (2013, October 13). Synergies in space: The case for an Indian Aerospace Command. *ORF Issue Brief, 59*.

Rajagopalan, R. P. (2016). *India's space program. Challenges, opportunities and strategic concerns*. Washington, DC: National Bureau of Asian Research.

Rao, U. (2015). Origins of satellite technology in India: The story of Aryabhata. In P. M. Rao (Ed.), *From fishing hamlet to red planet: India's space journey*. New Delhi: Harper Collins.

Sharma, G. (2011). *Space security: Indian perspective*. New Delhi: Vij Books.

Sheehan, M. (2007). *The international politics of space*. New York: Routledge.

Swaminathan, M. (2015). Genesis and growth of remote sensing applications in Indian agriculture. In P. M. Rao (Ed.), *From fishing hamlet to red planet: India's space journey*. New Delhi: Harper Collins.

United Nations. (1982, August 9–21). *Report of the second united nations conference on the exploration and peaceful uses of outer space*. Vienna. UN, A/Conf.101/10. United Nations.

Vasagam, R. (2015). APPLE in retrospect. In M. P. Rao (Ed.), *From fishing hamlet to red planet. India's space planet: India's space journey*. New Delhi: Harper Collins.

Chapter 3
Space Activities in India: The Current Landscape

This chapter provides an overview of the current landscape for space activities in India. It is comprised of four parts. In the first, attention is given to the institutional framework as it provides an overview of the organisational setup for space, of ISRO's mode of working and budgetary allocations and of ISRO's relationship with industry, academia and the military. The second part analyses India's space policy and regulatory framework, including the defining of India's long-term objectives and decision-making procedures, while the third part describes the programmes and the space, launch and ground infrastructures India has developed over the course of the last five decades. The final section gives a comparative overview of the Indian space programme.

3.1 Institutional Framework

3.1.1 Organisation of Space Activities

As discussed in Chap. 2, the Indian space programme was institutionalised in November 1969 with the formation of Indian Space Research Organisation (ISRO). Three years later, in June 1972, the Government of India passed a resolution setting up a supervisory body called the Space Commission and an administrative body called the Department of Space (DOS) and brought ISRO under the DOS in September 1972 (Indian Space Research Organisation – Deloitte 2010). Since this restructuring, the organisation of the Indian Space programme has been designed to ensure government support at the highest level, with the Space Commission, the apex policy body and the DOS, reporting direct to the prime minister (see Fig. 3.1).

Within this organisational setup, the Space Commission formulates policies and oversees the implementation of the Indian space programme to promote the development and application of space science and technology for the socio-economic

© Springer International Publishing AG 2018   47
M. Aliberti, *India in Space: Between Utility and Geopolitics*, Studies in Space Policy 14, https://doi.org/10.1007/978-3-319-71652-7_3

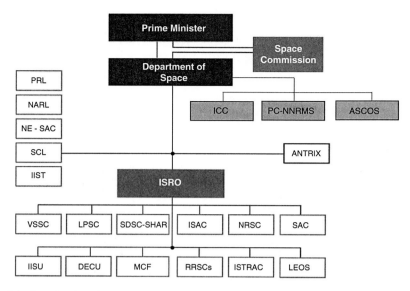

Fig. 3.1 Organisation of space activities in India (*Source*: ISRO)

benefit of the country (Sankar 2006), while the Department of Space, an independent department within the government, implements these programmes. The stated objectives of DOS are to:

- Provide national space infrastructure for the telecommunication needs of the country, including the required transponders and associated ground systems
- Provide satellite data required for weather forecasting, monitoring, etc.
- Provide satellite imagery and specific products and services required for the application of space technology for natural resource management/developmental purposes to the national government, state governments, NGOs and the private sector
- Promote research and development in space sciences and technology (Department of Space – Government of India 2016)

DOS implements the space programme mainly through the Indian Space Research Organisation (ISRO) and five major grant-in-aid institutions, namely, the Physical Research Laboratory (PRL), the National Atmospheric Research Laboratory (NARL), the North Eastern Space Applications Centre (NE-SAC), the Semi-Conductor Laboratory (SCL) and the Indian Institute of Space Science and Technology (IIST). The activities of these autonomous bodies supported by DOS are summarised in Table 3.1.

The main institution under DOS in charge of implementing the space programme is, however, ISRO, which is responsible for a variety of tasks, including:

- The design and development of launch vehicles and related technologies for providing access to space, including PSLV and GSLV

Table 3.1 ISRO/DOS establishments

Establishment	Location	Key work areas
Autonomous bodies		
Physical Research Laboratory (PRL)	Ahmedabad	Autonomous institution supported mainly by DOS
		Focuses on experimental and theoretical physics, astronomy and astrophysics and Earth, planetary and atmospheric sciences
National Atmospheric Research Laboratory (NARL)	Gadanki	Autonomous society supported by DOS
		Focuses on mesosphere, stratosphere, troposphere, lower atmospheric research
		Available for national and international scientists to conduct atmospheric research
North Eastern Space Applications Centre (NE-SAC)	Shillong	Joint initiative of DOS and the North Eastern Council for developmental support to the North Eastern region through space science and technology
		Provides developmental support by undertaking specific projects, utilising space technology inputs: remote sensing, satellite communication and space science
Semi-Conductor Laboratory (SCL)	Chandigarh	Research Institute supported by DOS
		R&D: very-large-scale integration (VLSI) devices and systems for telecommunication/ space sectors
Indian Institute of Space Science and Technology (IIST)	Thiruvananthapuram	Asia's first space university (established 2007)
		Bachelor's degree in space technology; specialisation in avionics and aerospace engineering
		Master's programme in applied sciences; emphasis on space-related subjects

- The design, development and realisation of satellites and related technologies for Earth observation, communication, navigation, meteorology and space science, including the Indian National Satellite (INSAT) programme, the Indian Remote Sensing (IRS) programme and the Indian Regional Navigation Satellite (IRSS) programme
- The development of space-based applications for societal improvement
- Research and development in space science and planetary exploration (Indian Space Research Organisation 2016a, b, c, d)

ISRO's activities are based on a network of centres and units, as well as industry and academia, that have been created in a phased manner over the years as centres of excellence, in place to meet needs in various fields of space technology, applications and sciences. As of 2016, ISRO had 13 centres and units spread over Bangalore, Thiruvananthapuram, Sriharikota, Ahmedabad and Hyderabad from where the programmes are executed. The main functions and work areas of ISRO/DOS centres and units are detailed in Table 3.2.

Table 3.2 ISRO centres and units

Establishment	Location	Key work areas
ISRO Centres		
Vikram Sarabhai Space Centre (VSSC)	Thiruvananthapuram	Design and development of launch vehicle technology
		Active R&D; core competence in aeronautics, avionics, materials, mechanisms, vehicle integration, chemicals, propulsion, space ordnance, structures, space physics, systems reliability, etc.
		Responsibilities for design, manufacturing, development and testing related to subsystems of various missions
		Programme planning and evaluation, human resource development, technology transfer, industry coordination
Liquid Propulsion Systems Centre (LPSC)	Valiamala, Bengaluru	Lead centre for development and execution of propulsion stages (launch vehicles and spacecraft)
		LPSC Valiamala: responsible for R&D, system design/engineering, delivery of liquid and cryogenic propulsion systems, control components and control power plants
		LPSC Bengaluru: design and development of satellite propulsion systems, production of transducers/sensors
Satish Dhawan Space Centre (SDSC-SHAR)	Sriharikota	Spaceport of India provides launch base infrastructure for the Indian space programme
		Responsibilities: (a) produce solid propellant boosters for ISRO's launch vehicles, (b) provide infrastructure for qualifying subsystems and solid rocket motors, (c) provide launch base infrastructure
		Two launch pads (main operational pads for PSLV and GSLV)
		Infrastructure augmentation underway to meet increased launch frequency requirements

(continued)

Table 3.2 (continued)

Establishment	Location	Key work areas
ISRO Satellite Centre (ISAC)	Bengaluru	ISRO's lead centre for design, development, fabrication and testing of Indian-made satellites
		Development of cutting-edge technologies of relevance to satellite building
		ISRO Satellite Integration and Test Establishment (ISITE): spacecraft integration and testing and assembly, integration and testing of all communication and navigation spacecraft
Space Applications Centre (SAC)	Ahmedabad	Focuses on payload development for societal application
		Development, realisation and qualification of communication, navigation, Earth observation and planetary payloads and related data processing in the areas of communications, broadcasting, remote sensing, disaster monitoring/mitigation, etc.
		Emphasis on outsourcing/indigenous development of technology and vendors
National Remote Sensing Centre (NRSC)	Hyderabad	Responsible for remote sensing satellite data acquisition and processing, data dissemination, aerial remote sensing and decision support for disaster management
		Regional Remote Sensing Centres (RRSCs) support remote sensing tasks and execution of application projects (mainly for natural resource management)
		Involved in software development, customisation and packaging specific to user requirements and conducting training programmes for users in geospatial technology, particularly digital image processing and GIS applications
ISRO Units		
ISRO Propulsion Complex (IPRC)	Mahendragiri	Assembly, integration and testing of earth-storable propellant engines, cryogenic engines and stages for launch vehicles, upper spacecraft thrusters, testing of its subsystems; production and supply of cryogenic propellants for Indian cryogenic rocket programme
		Responsible for supply of storable liquid propellants for ISRO's launch vehicles and satellite programmes

(continued)

Table 3.2 (continued)

Establishment	Location	Key work areas
ISRO Inertial Systems Unit (IISU)	Thiruvananthapuram	Centre of Excellence in Inertial Sensors and Systems
		Responsible for design and development of inertial systems for ISRO's launch vehicles and spacecraft
		Designs and develops actuators and mechanisms for spacecraft and allied applications
Development and Educational Communication Unit (DECU)	Ahmedabad	Unit involved in defining, planning, implementing and conducting socio-economic research and evaluation of various societal applications
		Major programmes supporting development, education and training: telemedicine (TM), tele-education (TE), disaster management system (DMS), Village Resource Centre (VRC)-related activities, etc.
Master Control Facility (MCF)	Hassan, Bhopal	Monitors and controls all ISRO's GEO satellites (INSAT, GSAT, KALPANA and IRNSS series)
		Responsible for orbit raising of satellites, in-orbit payload testing and on-orbit operations
		Interacts with user agencies for effective utilisation of satellite payloads and to minimise service disturbances during special operations
ISRO Telemetry, Tracking and Command Network (ISTRAC)	Bengaluru, Lucknow, Mauritius, Sriharikota, Port Blair, Thiruvananthapuram Brunei, Biak	Provides tracking support for all satellite and launch vehicle missions of ISRO
		Has a network of ground stations at various locations
Laboratory for Electro-Optics Systems (LEOS)	Bengaluru	Design, development and production of electro-optic sensors and camera optics for remote sensing and meteorological payloads
Indian Institute of Remote Sensing (IIRS)	Dehradun	Focuses on capacity building in remote sensing and geo-informatics and their applications through education and postgraduate training programmes
		Hosts and provides support to UN-affiliated Centre for Space Science and Technology Education in Asia and the Pacific

Source: ISRO

It is important to note that there is a very close connection between ISRO, DOS and the Space Commission, since the chairman of ISRO is also chairman of the Space Commission and the secretary of the Department of Space (DOS), as well as the head of the Antrix Corporation.

The Antrix Corporation Limited is a wholly owned Government of India company under the administrative control of the Department of Space (Antrix Corporation 2016). Antrix was incorporated in September 1992 as a marketing arm of ISRO for the promotion and commercial exploitation of space products, technical consultancy services and transfer of technologies developed by ISRO. Another major objective is to facilitate development of space-related industrial capabilities in India. Antrix also provides space products and services to international customers worldwide, such as Airbus, Intelsat, Avanti Group, etc. (Antrix Corporation 2016). The current business activities of Antrix include:

- Provisioning communication satellite transponders to various users
- Providing launch services for institutional and international customers
- Marketing data from Indian and foreign remote sensing satellites
- Building and marketing satellites as well as satellite subsystems
- Establishing ground infrastructure for space applications
- Providing mission support services for satellites

User-Driven Committees
A unique feature of the organisational structure of the Indian space programme lies in the special mechanisms created to provide effective user interfaces for the space systems. As the Indian space programme is predominantly application-driven, the creation of inter-linkages with other user departments within the Government of India as well as local governments has been key for ensuring the utilisation of Indian space assets in the most optimal manner. As explained by Sridhara Murthi et al., it was the experience gained with the Bhaskara and APPLE satellites, as well as with the SITE and STEP experiments, that paved the way for the creation of institutional frameworks designed to provide effective user interface for the space systems (Murthi et al. 2007). The three main bodies established for such purpose are the INSAT Coordination Committee (ICC), the Planning Committee on National Natural Resource Management System (PC-NNRMS) and the Advisory Committee on Space Research (ADCOS).

The ICC was first set up in November 1977 for coordinating and monitoring the implementation of the INSAT system as well as for planning its future development and advising on all technical issues related to the utilisation of ISAT. It comprises secretaries of the user departments and ministries of the INSAT system and "provides operational co-ordination, guidelines for planning of ISAT system, prioritisation of transponder requirements and other technical issues related to the utilisation of the INSAT system" (Department of Space – Government of India 2016).

The PC-NNRMS was constituted in 1992 for setting up an efficient management system of remote sensing data and other conventional data relevant for the management of natural resource in India. It is chaired by the member with the remit for science within India's Planning Commission (currently dismantled and replaced by the National Institution for Transforming India)[1] and includes secretaries of the Government of India departments – Space, Agriculture and Cooperation, Agricultural Research and Education, Environment and Forests, Earth Sciences, Finance, Mines, Planning Commission, Rural Development, Urban Development, Road Transport and Highways, Statistics, Science and Technology, Water Resources, Ocean Development and Biotechnology – and the chairman of the Central Water Commission as members.

ADCOS, consisting of national-level space scientists, coordinates space science research and supports the implementation of specific, nationally coordinated, multi-institutional, science payload instrumentation and science mission development projects in atmospheric and space sciences areas. All in all, these three special structures, adapted to meet user interface needs, have played a crucial role is sustaining various space endeavours (Murthi et al. 2007).

3.1.2 Mode of Working, Budget and Manpower

The organisational setup of the Indian space programme is predominantly project- and mission-oriented, and an important aspect in this respect is its so-called project/mission mode of working. Based on the long-term plans and demand for space-based services, "specific projects and programmes are conceived, undertaken and executed by DOS in a time-bound manner", meaning that once the objectives of the project are achieved, the project is closed and the resource re-deployed for other ongoing projects (Department of Space – Government of India 2016).[2]

The missions and projects are executed by the centres and units of ISRO – which are fixed entities responsible for technology, infrastructure and human resources required for execution of a given project – under a matrix management structure, which is designed to ensure optimum utilisation of resources and "to constantly review and monitor the progress and performance of the projects (Department of Space – Government of India 2016). A two-tier structure comprising a Project Management Council (PMC) and Project Management Board (PMB) provides the overall framework for the functioning of the project (see Fig. 3.2). The PMC,

---

[1] See Sect. 3.2 for more information on the recent institutional restructuring within the Indian government.

[2] The gestation period for space projects, i.e. development of satellites, launch vehicles and associated ground segments, is generally 3–5 years, while in some complex projects, it could extend up to 8–10 years. In the course of development, the project goes through various phases such as finalisation of configuration and detailed design, engineering and proto model development and qualification testing, fabrication of flight subsystem units and testing, assembly, integration and testing leading to launching of the satellite into orbit (Department of Space – Government of India 2016).

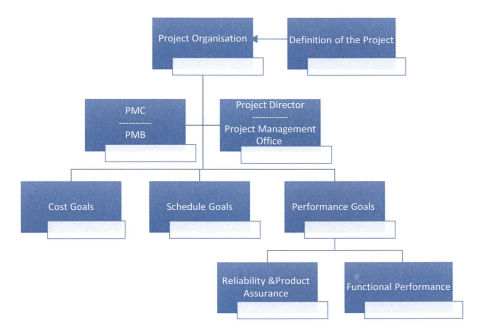

Fig. 3.2 ISRO's mode of working

chaired by the lead centre director, usually a distinguished scientist, provides overall direction for the project covering technical, managerial, financial and resource allocation aspects, while the PMB, chaired by a project director, holds regular meetings to discuss and tackle issues related to the progress of the project. These apart, the typical approach has been to carry out an in-depth and critical technical review of the project upon achieving specific project milestones such as preliminary design, detailed design, proto fabrication, flight readiness, etc. wherein experts outside ISRO also participate. It has been noted that such "project management structure for constant review and evaluation of activities during their implementation has been a key factor in achieving the programmatic targets in the space programme" (Murthi et al. 2007). Similarly, the process of "setting time and performance targets and working on a mission mode to achieve the targets has been a key feature of the Indian space programme that has ensured higher performance" (Ibid).

Concerning the actual implementation of space projects (e.g. development of satellites, launch vehicles and the associated ground systems), ISRO's centres and units are organised on the basis of their areas of specialisation/expertise. To illustrate, the Vikram Sarabhai Space Centre (VSSC) is the leading centre for launch vehicle projects, while ISRO's Satellite Centre (ISAC) is the leading centre for the development of space systems. The leading centre of the project has the primary responsibility for overall design, subsystem interface specifications, project management and coordination in addition to development of subsystems for which the lead centre has specialisation. The other centres/units of ISRO are in charge of realising specific subsystems/sub-assemblies for the project according to their expertise/specialisation. Tier-2-, Tier-3- and Tier-4-level work is also outsourced to private industry (see Sect. 3.1.3).

Typically, the output of the private industry and ISRO units (aside from the leading centre) are intermediate products which will then be integrated into the work of the leading centre into a physical output and eventually converted into the final outcome which the launch of the mission (Department of Space – Government of India 2016). This approach has been appropriately reflected in the Outcome Budget of DOS, where the output and outcome of a satellite or launch vehicle project during a given year are considered as the result of accumulated expenditure on the project during the previous years. Understandably, the "outlay for a satellite or launch vehicle project during a year does not necessarily result in output or outcome in the same year". While the deliverables and physical outputs are targeted and specified for each year for every project based on the development/realisation plans, the final outcome will accrue only upon the launch and operationalisation of the satellite.[3] The timeframes for such final outcomes are also specified in the Outcome Budget of DOS (Department of Space – Government of India 2016).

3.1.2.1 Budget Allocation and Expenditures

The budget proposals for DOS have been traditionally formulated in 5-year planning cycles by the Indian Planning Commission (see Sect. 3.2.1) under the framework of India's Five-Year Plans. The proposals are then approved on an annual basis by the Parliament before the beginning of each fiscal year (which runs from 1 April to 31 March) in accordance with the ceilings fixed by the Ministry of Finance.

Consistent with the "Project – Mission Mode of Working", the budget formulation process has evolved over the years with an emphasis on constantly reviewing the resource requirements so as to make financial management more effective. To this end, the Department has adopted a zero-based budgeting approach[4] and

---

[3] The outcome of a programme is largely dependent on the objectives of the programme. Given that the primary objective of the Indian space programme is to develop space technology, establish operational space systems in a self-reliant manner and demonstrate the potential applications of space systems for national development, "the nature of the Outcome of the Space Programmes will be in the form of (a) Indigenous capability to develop and realise complex space systems such as satellites and launch vehicles (b) Creating infrastructure in Space by launching and operationalisation of satellites including Space operations, which are utilised by various user agencies for national development (c) Capacity building in terms of critical technologies and ground technical infrastructure of relevance for future and (d) Benefits to the society arising from application of space technology/systems such as IRS satellites, INSAT satellites in various fronts. These have been appropriately reflected in the Outcome Budget against various programmes/schemes" (Department of Space – Government of India 2016).

[4] Zero-based budgeting (ZBB) is primarily concerned with making sure that the resources are distributed appropriately. ZBB is a process that can be repeated numerously to review every single dollar in the budget annually, establish a cost management culture and also manage financial performance monthly.

introduced a "mechanism of reviewing and monitoring the commitment and expenditure status of various programmes/projects approved in the annual budget and take appropriate action" (Department of Space – Government of India 2016). In line with the programmatic and financial guidelines of the department, multi-level budget reviews are carried out at the DOS/ISRO centres and at Project Management Boards/Councils.

The budget allocations approved during the 10th (2002–2007), 11th (2007–2012) and 12th (2012–2017) Five-Year Plans are provided in Fig. 3.3.

As the trend line in Fig. 3.3 clearly shows, the budget allocation has witnessed a steady increase over the past 15 years, passing from ₹2263 crores to ₹7509 crores. It is key to highlight, however, that budget allocations are not indicative of the real financial performance in a given year. The total amount of funds allocated, disbursed and actually spent during the last 5 financial years is provided in Table 3.3

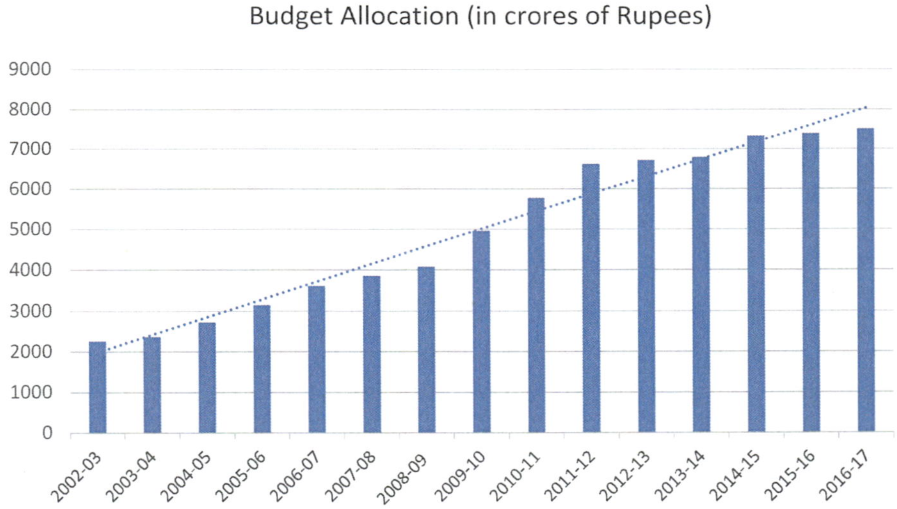

Budget Allocation (in crores of Rupees)

Fig. 3.3 ISRO's budget allocation (*Source*: IRSO)

Table 3.3 Recent expenditures (in crores of Rupees ₹) (*Source*: Department of Space)

Year	Budget allocation	Disbursement	Amount spent
2009–2010	4959.04	4167.04	4162.96
2010–2011	5778.04	4880.04	4482.23
2011–2012	6626.04	4432.04	3784.27
2012–2013	6715.04	4880.03	4856.28
2013–2014	6792.04	5172.04	5168.81
2014–2015	7328.00	5825.10	5798.81
2015–2016	7388.19	6959.44	
2016–2017	7509.14		

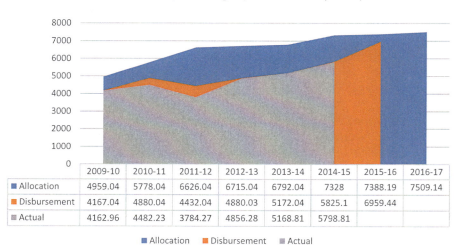

India's Space Budget (in crores of Rupees ₹)

	2009-10	2010-11	2011-12	2012-13	2013-14	2014-15	2015-16	2016-17
Allocation	4959.04	5778.04	6626.04	6715.04	6792.04	7328	7388.19	7509.14
Disbursement	4167.04	4880.04	4432.04	4880.03	5172.04	5825.1	6959.44	
Actual	4162.96	4482.23	3784.27	4856.28	5168.81	5798.81		

■ Allocation   ■ Disbursement   ■ Actual

Fig. 3.4 Space budget evolution (*Source*: Department of Space)

and in Fig. 3.4. For the years 2015–2016 and 2016–2017, the revised and actual expenditures are not yet available.

As Table 3.3 and Fig. 3.4 reveal, despite the steady increase in budget allocations, revised expenditures (disbursement) and actual expenditures have been generally lower than the initial allocation and in some cases even lower than the previous fiscal year. For instance, during the fiscal year 2011–2012, the budget allocation of ₹6226 crores was reduced to ₹4432 crores in compliance with the reduced ceilings fixed by the Ministry of Finance, and the actual expenditure during the year was ₹3784 crores, which is about 85.38% budget utilisation with respect to the initial allocation. This is consistent with the periodical reviews of the physical and financial performance of all the projects carried out as part of the planning and implementation strategy of DOS/ISRO.

It is also interesting to note that although India has witnessed a budgetary increase for its space activities, overall space expenditure as a percentage of its GDP has progressively decreased over the past 15 years, whereas in 2002 India invested about 0.092% of its GDP in space activities; that figure reduced to just 0.064% of GDP in 2013. As already analysed by N. Prasad, these can be taken as clear indicators of strong economic growth that allow the expansion of space activities with a progressively lower impact on the national GDP (Prasad 2016a, b, c) (Fig 3.5).

The assessment of the budget breakdown is also a very useful indicator. Considering that the establishment of operational space systems in a self-reliant manner has been a primary objective of India's space programme, it is not surprising that, since its inception, the greatest share of India's space budget has been devoted to the development of indigenous space technology capability. This item

Fig. 3.5 Space Expenditure in Comparison to GDP (*Source*: Prasad 2016b)

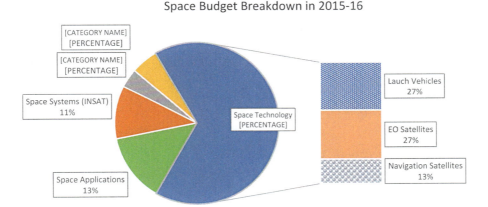

Fig. 3.6 ISRO's budget breakdown (*Source*: ISRO)

comprises the development of satellites, launch vehicles and associated ground segments. The budget breakdown for the fiscal year 2015–2016 is provided in Fig. 3.6.

With ₹4596.2 crores, space technology represents the first budget item for the year 2015–2016, followed by space systems (₹1320.95 crores) and space applications (₹962.32 crores). Space sciences and other programmes count for only a marginal share of the total budget.

3.1.2.2   Manpower

An important feature of the Indian space programme lies in the high level of manpower that ISRO has at its disposal. Interestingly, this has sensibly grown during the current Five-Year Plans (see Table 3.4). Today the total approved sanctioned strength of the department is 16,902 employees, of which 12,300 are in scientific and technical categories and 4602 are in administrative categories (see Fig. 3.7).

The majority of employees are concentrated in the VSSC (4443 employees as of 2016) followed by ISAC, SAC&DECU and NRSC. The distribution of the workforce in ISRO's main centres/units is provided in Fig. 3.8 (Indian Space Research Organisation 2016a, b, c, d).

Table 3.4 Evolution of manpower in ISRO

Year	2012	2013	2014	2015	2016
Employees	14,716	18,561	17,625	16,902	16,902

Source: ISRO

Fig. 3.7 Human resources
by category (*Source*:
ISRO)

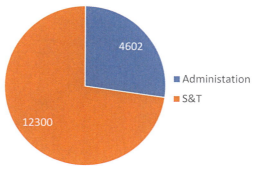

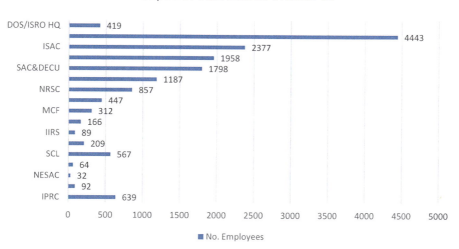

Fig. 3.8 ISRO's manpower (*Source*: ISRO)

3.1.3   ISRO and Industry

The remarkable manpower that DOS/ISRO has at its disposal is mainly due to the
fact that in both the upstream and downstream segments of India's space industry,
the role of the public actor is predominant, with ISRO acting as the main

manufacturer/provider of space products and services and Antrix as the main operator for the commercialisation of space-related assets, including remote sensing data imagery, transponder leases on Indian telecommunication satellites, ground station services, satellites launches and exports of satellite components and other products (see Sect. 4.5 for a more detailed assessment of India's commercialisation efforts).

The limited involvement of private industry is primarily a direct consequence of the limited industrial base India could rely on during its early stage of its space programme. Because of this situation, ISRO had to directly undertake the establishment of the required facilities and capabilities for the design, development, assembly, integration and testing of satellites and launch vehicles. Since the mid-1970s, ISRO has nonetheless started to support the creation of an industrial base for empowering domestic industries and offloading its activities through a policy of "technology transfer and buybacks" for the manufacturing of various products and services.

As reported by Murthi et al., thanks to the launch of this programme, "Industry started providing ISRO centres with a variety of hardware and software, including the propellants, structures for satellites and launch vehicle systems, electrical modules and subsystems, a variety of materials, components, ground systems and test equipment. The technology transfer programme was highly successful in farming out a substantial portion of manufacturing activities to industry and the industry derived several spin-offs including their involvement in markets related to space applications and imbibing the culture and practices associated with high degree of reliability and quality assurance" (Murthi et al. 2007). The overall impact of this programme was that over the course of four decades, an industrial base of 500 Indian enterprises, which are largely small and medium enterprises (SMEs), has been created and stimulated to contributing to the realisation of space, launch and ground infrastructure (Prasad 2016a, b, c).

Whereas ISRO now reports that 80% of the Polar Satellite Launch Vehicle (PSLV) production and almost 100% of certain applications systems are currently outsourced to industry (Indian Space Research Organisation 2016a, b, c, d), it is also important to stress that the current supply chain for space infrastructure development and exploitation continues to be centred around ISRO (see Fig. 3.9).

Within the current industry participation model, most of the private industry in India remains involved in the supply of Tier-2/Tier-3 products and services for ISRO (mainly subsystems components, as well as equipment and engineering services). There are a handful of Tier-1 space companies (delivery of complete subsystems) in the industrial ecosystem, a role that still belongs to ISRO. The final assembly, integration and testing (AIT) of spacecraft and launch vehicles are also conducted by ISRO (Prasad 2016a, b, c).

The major aerospace industries are organised in the Society of Indian Aerospace Technologies and Industries (SIATI), which counts around 300 industries from both the public and the private sector. Alongside the several private companies, some of the industries that work with ISRO include allied Government of India public sector units (PSU) such as Hindustan Aeronautics Limited (HAL), Bharat Electronics Limited (BEL) and Mishra Dhatu Nigam Limited (MIDHANI), to name a few. An overview of the main aerospace industries involved in the Indian space programme is provided in Table 3.5.

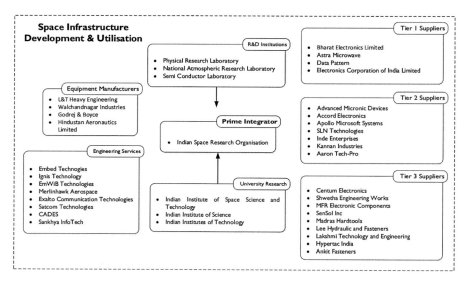

Fig. 3.9 The supply chain of the Indian space programme (*Source*: Prasad 2016a)

Table 3.5 Major Indian space industries

Name	Location	Contributions
Bharat Electronics Ltd.	Bangalore	Satcom payload items
Electronics Corp. of India	Hyderabad	Big and small Earth antennas and control systems
Godrej	Mumbai	Launcher engines shells
Hindustan Aircrafts Ltd.	Bangalore	Satellite bus for INSAT and IRS satellites
Kerala State Electronics Corp.	Thiruvananthapuram	Electronic parts for launchers
Larsen & Toubro	Mumbai	Launcher parts
Maehani Steel	Visakhapatnam	Supplies special steel for launchers
Prabhakar Products	Chennai	Road transportable carriers for satellite launcher segments and Earth station antennas
Walchand Industries	Walchandnagar	Launcher parts

Source: Confederation of Indian Industries

   Backed by the success of industry participation in the launch vehicle programme, ISRO has expressed a strong interest in enhancing the industry's role within the current ecosystem and in progressively letting the industry build satellites as well as launch vehicles, ultimately achieving a completely industry-built mission along the U.S. and European models.

   In order to reach this objective, in 2016 ISRO has set up two steering committees to delineate a comprehensive strategy for production with industry and is in the process of negotiating with industry players on the transitioning from solely vendors to integrators of the launch vehicle and satellites (The Economic Times 2016). In a question regarding industry participation in space programmes in the Lower House of Parliament, the strategy for expansion of industry participation and the role of the Steering Committees for stepping up launch capacity and satellite/payload capacity were articulated as follows:

> The terms of reference of the Steering Committee for stepping up the Launch Capacity include – (i) establishing the production profile of the launch vehicles; (ii) assessing the gaps in meeting the production requirement and (iii) arriving at a strategy for production including industry linkages, infrastructure build-up, technology sharing methods, quality assurance support.
>
> The terms of reference of the Steering Committee for stepping up the Satellite/Payload capacity include – (i) establishing the production profile of the Satellites and Payloads; (ii) assessing the gaps in meeting the production needs and (iii) creating a "Strategy Map" for production, industry linkages, joint ventures, infrastructure build-up, quality assurance support (Lok Sabha – Ministry of Space 2016).

Even though the typical supplier–system integrator relationship of SMEs and ISRO is now set to evolve, with industry gaining larger traction in the development and provision of turnkey product and services, it is clear that the dependence of ISRO on public sector companies, especially HAL, affects competitiveness in the marketplace, creating entry barriers for the private sector which is not assured of any customer other than ISRO due to export restrictions.

3.1.4   ISRO and Academia

Since the 1970s, ISRO has also nurtured the participation and contribution of academia in a variety of research and development activities related to space science, space technology and space applications. The main instrument to achieve this has been the RESPOND (Sponsored Research) programme.

The declared objective of the RESPOND programme is "to establish strong links with academic institutions in the country to carry out quality research and developmental projects which are of relevance to space and to derive useful outputs of such R&D to support ISRO programmes" (Indian Space Research Organisation 2016c, d). Through this programme, ISRO has provided financial support to research projects in a wide range of topics related to space technology, space science and application areas to universities/institutions. "In addition, conferences, workshops and publications, which are of relevance to the space programme, are also being supported" (Indian Space Research Organisation 2016c, d).

Apart from this, ISRO has also set up Space Technology Cells (STC) at premier institutions such as the Indian Institutes of Technology (IITs) – Bombay, Kanpur, Kharagpur and Madras – the Indian Institute of Science (IISc) in Bengaluru and the Joint Research Programme (JRP) with Savitribai Phule Pune University (SPPU) to carry out research activities in the areas of space technology and applications (Ibid).

In the 2015–2016 financial year, RESPOND supported 41 universities/colleges, 8 IIT institutes and 13 research centres to take up projects, both new and ongoing.

A large variety of projects was supported in all three broad areas, namely, space science (25), space applications (20) and space technology (35); in addition, 45 conferences/symposia/publications and other scientific/promotional activities were supported (see Figs. 3.10 and 3.11). During the year, 33 earlier sponsored projects were successfully completed.

Sponsored Research distribution by Institution (2015-16)

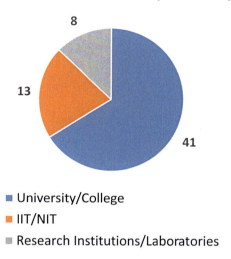

- University/College
- IIT/NIT
- Research Institutions/Laboratories

Fig. 3.10 Sponsored research by institutions (*Source*: ISRO)

Sponsored Research distribution by Area (2015-16)

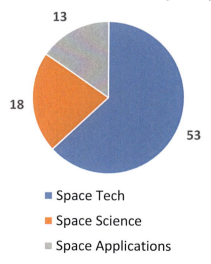

- Space Tech
- Space Science
- Space Applications

Fig. 3.11 Sponsored research by area (*Source*: ISRO)

Besides the RESPOND programme, ISRO's support for the development of academic research and the training of highly skilled human resources has been substantiated through additional tools. One such tool is the Indian Institute of Remote Sensing at Dehradun, a premier institute with the objective of capacity building in remote sensing and geo-informatics and their applications through education and training programmes at postgraduate level. The institute also hosts and provides support to the Centre for Space Science and Technology Education in Asia and the Pacific (CSSTE-AP), affiliated to the United Nations. The training and education programmes of the institute are designed to meet the requirements of various target/user groups, i.e. for professionals at working, middle and supervisory levels, fresh graduates, researchers, academia and decision-makers. In addition, IIRS also conducts special programmes for international and national participants on request from different organisations.

3.1.5   ISRO and the Military

India's national space programme has been characterised since its beginning by a strong civilian focus, rather than being driven by military interests, as has been historically the case of other spacefaring nations – with the sole exception of Japan. Hence, so far the involvement of the military has been rather marginal.

Since 2010, however, ISRO has become involved in the management of the Integrated Space Cell (ISC), the nodal agency within the Indian Ministry of Defence that oversees the security and utilisation of India's military and civilian space hardware systems. It is jointly operated by all the three services of the Indian Armed Forces (the Army, the Navy and Air Forces), the civilian Department of Space and ISRO.[5] The Integrated Space Cell is intended as a pivotal organisation in paving the way for an independent organisational structure for managing all the security- and defence-related aspects of the Indian space programme, while maintaining the civilian status of ISRO, which would otherwise run the risk of inhibiting its civilian counterparts in other spacefaring countries to enter into cooperative agreements with it.

Be this as it may, over the past decade, ISRO has launched four dual-use remote sensing missions with high-resolution optical and SAR capability, which have been used by the military to address some of their intelligence-gathering requirements. In addition, in August 2013, ISRO launched the first dedicated defence communication satellite (GSAT-7) for the Indian Navy (see Chap. 5).

On the same year, the Technology Perspective and Capability Roadmap (TPCR) was issued by the Headquarters Integrated Defence Staff of the Ministry of Defence in order to define the requirements of Indian Armed Forces, which include their space-based requirements. The roadmap identifies the utilisation of space assets for

---

[5]This move was intended to create the capabilities and necessary capacity to protect the country's space assets following the ASAT test by the Chinese in 2007. See Rao (2012) and Indian Armed Forces (2016).

meeting reconnaissance, surveillance, meteorological, navigation and communication requirements and the need for strategic forces to be supported by real-time information in the area of interest through sub-metric resolution imagery, aid in communications, all-day all-weather capability and signal intelligence, among others (Headquarters Integrated Defence Staff – Ministry of Defence 2013).

While the TPCR does not specify budget nor the organisation responsible for the implementation of the identified requirements, it appears that the remit is thus far entrusted to ISRO. At the same time, it should be highlighted that the civilian nature of ISRO creates inherent political hurdles in developing and deploying space assets for security and defence purposes (Prasad 2016a, b, c). Hence, the challenge for the government lies in creating an ad hoc organisational architecture with a dedicated budget that can enable the armed forces to integrate space capabilities into their operations and expand such capabilities, while allowing ISRO to effectively continue providing civilian applications (Ibid).

An independent organisational structure has been established in the area of the launch segment with the Defence Research and Development Organisation (DRDO) taking over the development of launchers for meeting the requirements of the armed forces. However, many policy-makers and commentators of the Indian space programme indicated the need to establish a dedicated aerospace command in the future to deal with all the various components of space security (Rajagopalan 2013).

3.2   Space Policies and Legislation

3.2.1   Space Policy Targets and Processes (Civilian Activities)

India has no declared, dedicated space policy that has been formalised into a single official document. The overall policy direction and goals of their space programme nonetheless can be derived from, and are enshrined in, a number of other key policy documents, including annual reports and the country's Five-Year Plans (which define the policy orientations of the space programme vis-à-vis access to space, EO, applications development, etc. for the subsequent 5 years; see below for specific policies per issue areas). In addition, since its inception, the space programme has been guided by so-called decade profiles, which set the goal and directions for long-lead-time technology development and investments. The major thrust areas identified for the decade 2010–2020 are summarised in Table 3.6.

Through the decade profiles, ISRO "commits itself to long term programmatic targets and strives to achieve these targets through Five Year Plans and Annual Plans" (Murthi et al. 2007). Such plans have been defined by the Planning Commission, the governmental body created in 1950 for managing India's economic planning processes.

In order to set out the space targets in its plans, the Planning Commission has traditionally formed a national-level S&T Steering Committee, within which a

Table 3.6 Thrust areas 2010–2020

2010–2020 thrust areas
Enhanced capabilities for space communications
Competitive and state-of-the-art space segment capacity augmentation in INSAT/GSAT system for national needs in the areas of communications, broadcasting and information infrastructure
Establishment of regional satellite navigation system and positioning services
Development of societal applications including tele-education, telemedicine and Village Resource Centres
Leadership in Earth observation
Enable EO infrastructure to meet the growing national imaging demand
Improve the National Natural Resource Management System, the Disaster Management Support System and the Weather and Ocean-State Forecast System
Further the development of EO-based developmental activities
Advancements in space transportation
Upgrading of the existing launch capabilities to meet the national launch needs
Realisation of a cost-effective two-stage-to-orbit (TSTO) vehicle by 2025
Critical technologies for human spaceflight
Development of space science
Undertake advanced space science endeavours
Execute planetary exploration missions (Moon and Mars)
Create an ecosystem for space activities
Enhance industry participation in realising space products and services
Promoting spinoffs and technology transfer
Human resource development
Strengthening the academia interface and forging international partnerships

Source: Planning Commission

working group on space is then constituted. Such a working group comprises all relevant stakeholders and is mandated with the following key tasks:

- To review the progress made during the previous Five-Year Plan
- To suggest plans and programmes for the next Five-Year Plan including thrust areas
- To suggest an optimum outlay for the plan

Based on an analysis of national development goals, including the overall resource position in the country as well as trends in space technology and other relevant aspects, a working document is produced by DOS and circulated among all the members of the working group, before submission to the Planning Commission (Indian Space Research Organisation 2006).

This participatory approach in the goal-setting process, as well as in the definition of specific projects, makes the public more committed and the goals less subject to political volatility. This has proved particularly important when the direct benefits and typical developmental rationales behind the initiation of new projects are not immediately straightforward. An interesting case in point is the initiative for robotic lunar exploration Chandrayaan-1, for which DOS/ISRO had to go through a complex and lengthy "process of consultation and justification with the scientific community, academia and political system and the public media before the project was given the go ahead" (Kasturirangan 2011). The process is summarised in Box 3.1.

Box 3.1: Decision-Making Process of Chandrayaan-1
Chandrayaan-1 mission's concept was presented in May 1999 by then ISRO Chairman K. Katsurirangan at a forum organised to celebrate National Technology Day. This was followed by presentations and discussions of the mission's scientific goals and challenges with the Indian Academy of Sciences and the Astronautical Society of India, respectively, in October 1999 and February 2000 (Kasturirangan, 2006). The two bodies strongly endorsed the proposed plans, which subsequently underwent a more detailed feasibility assessment. By April 2002, the positive political backing cutting across different party lines had led to the endorsement of the preliminary plans by a Standing Committee of the Parliament. The proposal was then further developed in further scientific and technical details by a competent task force and subsequently peer-reviewed by a national committee of experts in April 2003 (Ibid). The Space Commission gave the green light to the project in June 2003, and the announcement was eventually made by the Indian prime minister on 15 August 2003 (India's Independence Day). All in all, the decision-making process lasted 4 years, but it took only an additional 4 years to implement and launch the mission in 2008.

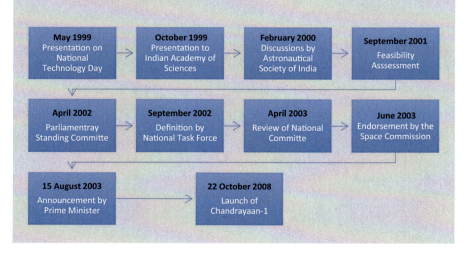

3.2.1.1   The 12th Five-Year Plan

The latest Five-Year Plan (the 12th) was launched in April 2012 and ends in March 2017. In terms of general goals and objectives, the focus of the current plan is on:

(a) The augmentation of the INSAT/GSAT capacity to meet demand for transponders.
(b) The establishment of the Indian Regional Navigational Satellite System over the Indian region.
(c) The continuation of established services with improved capabilities with thematic series of Indian EO satellites.
(d) Strengthening space-based Disaster Management Support.
(e) The expansion of satellite-based applications including societal applications.
(f) Strengthening of Polar Satellite Launch Vehicle (PSLV) and Geosynchronous Satellite Launch Vehicle (GSLV) as the workhorse vehicles.
(g) Realisation of developmental flights of next-generation launch vehicle (GSLV MK III) capable of launching 4T class INSAT satellites.
(h) Pursuance of semi-cryogenic engine development.
(i) Realisation of space science and planetary exploration missions, including Chandrayaan-2, Mars Orbiter Mission, Astrosat-1 and Aditya-1.
(j) The development of critical technologies for the pursuit of the human space-flight programme.

Overall, 58 missions have planned during the 12th Five-Year Plan period, including 33 satellite missions and 25 launch vehicle missions.

It is important to highlight that after the 12th Five-Year Plan, there will be no more plans because of the recent abolition of the Planning Commission and institutional changes introduced by the government of Narendra Modi (India Times 2015) (Box 3.2).

Box 3.2: What Happened to India's 13th Five-Year Plan?
In 2014, the Planning Commission was dismantled, and since January 2015, it was replaced by a new institution named NITI Aayog or the National Institution for Transforming India. The stated aim for NITI Aayog's creation is to foster involvement and participation in the economic policy-making process by the state governments of India. The emphasis is on bottom-up approach and on making the country to move towards cooperative federalism. It remains to be seen how this transformation will impact the Indian space programme in terms of budgetary allocation and space policy definition.

As an indication of the new approach, in 2015 the Cabinet secretary has asked all secretaries of all central ministries/departments to assess the current utilisation of space technology-based tools in their ministry/department and explore new potential application areas. Further, in order to hold proactive

(continued)

Box 3.2 (continued)
interactions with the central ministries, 18 expert teams were constituted within ISRO. These teams conducted one-to-one interactions with the ministries/departments, and joint action plan on "Effective Use of Space Technology" was prepared with 60 central ministries/departments. About 170 projects across various ministries/departments have emerged in the areas of natural resource management, energy and infrastructure, disaster and early warning, communication and navigation, e-governance and geospatial governance, societal services and support to flagship programmes. A one-day national meeting was organised by the NITI Aayog on 7 September 2015 in New Delhi to deliberate on the action plans of various ministries/departments (Government of India 2015).

3.2.2   Policy Framework of DOS

Besides the blueprint contained in the Five-Year Plan, the country's activities are embedded in a broader policy framework, which comprises a series of loosely defined policies (guidelines) and has the underlying objective of creating a cost-effective space infrastructure for the country in a self-reliant manner, ensuring its efficient utilisation for national development and enabling technology growth in multiple fields as spinoff benefits. This framework includes:

(a) Satcom Policy to enable the use of INSAT satellites by non-government sectors and to establish and operate private communication satellites
(b) Remote Sensing Data Policy to regulate the acquisition and distribution of satellite remote sensing data from Indian and foreign satellites for civilian users in India
(c) Industry Participation Policy to enhance the participation of Indian industries in national space endeavours
(d) Commercialisation Policy to extend the outreach of Indian space assets, products and services to the global market through Antrix Corporation
(e) International Cooperation Policy for promoting mutually beneficial bilateral and multilateral cooperative programmes
(f) Human Resource Development Policy to retain critical mass through the RESPOND programme
(g) User Participation Policy to involve the user agencies of the government of India in the planning and utilisation of space systems
(h) Technology Upgradation Policy to ensure state-of-the-art cost-effective space systems

Within this framework, the provisions of the Satcom Policy and Remote Sensing Data Policy deserve more attention, as these have the more elaborate provisions.

3.2.2.1   Remote Sensing Data Policy

The Remote Sensing Data Policy (RSDP) was adopted in 2001 and then amended in 2011. It contains guidelines for governing how remote sensing data from Indian and foreign satellites are to be acquired and distributed in India. In identifying remote sensing as a public good, the RSDP introduced the concept of "one window" access to any image through the National Remote Sensing Centre (NRSC), and that of "regulatory use-determination", whereby "images of resolution up to 1 m shall be distributed on a non-discriminatory basis", whereas "all data better than 1 m resolution shall be screened and cleared by the appropriate agency prior to distribution".

The policy also contemplates the possibility of "licensing", i.e. the possibility for future Indian private remote sensing satellites and private agencies to acquire/distribute any satellite images through "licensing". Although such an opening towards privatisation was envisaged as early as the 2001 RSDP, so far no licensing applications have in fact been encouraged, and NRSC has de facto remained the single monopolistic data provider. Such position is made even more monopolistic by the fact that the policy requires foreign satellite images to be routed through the NRSC, thereby denying Indian users higher-resolution images provided by foreign satellites. This has proven particularly burdensome for private users (Indian citizens and companies), as "India is unable to match the resolution quality of some commercial systems that have reached 0.3 m level in the global market" (Rao et al. 2015).

The restrictive approach enshrined in the 2011 NRSDP has not undergone significant changes with the National Geospatial Policy issued by the Department of Electronics and Information Technology in 2016. While the declared aim of this policy is to boost the emergence of a knowledge-based society, the licensing mechanisms for accessing geospatial data and information foresees onerous access requirements and procedures for the private users.

Overall, India's "current geospatial data and information distribution policies [remain] premised on the principle of "presumption of access denial", despite the National Data Sharing and Accessibility Policy released in 2012 clearly stating in its preamble that information and data "collected with the deployment of public funds should be made more readily available to all, for enabling rational debate, better decision making and use in meeting civil society needs" (Government of India 2012).

This policy approach is stark contrast to the geospatial data and distribution policies of the United States and Europe, which are based on the principle of "presumption of open access". While this is primarily the result of national security considerations, as argued by Ranjana Kaul, it also translates in "a self-imposed roadblock by the Government of India, which prevents the acceleration of new technology applications by indigenous innovations and start-up enterprises to assist in the timely achievement of national economic development goals" (Kaul 2017).

3.2.2.2   Satcom Policy

India Satellite Communication Policy was issued in June 1997, while the detailed norms, guidelines and procedures for implementation were approved by the Cabinet on 12 February 2000. The two documents address three major aspects: (a) the allocation and utilisation of INSAT capacity to non-governmental users (both Indian and foreign) on a commercial basis, (b) the establishment and operations of Indian satellite systems and (c) the utilisation of foreign satellites for the provisions of satcom services in India.

Concerning the first point, the policy authorises the leasing of INSAT capacity for the provision of services by the private sector. While most of this capacity is allotted for the use of the Department of Telecommunications, Doordarshan (a public service broadcaster owned by the Ministry of Information and Broadcasting) and All India Radio, clause 2.6 of the Norms specifies that a certain amount of capacity can be allotted for non-governmental users to offer communication services (including broadcasting) if authorised by the Department of Space. It also states that DOS "may enter into bilateral agreements with other agencies for marketing this capacity" and that in case the demand is more than the available supply capacity, DOS should adopt "suitable transparent procedures for allotting the capacity. This procedure may be in the form of auction, good faith negotiations, first come first served or any other equitable method" (Government of India 2000).

With respect to the second point, the policy framework envisages a "Committee for authorising the establishment and operation of Indian Satellite Systems (CAISS)" consisting of secretaries from the Department of Space, the Department of Telecommunications, Ministry of Information and Broadcasting, Ministry of Home Affairs, Ministry of Defence, Ministry of Industry (Department of Industrial Policy and Promotion) and Wireless Advisor to the Government of India. Such committee is responsible for reviewing applications and issuing licences. Clause 3.6 of the Norms specifies that "the application for notifying, registering and operating an Indian Satellite System/Network must be on behalf of a company registered in India". In fact, only Indian registered companies may be allowed to establish and operate an Indian satellite system. The foreign direct investment (FDI) in such a company shall not exceed 74%, although CAISS may license Indian registered companies with 100% FDI "to establish Indian Satellite Systems with the condition that over a period of five years after the issue of license for the establishment of the Satellite System the foreign direct investment should be brought down to the extent of 74% or less" (Government of India 2000).

Even though the Satcom Policy was put into place by the Government of India more than 15 years ago, the only licensee to date has been M/s Agrani Satellite Services Limited, which failed to kick off with Lockheed Martin as it failed to obtain an export licence due to the US export embargo on India (Harvey et al. 2010). Overall, foreign investments flowing into the Satcom field have remained untouched because of the several burdens and concerns inherently associated with the various norms and procedures. Besides the cumbersome and time-consuming process of applying for and securing licences,[6] the most critical factors identified by legal experts Ashok G.V. and Riddhi D'Souza include the fact that "the CAISS has been vested with tremendous discretion in reviewing applications seeking licences for operating and managing satellite systems. But the Satcom policy and the Norms fail to explicitly prescribe a framework for the exercise of this discretion" Since a key stakeholder in the CAISS is the Department of Space which also oversees ISRO, "there is justifiable apprehension of conflict of interest among the private sector enterprises who wish to attempt the application process for launching satellite systems. In addition, the Norms and the Policy are silent on key issues surrounding confidentiality of technology disclosures made to the CAISS by the applicants" (Ashok and D'Souza 2017).

Concerning the third point (the use of foreign satellites for operations in India), the provisions of the Satcom Policy are rather restrictive. In fact, the policy states that "operation from Indian soil with foreign satellites may be allowed only in special cases to be notified [and approved by the Indian government]. These may be in the case of overseas services using international inter-governmental systems, systems owned and operated by Indian Parties but registered in other countries before rules for registrations have been formulated in India, international private systems where there is a substantial Indian participation by way of equity or in kind contribution and where considered necessary reciprocal arrangements could be worked out with the country/countries of registration or ownership". In addition, in any case, "proposals envisaging use of the Indian satellites will be accorded preferential treatment" (Government of India 1997).

Unlike the RSDP, India's Satcom Policy has not changed since its introduction, although many commentators have highlighted the need to amend it in order to enable a more robust private investment in the country's space economy.

3.2.2.3 Non-proliferation Policies

In addition to these policies, mention must also be made of India's non-proliferation framework, since it has had huge impact on the development of its space programme. India is not part of the international non-proliferation regime, nor does India have a unified export control law, but it has an exemplary track record on non-proliferation of dual-use technologies. India's principles and policies of export control are embodied in a series of legal instruments that are in force, including (a) The Foreign Trade Act of 1992 and various policies issued pursuant to this act, (b) the Atomic Energy Act of 1962, (c) the Chemical Weapons Convention Act of 2000, (d)

---

[6]No deadlines are prescribed for clearances of such applications. In addition, the applicant is not only bound by the Satcom norms, but it must also secure licences for the specific services it proposes to offer through Indian satellite.

the Arms Act of 1959 and the Arms Rules of 1962, (e) the Customs Act of 1962 and
(f) the Weapons of Mass Destruction and Their Delivery Systems (Prohibition of
Unlawful Activities) Act of 2005.

3.2.3 Space Legislation

India does not have national space legislation, although discussions on the need for
and scope of a national space law were initiated as early as 2005. After 10 years,
ISRO has revitalised a process of formulating a National Space Act for regulating
space activities in India, facilitating enhanced levels of private sector participation
and offering more commercial opportunities. Towards this, a 2-day national workshop was held at ISRO Headquarters, Bangalore, in January 2015 with the participation of experts across the country. A draft version has been prepared, which is
now under consultation with experts, before possible approval by the Parliament.

Although within the current scenario the National Space Act remains far from
being approved any time soon, it is worth noting that space issues have been
receiving great attention within the two Houses of the Indian Parliament (Lok Sabha
and Rajya Sabha) during its three major sessions (budget session, monsoon session
and summer session). Such attention is demonstrated by the large number of
Parliamentary questions raised over the past years (see Fig. 3.12).

On average 75 questions have been raised each year in Parliament. The questions
have covered a plethora of topics, spanning from technology development and space
exploration to international cooperative ventures.

Fig. 3.12 Space in India's Parliament (*Source*: Government of India)

3.2.4 International Agreements

India is party to all five major international treaties on outer space activities, even if in the case of the Moon Agreement, India has signed but not ratified it. India has also been an active participant in a variety of multilateral fora, including the United Nations Committee on the Peaceful Uses of Outer Space (UN-COPUOS), the Committee on Space Research (COSPAR), the International Telecommunication Union (ITU), the Space Frequency Coordination Group (SFCG), the Committee on Earth Observation Satellites (CEOS), the Inter Agency Debris Coordination Committee (IADC), the Coordinating Group on Meteorological Satellites (CGMS), International Space Exploration Coordination Group (ISECG), International Global Observing Strategy (IGOS), the Asian Association for Remote Sensing (AARS), International COSPAS-SARSAT system for search and rescue operations, International Astronautical Federation (IAF), International Academy of Astronautics (IAA), International Institute of Space Law (IISL) and International Space University (ISU). In recent years, India hosted the 58th International Astronautical Congress (IAC) in Hyderabad, the 39th Scientific Assembly of COSPAR and the 26th Plenary of CEOS. In addition, "ISRO plays an active role in the provision of satellite data for the management of natural disasters through such multi-agency frameworks as the International Charter for Space and Major Disasters, UN-SPIDER and the Japan-led Sentinel Asia programme" (Indian Space Research Organisation 2016a, b, c, d).[7]

In terms of government-to-government cooperation, bilateral agreements have been signed with the space agencies of over 30 countries, including those of Argentina, Australia, Brazil, Brunei Darussalam, Bulgaria, Canada, Chile, China, Egypt, France, Germany, Hungary, Indonesia, Israel, Italy, Japan, Kazakhstan, Kuwait, Mauritius, Mexico, Mongolia, Myanmar, Norway, Peru, Republic of Korea, Russia, Saudi Arabia, Spain, Sweden, Syria, Thailand, the Netherlands, Ukraine, the United Kingdom, the United States and Venezuela, as well as with four multinational bodies, namely, the European Space Agency (ESA), the European Centre for Medium-Range Weather Forecasts (ECMWF), the European Organisation for Exploitation of Meteorological Satellites (EUMETSAT) and the South Asian Association for Regional Cooperation (SAARC) (Indian Space Research Organisation 2016a, b, c, d, p. 126). A more detailed analysis of India's cooperation undertakings will be provided in Chap. 6.

---

[7] The Centre for Space Science and Technology Education for Asia and the Pacific (CSSTE-AP) has been set up in India under the initiative of UN Office for Outer Space Affairs (UNOOSA) and offers 9-month postgraduate diploma courses in Remote Sensing and Geographic Information Systems (every year), Satellite Communication (every alternate year), Satellite Meteorology and Global Climate (every alternate year) and Space and Atmospheric Science (every alternate year). After completion of the course, students have the opportunity to carry out research in their own country for 1 year leading finally to the award of a master's degree from Andhra University (Indian Space Research Organisation 2016a, b, c, d).

3.3 Programmes and Infrastructure

3.3.1 Access to Space

Indian access to LEO and MEO currently relies on PSLV, while for GTO it relies on GSLV Mk II, which has a GTO capacity of approx. 2500 kg (see Table 3.7). In order to ensure independence from foreign sources and enable a domestic launch of all of its satellites, India is seeking to develop a launch vehicle able to lift its INSAT I-3K and I-4K satellite platforms into orbit. To satisfy its needs, ISRO is working in parallel on a possible Mk II evolution using an improved cryogenic upper stage and on the development of GSLV Mk III.

Development activities have however been progressing at a slower pace due to funding being bound by the procurement of foreign launch services for missions that cannot be launched on GSLV Mk II. Contrary to what the nomenclature may suggest, GSLV Mk III is an entirely different vehicle from Mk II and not just an evolution.

Following a positive sub-orbital test in 2014, and the first successful flight of the indigenous cryogenic stage in August 2015, the development of GSLV Mk III, a highly prioritised 4-tonne-class GTO launcher, is now scheduled for completion in 2017. Once GSLV Mk III becomes operational, it should replace GSLV Mk II.

PSLV and GSLV are launched from Sriharikota, Andhra Pradesh (Satish Dhawan Space Centre), where ISRO is currently operating two launch pads allowing for parallel operations of PSLV and GSLV following the inaugural launch from the Second Launch Pad (SLP) in 2005 (PSLV). Funding has been earmarked as of 2016 for the construction of a second vehicle assembly building.

PSLV production has been organised in batches of six in accordance with the initial intention of performing approximately two annual launches. As of 2016, annual launch rates have increased to 6 to allow one launch per year for foreign customers.

India's presence on the worldwide commercial launcher market has been very limited so far, mainly due to the decision to size its launcher vehicle production to satisfy just the needs of its domestic demand and the lack of a heavy-lift vehicle for large GEO telecommunication satellites (see also Chap. 4). Nevertheless, its PSLV launcher has achieved noticeable success in launching small satellites, often piggybacked on domestic LEO payloads. For example, several European satellites such as DLR-TUBSAT, Agile and Proba have been launched in this manner. It is worth noting that ISRO has stated its interest in privatising the small launcher PSLV within a 4-year timeframe, handling integration and launch of the rocket to an industrial consortium of Antrix Corporation (Laxman 2016). Such privatisation would enable an increase in the annual rate of launches to 12–18 (Ibid).

One of the primary objectives of the Indian space programme for the 2010–2020 decade is to complete its range of launch vehicles, as to avoid reliance on foreign launchers for its national payloads, particularly the multipurpose INSAT communication satellite series. In fact, the lack of a heavy-lift vehicle able to place large Indian telecommunication satellites in GEO orbit makes the completion of its launcher fleet a crucial priority. India also has ambitions to expand its space activities to human

Table 3.7 India's launch vehicles

Name	Stages					Performance (in kg)				Launch sites	Launches		
	Stage 0	Stage 1	Stage 2	Stage 3	Stage 4	LEO (400 km)	SSO	GTO	GEO		Maiden flight	Total launches	Total failures
PSLV-CA (core alone)	–	Core PS1 with S139 engine (solid propellant)	PS2 with L37.5H engine (UH25/NTO)	PS3 with S7 engine (solid propellant)	PS4 with L-2.5 engine (MMH/MON-3)	2100 52°	1,050,630 km	–	–	Sriharikota (FLP, SLP)	2007	10	1 + 1 partial
PSLV	6 solid propellant strap-ons (PSOM)					3500 52°	1,600,630 km	1050 1634 m/s	–		1993	11	0
PSLV-XL	6 extended PSOM-XL strap-ons (solid)					3700 52°	1,750,630 km	1100–1500 1634 m/s	–		2008	7	0
GSLV Mk I	4 strap-ons (L40H) (UH25/NTO)	GS1 with S139 engine (solid boosters)	GS2 with L37.5 engine (UH25/NTO)	GS3 with C12 cryogenic upper stage (LOX/LH2)	–	6000 49°	2,300,835 km	2000 1655 m/s	–		2001	6	2 + 2 partial
GSLV Mk II				GS3 with CUS indigenous cryogenic upper stage (LOX/LH2)		> 6000 49°	> 2,300,835 km	2500 1655 m/s	–		2010	2	1

(continued)

Table 3.7 (continued)

Name	Stages					Performance (in kg)				Launch sites	Launches		
	Stage 0	Stage 1	Stage 2	Stage 3	Stage 4	LEO (400 km)	SSO	GTO	GEO		Maiden flight	Total launches	Total failures
GSLV Mk III	2 S200 strap-ons (solid)	L 110 core stage with 2 Vikas engines (NTO/UDMH)	C25 stage with 1 CE-20 engine (LOX/LH2)	–	–	10,000	–	4500–5000 1655 m/s?	–	–	–	–	–

Source: ISRO

spaceflight in the longer term. Albeit with a relatively small budget allocated so far, India has pursued the development and testing of re-entry and landing prototypes and foresees a fully autonomous manned space vehicle to LEO by 2020.

With more than one-third of its overall space budget devoted to the development of launch vehicles, it is expected that once the GSLV Mk III is ready for deployment in mid-2017, ISRO will pursue the development of other elements of its space programme, including human spaceflight, but also further development of cryogenic technology to improve GSVL Mk III performance, as well as possibly development of a fully reusable two-stage-to-orbit launcher. From a reusability perspective, India is still at a very fundamental stage in the development of reusable rockets. Given the development and test cycles involved in achieving this challenge, India might reach this milestone only by 2030 (Rajwi 2016).

Having developed a complete family of launchers able to satisfy domestic demand, at the beginning of the 2020s, India could then look to substantially increase its share on the worldwide launcher market by offering increasingly reliable launch solutions, accompanied by moderately cheap prices.

ISRO has recently proposed establishing a third launch pad at the Satish Dhawan Space Centre, Sriharikota, to support increased launch frequency, provide active redundancy to existing launch pads and support the launching requirements of advanced launch vehicles. Further work on the design of the launch pad will be taken up at an appropriate time after finalising the configuration of the advanced launch vehicle, operationalisation of GSLV MIII, programmatic requirements and having resource availability.

3.3.2 Satellite Programmes

ISRO's satellite programmes mainly include communication and Earth observation satellites. Recent years, however, have witnessed a remarkable growth also in terms of navigation, space science and exploration and small satellite programmes.

3.3.2.1 Communication Satellites

The Indian National Satellite (INSAT) system, established in 1983, is "the largest domestic communication satellite system in the Asia Pacific Region" (Indian Space Research Organisation 2016a, b, c, d). It has several communication satellites in operation including commercial communication satellites such as INSAT-3A, INSAT-3C, INSAT-4A, INSAT-4B, INSAT-4CR, GSAT-6, GSAT-8, GSAT-10, GSAT-12, GSAT-14, GSAT-16 and GSAT-15. The overall coordination and management of the INSAT system rest with the INSAT Coordination Committee (Ibid). Table 3.8 page provides an overview of these satellites.

Table 3.8 Telecommunication satellites

Name	Launch date	Launch mass (in kg)	Power (in W)	Launch vehicle	Orbit type	Additional information
GSAT-18	06.10.2016	3404	6474[a]	Ariane-5 VA-231	GSO	48 transponders: normal and extended C-band, Ku-band
GSAT-15	11.11.2015	3164	6200[b]	Ariane-5 VA-227	GSO	24 transponders: Ku-band, two channel GAGAN
GSAT-6	27.08.2015	2117	3100	GSLV-D6	GTO	High-power S-band, CxS and SxC transponders
GSAT-16	07.12.2014	3181,6	6000[c]	Ariane-5 VA-221	GSO	48 transponders: normal and extended C-band, Ku-band
GSAT-14	05.01.2014	1982	2600	GSLV-D5	GSO	14 transponders: extended C-band, Ku-band, Ka-band beacons
GSAT-7	30.08.2013	2650	3000	Ariane-5 VA-215	GSO	Ku-band, C-band
GSAT-10	29.09.2012	3400	6474	Ariane-5 VA-209	GSO	30 transponders: normal and extended C-band, Ku-Band, GAGAN
GSAT-12	15.07.2011	1410	1430	PSLV-C17	GSO	12 transponders: extended C-band SSPA
GSAT-8	21.05.2011	3093	6242	Ariane-5 VA-202	GSO	24 transponders: Ku-band, GAGAN
INSAT-4CR	02.09.2007	2130	3000	GSLV-F04	GSO	12 transponders: Ku-band
INSAT-4B	12.03.2007	3025	5859	Ariane-5	GSO	24 transponders: Ku-band, C-band
INSAT-4A	22.12.2005	3081	5922	Ariane-5 V169	GSO	24 transponders: Ku-band, C-band
INSAT-3A	10.04.2003	2950	3100	Ariane-5 V160	GSO	25 transponders: normal and extended C-band, Ku-band, SAS&R
INSAT-3C	24.01.2002	2650	2765	Ariane-5 V147	GSO	32 transponders: normal and extended C-band, MSS

Source: ISRO
[a]Solar array providing 6474 Watts and two 144 AH lithium-ion batteries
[b]Solar array providing 6200 Watts and three 100 AH lithium-ion batteries
[c]Solar array providing 6000 Watts and two 180 AH lithium-ion batteries

3.3.2.2 EO Satellites

The IRS system is the second major satellite system established by ISRO. As of June 2017, the IRS constellation comprises 13 operational satellite SSO, namely, Resourcesat-1, Resourcesat-2, Resourcesat-2A, Cartosat-1, Cartosat-2, Cartosat-2A, Cartosat-2B, RISAT-1, RISAT-2, Oceansat-2, Megha-Tropiques, SARAL and SCATSAT-1, and four satellites in GEO, namely, INSAT-3A, INSAT-3D, KALPANA and INSAT-3DR (Indian Space Research Organisation 2016a, b, c, d). Table 3.9 provides an overview of these satellites.

3.3.2.3 Navigation Satellites

Satellite navigation has been identified as one of the most important programmes of the Department of Space for the 2012–2017 period. It includes activities such as GAGAN and the Indian Regional Navigation Satellite System (IRNSS) (Indian Space Research Organisation 2016a, b, c, d).

GAGAN is the first satellite-based augmentation navigation system over the Indian region. Its signals augment those of GPS in order to provide users with more precise positioning and reliability of GPS signals in terms of "integrity parameters". GAGAN is a joint project of ISRO and the Airport Authority of India. The main objectives of GAGAN are to provide satellite-based navigation services with accuracy and integrity required for civil aviation applications and to provide better air traffic management over Indian airspace.

The GAGAN architecture consists of three segments, namely, the ground segment, space segment and user segment. The ground segment consists of 15 Indian Reference Stations (INRES), 2 Indian Master Control Centres (INMCC) and 3 Indian Land Uplink Stations (INLUS). The GAGAN signal is being broadcast through two geostationary Earth orbit (GEO) satellites – GSAT-8 and GSAT-10 – covering whole Indian Flight Information Region (FIR) and beyond. An on-orbit spare GAGAN transponder is flown on GSAT-15. The availability of the GAGAN signal in space will bridge the gap between European Union's EGNOS and Japan's MSAS coverage areas, thereby offering seamless navigation to the aviation industry (Ibid).

As for the Indian Regional Navigation Satellite System (IRNSS), renamed NAVIC by Prime Minister Modi, it is an independent regional navigation satellite system designed to provide accurate position information service to users in India as well as the region extending up to 1500 km from its borders, which is its primary service area. IRNSS will provide two types of services, namely, Standard Positioning Service (SPS) and Restricted Service (RS), and is expected to provide a position accuracy of better than 20 m in the primary service area (GPS Daily 2016).

The IRNSS system mainly consists of a space segment, ground segment and user segment. The space segment is comprised of seven satellites: three satellites in GEO and four satellites in inclined GSO. The satellites, together with some relevant information, are listed in Table 3.10.

The whole IRNSS constellation became fully operational on 1 December 2016.

Table 3.9 EO satellites

Name	Launch date	Launch mass (in kg)	Power (in W)	Launch vehicle	Orbit type	Additional information
Resourcesat-2A	07.12.2016	1235	1250	PSLV-C36	SSPO	3 payloads: LISS-4, LISS-3, AWiFS
SCATSAT-1	26.09.2016	371	750	PSLV-C35	SSPO	Ku-band scatterometer
INSAT-3DR	08.09.2016	2211	1700	GSLV-F05	GSO	Middle infrared band and thermal infrared band imaging, higher spatial resolution in visible and thermal infrared bands
Cartosat-2 SS	22.06.2016	7375	986	PSLV-C34	SSPO	Regular remote sensing services using panchromatic and multispectral cameras
INSAT-3D	26.07.2013	2060	1164	Ariane-5 VA-214	GSO	4 payloads: 6-channel multispectral imager, 19-channel sounder, data relay transponder, search and rescue transponder
SARAL	25.02.2013	407	906	PSLV-C20	SSPO	ISRO-CNES joint mission; 3 payloads: Ka-band altimeter (Altika), Argos Data Collection System, solid state C-band transponder
RISAT-1	26.04.2012	1858	2200	PSLV-C19	SSPO	Carries synthetic aperture radar operating in C-band
Megha-Tropiques	12.10.2011	1000	1325	PSLV-C18	SSPO	ISRO-CNES joint mission, 4 payloads: MADRAS imaging radiometer, SAPHIR, ScaRaB, ROSA
Resourcesat-2	20.04.2011	1206	1250	PSLV-C16	SSPO	LISS-3, LISS-4 and AWiFS enhancements + Automatic Identification System (AIS) + 2 solid state recorders (200 GB each)
Oceansat-2	23.09.2009	960	1360	PSLV-C14	SSPO	3 payloads: OCM, ROSA, Ku-band pencil beam scatterometer (SCAT)
RISAT-2	20.04.2009	300		PSLV-C12	SSPO	Radar imaging satellite with all-weather capability
Cartosat-2	10.01.2007	650	900	PSLV-C7	SSPO	Panchromatic camera; frequent imaging
Cartosat-1	05.05.2005	1560	1100	PSLV-C6	SSPO	2 panchromatic cameras; cartographic applications
INSAT-3A	10.04.2003	2950	3100	Ariane-5V160	GSO	Very high-resolution radiometer and charge-coupled device
KALPANA-1	12.09.2002	1060	550	PSLV-C4	GSO	VHRR and DRT payloads

Source: ISRO

Table 3.10 IRNSS constellation

Name	Launch date	Launch mass (in kg)	Power (in W)	Launch vehicle	Orbit type	Additional information
IRNSS-1G	28.04.2016	1425	1660	PSLV-C33/ IRNSS-1G	GTO	Navigation payload: L5-band, S-band, rubidium atomic clock; ranging payload, C-band transponder; together the 7 satellites constitute the IRNSS space segment
IRNSS-1F	10.03.2016	1425	1660	PSLV-C32/ IRNSS-1F	GTO	
IRNSS-1E	20.01.2016	1425	1660	PSLV-C31/ IRNSS-1E	GTO	
IRNSS-1D	28.03.2015	1425	1660	PSLV-C27/ IRNSS-1D	GTO	
IRNSS-1C	16.10.2014	1425,4	1660	PSLV-C26	GTO	
IRNSS-1B	04.04.2014	1432	1660	PSLV-C24/ IRNSS-1B	GTO	
IRNSS-1A[a]	01.07.2013	1425	1660	PSLV-C22/ IRNSS-1A	GTO	

Source: ISRO
[a]Pending replacement due to failure of atomic clocks (as of 2/17)

3.3.2.4 Space Science and Exploration

India is a relatively recent player in the field of space science and exploration. For decades, ISRO's purely utilitarian approach has limited ISRO's space science missions to the Stretched Rohini Satellite Series (SROSS), which was launched between 1987 and 1994 (Prasad 2016a, b, c). While the four satellites were primarily intended to test development of the Augmented Satellite Launch Vehicle (ASLV), they were also used to conduct experiments in the fields of gamma-ray astronomy and research in astrophysics and upper atmospheric monitoring. In 1996, ISRO also launched the Indian X-ray Astronomy Experiment (IXAE) aboard the newly introduced PSLV to study periodic and aperiodic variability of X-ray binaries (Agrawal et al. 2007). Pursuant the introduction of the PSLV, ISRO's understandably focused its efforts on the deployment of its IRS system, and space science missions were put on the back-burner until 2003, when India's Chandrayaan-1 lunar exploration mission was approved by the government. The eventual success and international resonance of this mission – launched in 2008 – then catalysed an increased interest in the field of space science, as demonstrated by the list of missions approved during the 11th and 12th Five-Year Plans. Table 3.11 provides a list of missions pursued by ISRO during the 12th Five-Year Plan as well as those planned over the next few years.

   In addition to these missions, activities in space sciences undertaken by ISRO currently include research in:

- Astronomy and astrophysics
- Atmospheric sciences
- Planetary science and exploration (PLANEX)
- Space sciences promotion

Table 3.11 Space science and exploration missions

Satellite	Launch date	Launch vehicle	Notes
Mars Orbiter Mission	5.11.2013	PSLV-C25	ISRO's first interplanetary mission to Mars making India the first nation in the world to have successfully entered the Mars orbit on the first attempt
ASTROSAT	28.09.2015	PSLV-C30	India's first multi-wavelength astronomy mission covering soft-X rays, hard-X rays, near and far ultraviolet bands and visible band. It is designed for studies of time variability phenomena, like pulsations, flaring activity, etc.
Chandrayaan-2	2018	PSLV-XL	India's first attempt to develop a lander and a rover indigenously, though the mission was originally intended as a joint India–Russia mission
Aditya-1	2020	PSLV-XL	India's first dedicated scientific mission to study the sun and understand the physical processes that heat the solar corona, accelerate the solar wind and produce coronal mass ejections.

Source: ISRO

3.3.2.5   Small Satellites

ISRO's small satellite project aims to provide a platform for stand-alone payloads for Earth imaging and science missions within a quick turnaround time. With the purpose of creating a versatile platform for different kinds of payloads, two kinds of buses have been configured and developed.

The Indian Mini Satellite-1 (IMS-1) bus is conceived as a versatile bus in the 100-kg class which includes a payload capability of around 30 kg. The bus has been developed using various miniaturisation techniques. The first mission of the IMS-1 series was launched successfully on 28 April 2008 as a co-passenger along with Cartosat-2A. YouthSat is the second mission in this series and was successfully launched along with Resourcesat-2 on 20 April 2011.

The IMS-2 bus has evolved as a standard bus of the 400-kg class which includes a payload capability of around 200 kg. IMS-2 development is an important milestone as it is envisaged to be a work horse for different types of remote sensing applications. The first mission of IMS-2 is SARAL. SARAL is a cooperative mission between ISRO and CNES with payloads from CNES and the spacecraft bus from ISRO (Indian Space Research Organisation 2016a, b, c, d).

3.3.3   Ground Infrastructure

3.3.3.1   Launch Centre

Satish Dhawan Space Centre (SDSC) in Sriharikota – also referred to as the Spaceport of India – provides India with the necessary launch base infrastructure. Facilities for solid propellant processing, static testing of solid motors, launch vehicle integration and launch operations, range operations including telemetry, tracking and command network and mission control centre are found at SDSC. Two launch pads are utilised for PSLV and GSLV launches. The centre bears the paramount responsibilities of (a) producing solid propellant boosters for ISRO's launch vehicle programmes; (b) providing an infrastructure for the qualification of subsystems and solid rocket motors, as well as conducting the necessary tests; and (c) providing the vital launch base infrastructure. In preparations for expected increases in launch activity, the centre is undergoing augmentations.

While in the Vehicle Assembly Building (VAB), India's launch vehicles PSLV, GSLV and GSLV Mk III are being integrated for launch from the second pad, and the second Vehicle Assembly Building (SVAB) serves the role of backing up VAB and bolstering ISRO's infrastructure.

Sounding rockets and payloads along with their launch infrastructure are assembled, integrated and launched from a different location within SDSC SHAR (Indian Space Research Organisation 2016a, b, c, d).

3.3.3.2   ISTRACK

Located in Bengaluru, the ISRO Telemetry, Tracking and Command Network (ISTRAC) provides tracking support to all ISRO launch vehicle and satellite missions. Common responsibilities include the estimation of satellites' preliminary orbits as they are launched into space, mission support to all remote sensing and scientific satellites operational in normal phase, the operation and maintenance of the IRNSS ground segment and radar system development for meteorological and launch vehicle tracking purposes. ISTRAC's manifold responsibilities do not end there; it also plays important roles in ISRO's deep space missions, search and rescue, disaster management and various civil applications such as TM and TE. With the Bengaluru Mission Operations Complex (MOX) connected to a network of nearly a dozen stations spread across the country and region, ISTRAC manages to operate on a broad scale. Especially noteworthy was ISTRAC's establishment of 18 vital IRNSS ground stations in support of the navigation system (Indian Space Research Organisation 2016a, b, c, d).

3.4   Indian Space Programme in a Comparative Perspective

In order to assess the *relative* position that the Indian space programme has in a comparative international perspective, a number of tools can be deployed. Certainly, it would be useful to compare the space-related expenditures of the major spacefaring nations and their respective global ranking. Comparing space budgets is, however, inherently misleading and not necessarily indicative of the real performance of a given space programme. Considering the several complications involved in this type of comparison (including the economic reality of India), this comparison should be intended as offering an order of magnitude only. Figure 3.13 provides a comparison of space budgets.

At $912 million, ISRO maintains a relatively cost-effective annual budget considering its various missions and ongoing projects. The highest spender, the United States, allocates more than 22 times the amount of money ISRO does for its civil space applications, while the ESA's budget exceeds that of ISRO by more than 400%. Yet, it is necessary to put this figure into perspective, given the economic

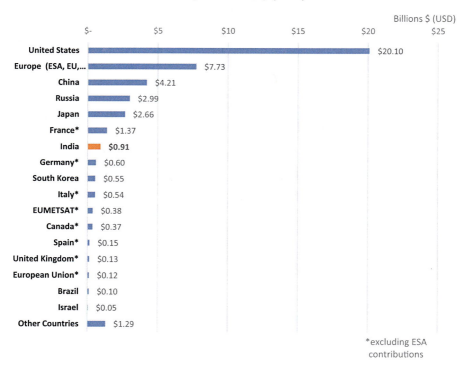

Fig. 3.13 Civil space budgets 2015 (*Source*: Space Foundation 2016)

reality of India. It is undeniable that ISRO manages to put its funds to work efficiently and effectively, sending spacecraft into Martian orbit, developing GSLV Mk III, providing globally respected commercial launch services which have resulted in ISRO's record-breaking single launch of 104 satellites, establishing the Indian Regional Navigation Satellite System (IRNSS), steadily improving communication and EO satellites and progressively contributing to planetary and space sciences. In this sense, as mentioned earlier, ranking nations by civil space spending and attempting to derive a nation's space competencies, capacity or overall potential from this information alone is misleading and inconclusive.

Further putting space spending into perspective and neutralising currency conversion bias, Fig. 3.14 compares India's capital outflow in the space sector to that of other spacefaring nations as a percentage of gross GDP for each respective nation:

India's space expenditure relative to GDP places it well below big spenders such as Russia and the United States. However, the spending average for spacefaring nations excluding Russia and the United States stands at 0,038% of GDP, thus barely placing India above this average. This is rather remarkable given the extraordinary achievements made by ISRO over the past years and decades.

In summary, India is without doubt enjoying a growth curve as far as space activity at large is concerned. Their sub-billion-dollar 2015 space budget perfectly reflects the very nature of ISRO, which essentially specialises in highly economic operations and manages to make big moves on a global scale with moderate expenditure, both in absolute terms and relative to GDP.

In order to better capture India's relative position among the leading spacefaring nations, additional comparative methods must be included. Accordingly, Fig. 3.15 provides an overview of the workforce of the major space agencies worldwide.

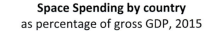

Space Spending by country
as percentage of gross GDP, 2015

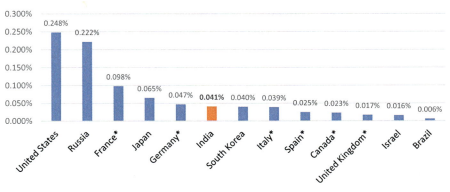

*including ESA contributions

Fig. 3.14 Space expenditures as percentage of GDP 2015 (*Source*: Space Foundation 2016)

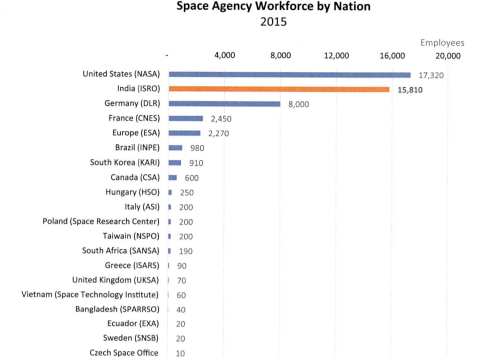

Fig. 3.15 Space Agency Workforce (*Source*: Space Foundation 2016)

Another meaningful comparative method is given by the statistics comparing the number of launches. Figure 3.16 displays the number of launches performed by the major spacefaring nations over the period 1957–2016.

As the figure shows, India accounts for only a tiny portion of world launches since 1957. In a historical context, India's launch activity not only falls far behind that of superpowers Russia and the United States but remains miniscule even in comparison to Japan, for instance, which has carried out more than double the launches of India. Russia clearly leads the pack, having recorded over 3000 orbital launches since the genesis of the space era six decades ago. This means that Russia has carried out an average of 51 orbital launches a year, or just about one a week, for the past 60 years. India, in stark contrast, has averaged only a single launch every 16 months over the same time period and has cumulatively conducted only 1.4% of the launches of Russia. The United States trails in a distant second place with just under half of Russia's launches, followed by yet another substantial gap, on the other side of which Europe and China stand in the mid-2000s. In essence, uncontested leaders Russia and the United States far exceed what can be considered a second tier – consisting of Europe and China – and a third tier made up of emerging Asian powers, China and India, in regard to orbital launch history.

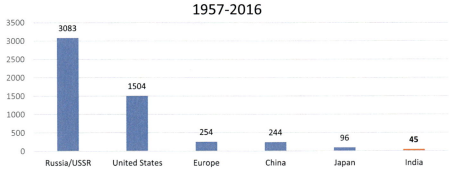

Fig. 3.16 Worldwide number of launches 1957–2015 (*Source*: ESPI)

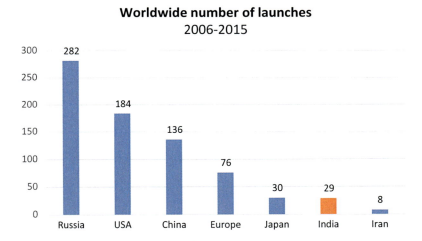

Fig. 3.17 Number of launches 2006–2015 (*Source*: FAA)

India's small share in global launch activity has, however, moderately increased over the past 10 years. Evaluating India's current standing in the global space community requires us to focus on a more recent timeframe (see Fig. 3.17).

In light of such numbers, one would imagine ISRO to be but a blip on the map of global space activity. Upon closer examination of more recent launching data, however, India is increasingly sending payloads into orbit as these grew from two launches in 2012, to five in 2015, to seven orbital launches in 2016. This average of four annual launches over the past five years of course is still just a fraction of the 26 Russian, 18 Chinese and 17 American average annual launches over the same timeframe. Having said that, India is on a strong upward trend regarding launch frequency, having increased its orbital launches by 40% from 2015 to 2016, while

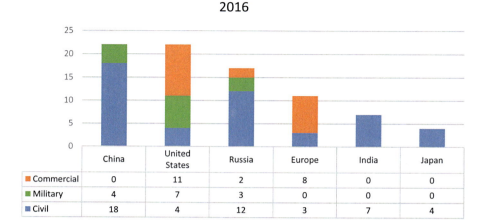

Fig. 3.18 Orbital launches 2016 (*Source*: FAA)

China's launches have increased by just 16% and the United States' by 10%, Europe's have stayed constant and Russia has seen a 35% reduction. Coupling this recent increase in orbital launch frequency with the steady capability advancement of the Indian space programme certainly speaks for a stronger Indian space presence in the future. In yet another narrowing of the scope, Fig. 3.18 breaks down global launches by application category.

With all of ISRO's orbital launches conducted in 2016 serving civil purposes, the space agency's focus is clear: when ISRO's launch vehicles blasted off from Indian soil last year, their application aimed in one way or another to directly improve life quality for the Indian people. A similar approach was taken by China, 82% of whose launches were part of civil space missions. The United States, by contrast, only undertook 18% civil launches and rather focused on commercial launch services which accounted for half of all their total orbital launches last year.

Furthermore, the total payloads delivered in last year's orbital launches on a global scale are displayed in Fig. 3.19, followed by Fig. 3.20 showing worldwide operational satellites by nation.

As we have now seen time after time, India's most basic space activity simply is no match for the sheer volume of the United States or Russia and more recently China. The great disparity in space activity is at least partially due to India's historically one-dimensional civil space application focus and long-time dependence on space superpowers before reaching greater self-sufficiency.

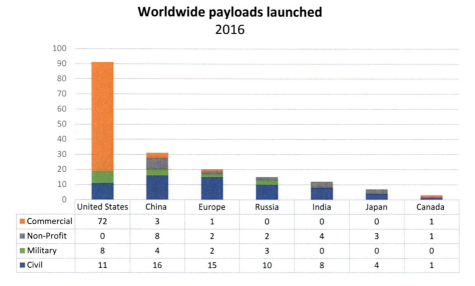

	United States	China	Europe	Russia	India	Japan	Canada
■ Commercial	72	3	1	0	0	0	1
■ Non-Profit	0	8	2	2	4	3	1
■ Military	8	4	2	3	0	0	0
■ Civil	11	16	15	10	8	4	1

Fig. 3.19 Payloads launched 2016 (*Source*: FAA)

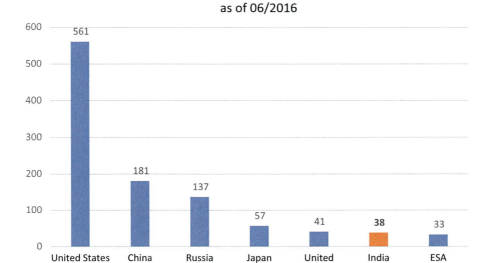

Fig. 3.20 Worldwide operating satellites (*Source*: Union of Concerned Scientists 2016)

References

Agrawal, P., Sreekantan, B., & Bhandari, N. (2007, December). Space astronomy and interplanetary exploration. *Current Science, 3*(12), 1767–1778.

Antrix Corporation. (2016). *About us.* Retrieved September 10, 2016, from Antrix Corporation Limited. http://www.antrix.gov.in/about-us

Ashok, G. V., & D'Souza, R. (2017). SATCOM policy: Bridging the present and the future. In R. P. Rajagopalan & N. Prasad (Eds.), *Space India 2.0 commerce, policy, security and governance perspectives* (pp. 119–139). New Delhi: Observer Research Foundation.

Department of Space – Government of India. (2016). *Outcome budget 2016–17.* New Delhi: Department of Space.

Government of India. (1997). *Policy framework for satellite communications in India.* New Delhi: Department of Space.

Government of India. (2000). *Norms, guidelines and procedures for implementation of the policy frame-work for satellite communications in India.* New Delhi: Government of India.

Government of India. (2012). *National data sharing and accessibility policy.* New Delhi: Government of India.

Government of India. (2015, September 7). *PM Modi to address national meet on promoting space technology on 7 September 2015.* Retrieved September 10, 2016, from Narendra Modi: http://www.narendramodi.in/pm-toaddress-the-national-meet-on-promoting-space-technology-based-tools-and-applications-in-governance-anddevelopment-at-vigyan-bhavan-on-7th-september-2015-291009

GPS Daily. (2016, May 4). *Operation of Indian GPS will take some more time: ISRO.* Retrieved September 21, 2016, from GPS Daily http://www.gpsdaily.com/reports/Operation_of_Indian_GPS_will_take_some_more_time_ISRO_999.html

Harvey, B., Smid, H. H., & Pirard, T. (2010). *Emerging space powers.* Chichester: Springer – Praxis.

Headquarters Integrated Defence Staff – Ministry of Defence. (2013). *The technology perspective and capability roadmap.* Retrieved September 10, 2016, from Ministry of Defence. http://mod.gov.in/writereaddata/TPCR13.pdf

India Times. (2015). *15 year development agenda to replace five year plans.* Retrieved from India Times http://economictimes.indiatimes.com/news/economy/policy/15-year-development-agenda-to-replace-five-year-plans-to-include-internal-security-defence/articleshow/52247186.cms

Indian Armed Forces. (2016). *Integrated space cell.* Retrieved from Indian Armed Forces: http://indianarmedforces.weebly.com/integrated-space-cell.html

Indian Space Research Organisation. (2006). *Eleventh five year proposal.* Bangalore: ISRO.

Indian Space Research Organisation. (2016a). *Small satellites.* Retrieved September 7, 2016, from ISRO http://www.isro.gov.in/spacecraft/small-satellites

Indian Space Research Organisation. (2016b). *International Cooperation.* Retrieved September 20, 2016, from Indian Space Research Organisation http://www.isro.gov.in/international-cooperation

Indian Space Research Organisation. (2016c). *ISRO annual report 2015–2016.* Bangalore: ISRO.

Indian Space Research Organisation. (2016d). *Sponsored research – ISRO.* Retrieved 2016, from ISRO http://www.isro.gov.in/sponsored-research-respond

Indian Space Research Organisation – Deloitte. (2010). *Overview of the Indian space sector.* Bangalore: Deloitte.

Kasturirangan, K. (2006). *India's space enterprise: A case study in strategic thinking and planning.* Canberra: Australia South Asia Research Centre.

Kasturirangan, K. (2011). *Relating science to society.* Mysore: University of Misore.

Kaul, R. (2017). A review of India's geospatial policy. In R. P. Rajagopalan & N. Prasad (Eds.), *Space India 2.0 commerce, policy, security and governance perspectives* (pp. 141–150). New Delhi: Observer Research Foundation.

Laxman, S. (2016, February 15). *Plan to largely privatize PSLV operations by 2020: ISRO chief.* Retrieved September 20, 2016, from The Times of India http://timesofindia.indiatimes.com/india/Plan-to-largely-privatize-PSLV-operations-by-2020-Isro-chief/articleshow/50990145.cms

Lok Sabha – Ministry of Space. (2016). *"Questions: Lok Sabha," Government of India.* Retrieved September 2, 2016, from Lok Sabha http://164.100.47.192/Loksabha/Questions/QResult15.aspx?qref=35204&lsno=16

Murthi, K. S., Sankar, U., & Madhusudhan, H. (2007, December). Organisational systems, commercialisation and cost-benefit analysis of Indian space programme. *Current Science, 93*(12), 1812.

Prasad, N. (2016a, March). Demystifying space business in India and issues for the development of a globally competitive private space industry. *Space Policy, 36*, 1–11.

Prasad, N. (2016b). Diversification of the Indian space programme in the past decade: Perspectives on implications and challenges. *Space Policy, 36*, 38–45.

Prasad, N. (2016c). Industry participation in India's space programme: Current trends & perspectives for the future. *Astropolitics: The International Journal of Space Politics and Policy, 4*(4).

Rajagopalan, R. P. (2013, October 13). Synergies in space: The case for an Indian aerospace command. *ORF Issue Brief, 59.*

Rajwi, T. (2016, March 29). *RLV-TD likely to touch record Mach 5 speeds: ISRO officials.* Retrieved from The New Indian Express http://www.newindianexpress.com/cities/thiruvananthapuram/RLV-TD-Likely-to-Touch-Record-Mach-5-Speeds-ISRO-Officials/2016/03/29/article3351385.ece

Rao, R. (2012, February 1). *Why does India need an aerospace command?* IPCS(3570).

Rao, M. K., Murthi, S., Raj, B. (2015). Indian Space - Towards a "National Eco-System" for Future Space Activities. Jerusalem: 66th International Astronautical Federation.

Sankar, U. (2006). *The economics of India's space program. An exploratory analysis.* New Delhi: Oxford University Press.

Space Foundation. (2016). *The space report 2016: The authoritative guide to global space activity.* Colorado Springs: Space Foundation.

The Economic Times. (2016, May 12). ISRO has set up committees for production of satellite launch vehicles with private sector. *The Economic Times, 2016.*

Union of Concerned Scientists. (2016). *UCS satellite database.* Retrieved April 6, 2017, from Union of Concerned Scientists: https://www.ucsusa.org/nuclear-weapons/space-weapons/satellite-database#.WlR3WyOZPOQ

Chapter 4
Understanding India's Uses of Space: Space in India's Socio-Economic Development

4.1 Introduction: Embedding Space in the Broader Context

In the story of "the blind men and the elephant" that originated on the Indian sub-continent, six blind men are asked to determine what an elephant looks like by each feeling a different part of the elephant's body. The blind man who feels a leg says the elephant is like a pillar; the one who feels the tail says the elephant is like a rope; the one who feels the trunk says the elephant is like a tree branch; the one who feels the ear says the elephant is like a hand fan; the one who feels the belly says the elephant is like a wall; and the one who feels the tusk says the elephant is like a solid pipe. A king explains to them: "All of you are right. The reason every one of you is telling it differently is because each one of you touched the different part of the elephant. So, actually the elephant has all the features you mentioned".

The ancient Jain text uses the story of the blind men to explain the concepts of *anekāntavāda*, or non-one-sidedness, which acknowledges the concepts of perspective-based truth and highlights the importance of considering all viewpoints in obtaining a full picture of reality. The doctrine of Anekantavada holds that it is "impossible to properly understand an entity consisting of various properties without including all viewpoints, since it will otherwise lead to a situation of seizing a superficial, inade-quate cognition, on the maxim of the blind men and the elephant" (Mallisena 1933).

Similarly, the Indian space programme presents different features that cannot be fully appreciated if taken in isolation from the context in which it has developed, particularly when looking at the uses policy-makers have made of it. Indeed, while the development of indigenous technological capability in a self-reliant manner is generally identified as the primary objective of the Indian space programme, tech-nology was never "visualised as being independent and separate from the social and political contexts in which they were happening". Quite to the contrary, it was understood as a tool for achieving broader objectives and addressing many of the societal issues facing this huge and developing country.

© Springer International Publishing AG 2018

M. Aliberti, *India in Space: Between Utility and Geopolitics*, Studies in Space Policy 14, https://doi.org/10.1007/978-3-319-71652-7_4

Socio-economic development itself presents different facets: there are not only drivers related to societal needs and the alleviation of poverty but also macroeconomic goals as well as scientific, commercial and technologically innovative issues. All these elements must be disentangled to better seize the current dynamics of the Indian space programme. In addition, although ISRO has traditionally taken a very utilitarian approach to space, the programme is undergoing diversification, meaning that this narrow, utilitarian focus is expanding. This is not surprising. India is both a developing country and a developing power. By necessity, it needs to maintain a coordinated and comprehensive approach to development, thus simultaneously leveraging the different facets.

By the beginning of the twenty-first century, with the imminent "rise of India" and the vast progress of its space programme, the societal uses of space of the early decades witnessed an expansion. Space has progressively become an active enabler of economic expansion also in terms of macroeconomic production functions, particularly through technological innovation and industrial upgrading and commercialisation, as explained in the following sections.

In addition, alongside the socio-economic development facet, there is a new geopolitical dimension that is currently taking root. Space is gradually becoming a source of prestige and a tool in India's geopolitical calculus and action on the international stage. Hence, when trying to disentangle and assess India's uses of space, it is essential to embed space activities in the broader social, economic and political context. This is the objective of this chapter and of the following one. The chapter in hand discusses India's uses of space as a tool of its economic emersion strategy and a way to leapfrog developmental stages; the second analyses the emerging uses related to India's diplomacy and security.

4.2   India's Emergence Amidst Development Challenges

For most of its independent history, India has followed a socialist-inspired development model. In stark contrast to the bottom-up development model initially proposed by Gandhi – characterised by a weak central government with self-sustaining villages being the actual centre of economic production – the vision for India of Prime Minister Jawaharlal Nehru was based on a top-down, Soviet-style, state-led industrialisation.[1] Nehru, along with the statistician Prasanta Chandra Mahalanobis, believed a large government-run public sector, import substitution industrialisation (ISI), high regulations and central planning would be the engine of India's future

---

[1] Broadly speaking, post-independent India witnessed two broad development discourses. Although the objective was the same, the approach greatly varied as two different paths emerged: while the Gandhian model was characterised as a bottom-up model with epistemological roots of inclusive innovation clearly traced to this model, and the Nehruvian model was often characterised as a top-down model, as reflected in the approach followed by space, atomic, defence and science agencies such as CSIR.

emergence. But this vision ultimately failed the reality check. Neither the economy nor industrial manufacturing took off, and the growth rate barely matched targets set by India's Five-Year Plans until the 1980s. Indian economist Raji Krishna coined the expression "Hindu rate of growth" to contrast the stagnant economic performance under the License Raj system with that of Confucian Asia, especially the Asian Tigers (Staley 2006).

4.2.1   India Rising

Departure from the economic policies of Nehruvian socialism began in the early 1980s and eventually took root in the economic liberalisation of 1991. Since then, India's economic performance has become impressive. A simple look at the curve of India's GDP growth pre- and post-liberalisation highlights this vertical take-off in India's economy (see Fig. 4.1).

As seen in Fig. 4.2, India had been on a steady upward growing trend leading up to economic liberalisation in the early 1990s. After recording negative annual GDP growth rates in 1989, 1991 and 1993, the Indian economy grew rapidly as of 1994, when a 4-year growth spurt boosted India's economy by 27%. Explosive growth continued through the 1990s and into the early 2000s when India recorded yearly

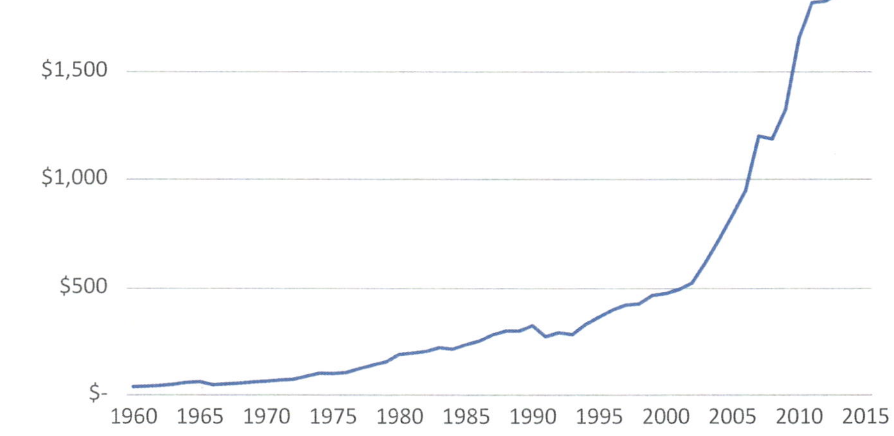

Fig. 4.1 The take-off of India's economy (*Source*: World Bank)

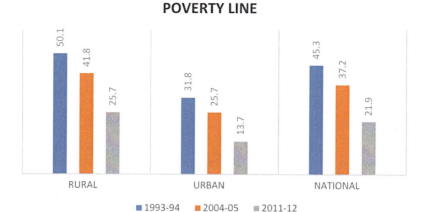

Fig. 4.2 Poverty rate (1993–2012) (The World Bank 2016a, b, c, d)

GDP growth high points of 26.5% in 2007 and 25.1% in 2010. The global financial crisis took a toll on India when the annual growth of India's economy went from the record 26.5% in 2007 to −1.18% in 2008. This was the first time in 14 years that India's economic growth halted and dropped below zero and only one of 6 years in which this had occurred since 1960. As the financial crisis and subsequent great recession sent numerous economies around the world into chaos, India proceeded to recover swiftly, leaving no doubt about the global economic force it had become over the past few decades. In 2009, 2010 and 2011, GDP growth helped reinvigorate the economy at rates of 11.54%, 25.13% and 10.05% annual growth, respectively. In the last 5 years, India's economy has been growing at single digit rates and remains among the world's fastest growing economies (The World Bank 2016a, b, c, d).

With an average annual growth of 7.5%, India is now among the top 10 percentile of fast-growing nations and has become a prominent global voice, heightening the interest of investors, companies and political leaders from around the world. India's impressive economic performance has undoubtedly been the main driver for change in the country and has enabled the country to attain one of the most significant achievements of recent times in terms of human development.

Since the early 1990s, India's poverty rate has fallen from about 45.3% to 21.9% in 2012, hence allowing the country to outstrip the UN Millennium Development Goal of halving the share of people living under the poverty line of $1.90 per day (see Fig. 4.2).

Likewise, achievements in human development have been significant: it suffices to note that life expectancy more than doubled from 31 years in the years after independence to 65 years in 2012 and adult literacy more than quadrupled from 18% in 1951 to 72% in 2015 (The World Bank 2016a, b, c, d).

In parallel to this significant progress in terms of human development, the miraculous economic performance of the past 15 years has enabled the country to catch up with other major Western economies and become more and more competitive

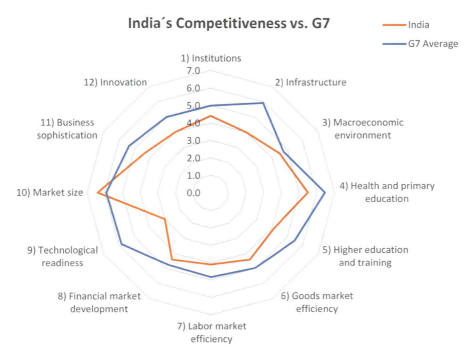

Fig. 4.3 India's performance across 12 pillars of competitiveness (*Source*: World Economic Forum)

internationally (see Fig. 4.3). In its global competitiveness index, the World Economic Forum positions India as 39th out of 138 across all indicators of performance, jumping 16 places in 2 years.

In line with these transformations, India has now emerged as the world's seventh largest economy in nominal terms and the third largest in PPP terms. Its share of world's GDP has increased from 1.8% in 2005 to approximately 3% in 2015. However, when considering its population being 18% of the world, that share remains proportionally very small and contributes to making India a middle–low-income economy with a per capita GDP as low as $1500.[2]

4.2.2   Persisting Development Challenges

This apparent contradiction is particularly evident at international negotiating tables, where India simultaneously represents one of the emerging economic giants and one of the poorest countries on Earth facing huge development challenges.

---

[2] To illustrate, India's overall GDP is about eight times smaller than the United States'.

To be sure, the country has come a long way in reducing endemic poverty and improving the living conditions of its population. According to the World Bank, India's progress in economic and human development is one of the most significant global achievements of recent times (The World Bank 2016a, b, c, d). Yet, in spite of its praiseworthy accomplishments, India remains the country with the highest number of poor globally – 270 million people in India live under the poverty line of $1.90 (The World Bank 2016a, b, c, d) – and continues to lag behind regional and international averages in terms of human development outcomes. In 2015, India ranked 130th out of 188 countries in the UN Human Development Index (United Nations 2015). As the index and other data sources reveal, India's malnutrition rates are the highest in South Asia (almost 40% of children aged 5 or lower are malnourished) and its life expectancy at birth (66.2 years remains the second lowest in the region). The country still accounts for over one-fourth (650 million) of the people living without access to sanitation facilities, and about one-third (almost 300 million) of the world's illiterate people live in India (The World Bank 2016c).[3] In addition, even though gains on the welfare and poverty reduction fronts have generally been widespread, several segments of society (in particular scheduled castes and tribes) and regions (in particular the low-income states) have been falling further behind. The World Bank estimated that "scheduled tribes in 2012 still experienced levels of poverty (43%) seen in the general population 20 years earlier in 1994 (45.5%)" (Ibid). It also notes that "although poverty has declined everywhere, it is increasingly concentrated in the poorest states – also the most populous – which have lagged in both growth and responsiveness of poverty to growth. The low-income states (LIS), as a group, have a poverty rate that is twice that of other states, and are home to a disproportionate share (61.5%) of India's poor people" (see Table 4.1).

These outstanding levels of poverty among several segments of society not only explain why India's per capita income is much lower than in other emerging economies but also shed light on the fact that the country's wealth is unevenly distributed, with severe income disparities increasingly making India one of the most unequal countries in the world.

Table 4.1 Poverty rates in low-income Indian states

	Share of population	Share of poor	Poverty rate
Bihar	8.6	13.3	33.7
Chhattisgarh	2.1	3.9	39.9
Jharkhand	2.7	4.6	37.0
Madhya Pradesh	6.0	8.7	31.7
Odisha	3.5	5.1	32.6
Rajasthan	5.7	3.8	14.7
Uttar Pradesh	16.5	22.2	29.4
Total for LIS	45.1	61.5	29.8

Source: World Bank 2016c

---

[3] Access to education is indeed an area of great concern in India. More information and data on this issue are offered in subsequent chapters.

4.2.3 A Rising Inequality and an Elusive Middle Class

The rise of inequality is well evidenced by the progressive increase of India's Gini coefficient (a measure of how income deviates from a perfectly equal distribution), which is now the highest among low-income economies (Agraval 2016; Corrigan and Di Battista 2015). Building on the research conducted by French economist Thomas Piketty, the India's Economic Survey 2015–2016 indeed confirms that over the past few years, there has been a growing concentration of income at the top. The top 1%, the top 0.5% and the top 0.1% of people in the overall income distribution accounted for 12.4%, 9.4% and 5% of the income of the entire Indian economy, respectively (Government of India 2016a, b).

The Credit Suisse provides an even more disconsolate picture. According to its Global Wealth Databook 2015, India's richest 1% accounts for over 53% of the country's wealth, the richest 5% owns 68.6% and the richest 10% owns 76.3% (Credit Suisse 2015). In addition, whereas several middle-income countries utilise redistribution to reduce inequality, this is not the case of India. As the World Economic Forum (WEF) points out, "tax revenues are extremely low and India's tax code is regressive, meaning that the poor bear a heavier burden than the rich, [a burden that] is not offset by social spending. The country spends only 2.5% of GDP on social protection compared with over 6% in many peer countries" (Corrigan and Di Battista 2015).

Primarily owing to this concentration, the size of India's middle class remains pitifully small, though there is still no consensus among economists on what income or consumption thresholds are most appropriate to define middle class. Owing to the different datasets and benchmarks that recent studies have deployed, the size of India's middle class has been estimated to be anywhere from 24 million people to 153 million people.[4]

Among these studies, of particular relevance is that from the Pew Research Center, since it is a massive study showing the evolution from 2001 to 2011 and including data from 111 countries (Kochhar 2015a, b). The study, which makes use of 2011 PPP prices, divides India's population into five income groups: poor (a daily per capita income of $2 or less), low income ($2.01–10), middle income ($10.01–20), upper-middle income ($20.01–50) and high income (more than $50) (Fig. 4.4).

Using these benchmarks, the study shows that the share of Indians who are middle income adds up to less than 3% of its population, the vast majority of which belongs to the low-income group (see Table 4.2 and Fig. 4.5). It also shows that while the country was very successful in cutting the poverty rate, which moved 133 million Indian out of poverty and fell from 35% in 2001 to less than 20% in 2011, this reduction of poverty mainly resulted in a transition towards the low-income

---

[4] For instance, the Credit Suisse estimated India's middle class at 24 million people; Pew Research Center projected 32 million, and McKinsey Global Institute report estimated 50 million, while NCAER projected 153 million. This could mean that anything from 3% to 24% of India's population can be considered middle class and shows how the parameter used in determining middle class status can produce the most different outcomes.

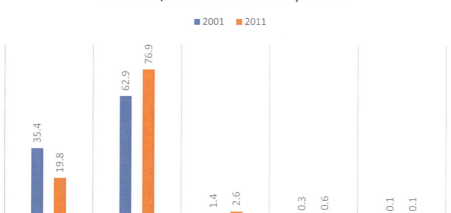

Fig. 4.4 India's population by income distribution (*Source*: Pew Research Center 2015)

Table 4.2 India's population by income distribution in 2001 and 2011

Income distribution 2011	2001 (%)	2011 (%)	Variation
Poor	35.4	19.8	−15.6
Low income	62.9	76.9	+14
Middle income	1.4	2.6	+1.2
Upper-middle income	0.3	0.6	+0.3
High income	0.0	0.1	+0.1

Source: Pew Research Center 2015

group, whose share rose from 63% in 2001 to 77% in 2011 (Kochhar 2015a, b). By contrast, gains for the middle-income group have by any measure remained modest. Their share did barely improve, passing from 1% in 2001 to less than 3% in 2011. And for the upper-middle and high-income groups, gains have been even smaller, less than 0.5%. It is also interesting to note that India has more poor and low-income people and fewer middle-, upper-middle- and high-income people when compared to other emerging economies and global averages (Pew Research Center 2015).

Other studies on the size of India's middle class have come to different findings, suggesting that it may account for between 5% and 10% of the population (Kochhar 2015a, b).[5] Although that share remains in any case small, it is nonetheless quite possible that these studies actually understate household well-being as might be measured using income only. As Pew also notes "researchers who adjust household

---

[5] However, these studies draw on different data sources and extend the middle-class status to encompass people living on up to $50 per day.

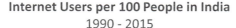

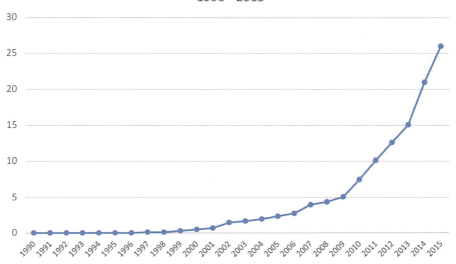

Fig. 4.5 Internet users in India (*Source*: The World Bank – ITU)

consumption data to account for the gap vis-à-vis estimates of income find that a somewhat larger share of India's population is middle income".[6]

4.2.4 What Indians Own and How They Live

In order to have a better picture of India's living conditions and associated socio-economic challenges besides the different and often conflicting estimates based on national income accounts, a commodity–/consumption-based approach could be more appropriate.

In this respect, statistics released by the India Census, the last being that of 2011, are a very useful tool (Ministry of Home Affairs – Government of India 2011). They provide a revealing account on how Indians actually live and shed new light on "a country in the throes of a complex transition, where millions have access to state-of-the-art technologies and consumer goods – but a larger number lacks access to the most rudimentary facilities" (The Hindu 2012). Table 4.3 provides an overview of India's household assets and access to various amenities.

---

[6]For instance, Birdsall finds that 70 million Indians, or 6% of the population, lived on $10–50 daily in 2010, while Kharas estimates that 5–10% of India's population earned $10–100 daily in 2010. Both estimates are based on 2005 Purchasing Power Parity (Kochhar 2015a, b).

Table 4.3 Household assets and access to amenities

	2001			2011			2001–2011 % age change		
	All-India	Rural	Urban	All-India	Rural	Urban	All-India	Rural	Urban
Total households	191.963.935	138.271.559	53.692.376	246.692.667	167.826.730	78.865.937	28.5%	21.4%	46.9%
No exclusive room households	3.1%	3.4%	2.3%	3.9%	4.3%	3.1%	0.8%	0.9%	0.8%
1 Room households	38.5%	39.8%	35.1%	37.1%	39.4%	32.1%	−1.4%	−0.4%	−3.0%
2 Room households	30%	30.2%	29.5%	31.7%	32.2%	30.6%	1.7%	2.0%	1.1%
3 Room households	14.3%	13.3%	17.1%	14.5%	12.7%	18.4%	0.2%	−0.6%	1.3%
4+ Room households	14.1%	13.4%	16%	12.9%	11.4%	15.9%	−1.2%	−2.0%	−0.1%
Tap water	36.7%	24.3%	68.7%	43.5%	30.8%	70.6%	6.8%	6.5%	1.9%
Bathing facility in house	36.1%	22.8%	70.4%	58.4%	45%	87%	22.3%	22.2%	16.6%
Toilet in house	35.8%	21.9%	73.7%	46.9%	30.7%	81.4%	11.1%	8.8%	7.7%
Kitchen in house	64%	59.4%	76%	55.8%	45.4%	77.8%	−8.2%	−14.0%	1.8%
Firewood for cooking	52.5%	64.1%	22.7%	49%	62.5%	20.1%	−3.5%	−1.6%	−2.6%
LPG/PNG for cooking	17.5%	5.7%	48%	28.5%	11.4%	65%	11.0%	5.7%	17.0%
Electricity for lighting	55.8%	43.5%	87.6%	67.2%	55.3%	92.7%	11.4%	11.8%	5.1%
Kerosene for lighting	43.3%	55.6%	11.6%	31.4%	43.2%	6.5%	−11.9%	−12.4%	−5.1%
Television	31.6%	18.9%	64.3%	47.2%	33.4%	76.7%	15.6%	14.5%	12.4%
Telephone (any)	9.1%	3.8%	23%	63.2%	54.3%	82%	54.1%	50.5%	59.0%
Bicycle	43.7%	42.8%	46%	44.8%	46.2%	41.9%	1.1%	3.4%	−4.1%
Scooter/other two-wheeler	11.7%	6.7%	24.7%	21%	14.3%	35.2%	9.3%	7.6%	10.5%
Availing banking services	31.0%	30.4%	33.1%	58.7%	54.4%	67.8%	27.7%	24.0%	34.7%

Source: Ministry of Home Affairs

The first thing the census data shows is that India's society is principally structured around nuclear families residing in rural areas.[7] The data reveal that the vast
majority of households live in one or two room houses, while less than 15% live in
houses with more than three rooms. In addition, there are still 10 million households
(3.91% of the total) without any exclusive room for living, and a third of urban
households live in slums (Ministry of Home Affairs – Government of India 2011).

The figures on access to key services such as water, electricity and improved
sanitation are a clear reminder that India's development agenda remains a complex
work in progress. The percentage of houses with an indoor toilet is still less than
50%, and there is a huge gap between urban and rural households, 67% of which
relieve themselves in the open area (Office of the Register General and Census
Commissioner 2011). One-third of the population still does not have access to electricity (only 55% of rural households compared to 93% of urban households), and
nearly half of the households still use firewood for cooking, a figure that has barely
changed since the census of 2001. Similarly, only a small portion of the population
(32% of all households) enjoys access to tap water from a treated source, and access
remains unevenly distributed.

By contrast, the figures on the ownership of certain amenities such as televisions,
computers and telephones stand out. Nearly half of the households own a television,
and more than 60% have at least one telephone, more than half in rural areas and
82% in urban areas. In terms of vehicle ownership, however, the large majority of
Indians continues to rely on cycles and two-wheelers for mobility. If car ownership
was used as an indicator of middle-class status, only 4.7% of Indian households
would enjoy that status (Haub 2012).

What is also remarkable to note is that Indians entering the Internet age are doing
so through mobile devices rather than desktop computers. The large majority of
Indians have never had a desktop and probably never will: India's Internet is increasingly built around smartphones (IndoGenius 2016b, c). Using the data compiled by
the ITU and the World Bank, Fig. 4.5 provides an overview of the Internet penetration in India, while Figs. 4.6 and 4.7 compare the number of fixed broadband subscriptions versus mobile cellular subscriptions. All in all, despite that Internet
penetration in the country remains limited, representing less than a third of the
population, with it 375 million people connected to the Internet, India already has
the second highest number of Internet users in the world, after China (The World
Bank 2016a, b, c, d).

---

[7] India has a population of 1.3 billion but only 246 million households, compared that to the USA
with a population one-fourth that of India but with 120 million households, nearly half that of
India.

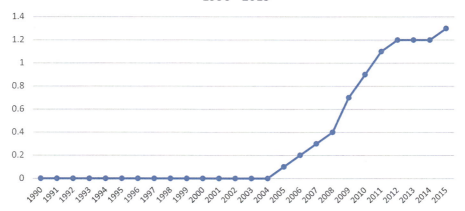

Fig. 4.6 Fixed broadband subscriptions in India (*Source*: The World Bank – ITU)

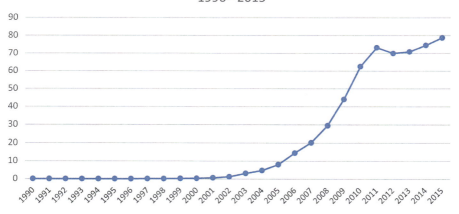

Fig. 4.7 Mobile cellular subscriptions in India (*Source*: The World Bank – ITU)

4.2.5   Urban–Rural Transformations and Impact

As all the previous figures show, there is a large divide between urban and rural India, particularly with regard to certain assets such as access to sanitation and electricity. It is hence useful to have a deeper look, though a similar variety of data can be seen among different regions and among different strata of society (tribes) within the same regions, cities or villages.

According to the World Bank, 67.2% (881 million people) of India's population lives in rural areas, while India's towns and cities are home to 429 million people or 32.8% of the population (The World Bank 2016a, b, c, d). Although India's population remains by far concentrated in rural areas, the country is nonetheless undergoing a profound urban transformation, and since the last decade, it has been witnessing greater absolute growth in its urban rather than rural population (see Table 4.4). Over the past 15 years, the urban population has increased by 137.9 million, passing from 27.6% to 32.8% of the total.

In the same period, the number of towns with a population of 5000 people or more has grown from about 5.000 to more than 8000, and 53 cities have now a population greater than one million. Eight of these cities have populations of more than five million people. And four of those cities (Delhi, Mumbai, Kolkata and Bangalore) have populations exceeding ten million people (IndoGenius 2016c). New Delhi, according to the UN, is projected to become the largest city in the world, overtaking Tokyo sometime in the 2030s, and India's urban population is expected to rise by 50% in the next 20 years and reach 600 million, or 40% of the whole population, by 2031, according to the UN World Urbanisation Prospects (United Nations 2014).

This accelerating urbanisation is thought to be key to India's development and poverty reduction. However, it would be erroneous to think that the country's wealth is concentrated in its richer urban areas and that there is a single type of rural India in terms of income and level of development. In fact, research demonstrates that it is more appropriate to think of three broad types of rural India, namely, a developed rural India, an emerging rural India and an underdeveloped rural India (People Reserach on Indian Consumer Economy 2014). Out of the 179 million rural households, "29 million can be classified as developed rural, 53 million as emerging rural and 97 million as underdeveloped" (IndoGenius 2016c). It is furthermore interesting to note that about half of "the 29 million households in developed rural India are actually in the top 20% of all the incomes in the country and that a further 17% of the households in emerging India are also in that top 20%" (IndoGenius 2016c). It is also noteworthy that around 40% of total household income in rural India comes

Table 4.4 Urban and rural population in 2000 and 2015 (The World Bank 2016b)

	2000	2015	Difference
Urban	291.4	429.3	137.9
Rural	762.0	881.7	119.7
All-India	1053	1311	258

from nonagricultural sources (People Reserach on Indian Consumer Economy 2014) and that consumption levels in rural India are also relatively high. In fact, 56% of India's entire consumer spending is rural, not urban (IndoGenius 2016c).[8]

In short, what these figures all underline is that generalisations about the level of wealth in rural versus urban India can prove deceiving, because there are many different income levels within both urban and rural areas. Similarly, it would be misleading to think that the ongoing urbanisation processes are exclusively taking place in India's large metro cities. In fact, these processes "are taking place in thousands of towns and villages right across the country, all of which have ever-increasing demand for amenities and services like education and healthcare, in addition, of course, to housing and infrastructure" (IndoGenius 2016c).

What on the contrary appears more evident is that these profound urban–rural transformations are generating far-reaching impacts on many issue-areas, not least on the environment. The urbanisation processes, combined with faster economic growth and private sector development, have indeed "accelerated environmental degradation and depletion of scarce natural resources essential for sustaining growth and eliminating poverty". As stressed by the World Bank, "India's long-term growth is predicated on its own ability to address environmental problems such as soil erosion, water and air pollution growing water scarcity, and the declining quality of forests". The cost of environmental degradation is estimated to be 6.6% of India's GDP (The World Bank 2016b, c).

Moreover, India is much more vulnerable to climate change because of high levels of poverty, high population density, heavy reliance on natural resources and an environment already under stress. With an increase in mean annual temperatures of 1.1 to 2.3 degrees Celsius (a moderate climate change scenario), the risk of increased frequency and severity of natural hazards are likely to increase, and densely populated cities will be at extreme risk. Kolkata is among the six fastest-growing cities worldwide that are classified to be at extreme risk, whereas Mumbai, Delhi and Chennai are among the ten that are classified as high risk. Overall, India is ranked the second most vulnerable country in the world. Institutions and mechanisms for enhanced disaster risk management and climate resilience, especially in agriculture and water-intensive sectors, are, however, either weak or nonexistent (The World Bank 2016c).

4.2.6   Summary

All in all, although India's recent progress on economic and human development is by any measure impressive, relevant socio-economic challenges persist. As stressed by several institutions, the challenges are many, cutting across all sectors and all 35

---

[8] It is important to note that farming and agriculture are not the only drivers of the rural economy, which on the contrary includes a variety of income-producing activities, including education, construction and retail.

union states and territories, across urban and rural areas, and impacting the lives of 1.2 billion people. Addressing these challenges is central to the goal of reducing poverty and boosting shared prosperity, but requires India to have strategies that appreciate its great diversity of development and to devise apt instruments that can accelerate – and even leapfrog – its development path.

The digital infrastructure India is currently building is certainly one such instrument to leapfrog (see next sections). But space has proved to be an equally powerful tool for addressing these challenges while supporting and accelerating the pace of national development. And this is so not only because it is an indispensable pillar supporting digital infrastructure. Space has been a tool to address the many developmental issues faced by India.

When looking specifically at the development needs and priorities of developing India, it is not difficult to see how an appropriate application of space technology and creation of services based on space technology is highly relevant in a variety of fields, spanning from agriculture and use of natural resources to education and health (Kasturirangan 2011). The next section provides a summary of the applications used for meeting developmental challenges and supporting the pace of national development in India.

4.3 Space Applications for Societal Empowerment

Given the formidable development challenges post-independence India has been facing and continues to face nowadays, it is understandable that, since its inception, the country's policy-makers in India have linked the space programme to the alleviation of poverty and the progressive empowerment of its citizen. As already noted in earlier chapters, they have always insisted on making it socially relevant by putting societal needs at the forefront of their action and hence reversing the perception from "why should poor India have a space programme?" to "India should have a space programme precisely because it is poor" (Lele 2013). Towards this, the apt utilisation of space applications through a distinctive approach has proved to be of paramount importance in a variety of fields.

4.3.1 India's Unique Approach

The use of space applications for supporting and accelerating the pace of development began in the early 1970s, with a series of pioneering experiments to detect coconut wilt diseases in the field of remote sensing and with the Satellite Instructional Television Experiment (SITE) in the field of communication.

The newly established ISRO has since then refined a unique approach based on societal pull rather than technology push. Building on specific user requests and on their needs assessment, this process has been structured along three phases, namely,

(a) joint experimentation, (b) technology/technique development and transfer and (c) integration and institutionalisation.

The first step in the implementation of any application has been joint experimentation with the user agency. As reported by B.S. Bhatia, former director of the Development and Educational Communication Unit of ISRO, "the unique feature of this joint experimentation has been that these are conducted jointly from 'end to end'. This means that the space agency does not limit its role to providing the space segment, but it gets involved in each and every step of utilising the technology and evaluating it to see how a user can use it most effectively. This requires that the space agency moves beyond providing space infrastructure and actually gets involved in every step the user makes for data collection, processing, technique development for specific applications, evaluating effectiveness and providing feedback to both the space segment and the user agency" (Bhatia 2016).

Such an intensive and involved process of joint experimentation and evaluation helps in demonstrating the strengths and advantages of space inputs in the work of the user agency and in developing techniques/methods that are acceptable to the end user. This in turn helps in integrating the space inputs into the routine operations of the user agency. If it is deemed necessary by the user agency, especially if multiple agencies are involved, a separate institution is created for a more effective utilisation of the space inputs. Thus, new institutions are created in the user agency for effective use of space inputs and its future growth. The space organisation continues its involvement in a support role for the smooth functioning of the new institution. The results of such joint working are presented periodically to decision-makers and funding agencies. This enables the user agency to receive funding to accept, adapt and internalise the space inputs (Ibid).

To illustrate, in the early 1970s, ISRO was requested to try and study the area under paddy cultivation in a district in Andhra Pradesh and a district in Punjab. Joint experiments called ARISE (Agriculture Resource Inventory and Survey Experiment) and CAPE (Crop Acreage and Production Estimation) were undertaken with the Indian Agricultural Research Institute (IARI). The procedures put in place for CAPE were continuously revised and upgraded (phase 3: technology development and transfer) to improve the accuracy and timeliness of crop estimates. The wide swath coverage and the ability to quantify different crops subsequently led to the National Crop Forecasting programme for a number of crops including sugar cane, potato, cotton, jute, mustard, etc (Ibid).

Eventually, the recognition that remote sensing alone cannot provide a system for making multiple and reliable forecasts led to the institutionalisation of a programme called FASAL (Forecasting Agricultural Output Using Space, Agrometeorology and Land-based observations), which ultimately led to the establishment of the Mahalanobis National Crop Forecast Centre at the IARI (see Fig. 4.8).

Looking now specifically to the development needs and priorities of developing India, it is not difficult to realise that an appropriate application of space technology and creation of services based on space technology has been highly relevant in a variety of fields, spanning from agriculture and use of natural resources to education

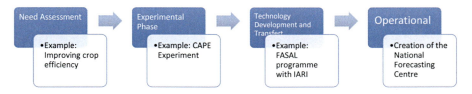

Fig. 4.8 India's societal pull approach to space applications

and health. The rest of this section provides a summary of the applications used for meeting developmental challenges and supporting and accelerating the pace of national development in India.

4.3.2 Natural Resource Management

Because of the large size of Indian population, the efficient use of natural resources such as crops, land and water has always been of utmost importance.

As noted above, the utilisation of space applications to support agriculture began with small experiments in the field of remote sensing and went on to more ambitious experiments covering a variety of areas, which have led to the development of several applications and their integration in user agencies.

Eventually, in order to make information regularly available to decision-makers at various levels, it was felt necessary to have a unified system of resource management integrating space technologies with conventional techniques. To this end, the Natural Resource Management System (NRMS) was established, with DOS acting as its nodal agency and nine standing committees addressing the user/application needs in all resource-related areas (see Fig. 4.9 for the current framework). These areas include agriculture, forestry, water resources, mineral resources, urban planning, rural development and ocean and atmospheric sciences. The needs of the major areas and select examples are presented below.

Agriculture and Rural Development
Agriculture represents 70% of rural employment and approximately 18% of India's GDP. Despite the "Green Revolution" and the increase of agricultural production from 50 million tonnes in 1950 to 240 million tonnes in 2012 (Navalgund 2015), India still faces many challenges including stagnation of productivity in some of the Green Revolution regions, low productivity in the eastern region of the country, high uncertainty in rain-fed agriculture, soil erosion, waterlogging, etc. The current conventional agricultural statistics system fails to provide timely and reliable forecasts necessary for taking many policy decisions. Increasing agricultural production to 350 million tonnes by 2025 to meet the country's demand requires increasing the area under agriculture by identifying cultivable wastelands/marginal lands, increasing cropping intensity by cultivating in suitable fallows, increasing productivity by soil and moisture conservation, improving soil fertility, using high-yielding

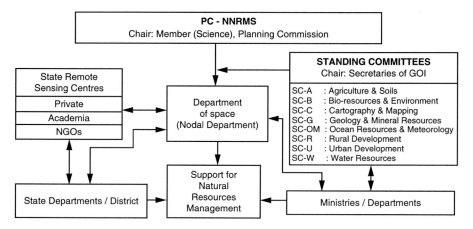

Fig. 4.9 The natural resource management system (*Source*: ISRO)

varieties, ensuring the use of irrigation potential created, improving rain-fed agriculture and establishing a robust system for generating crop production statistics and a forecast system.

In this context, applications of remote sensing in agriculture have been very demanding, and in fact, space systems have been specifically designed to meet information needs of this sector. The main application projects/programmes carried out in this context include (a) crop production forecasting; (b) crop intensification, rotation and cropping system analysis; (c) mapping wastelands and salt-affected soils; and (d) horticulture (Navalgund 2015).

Ocean Resources and Fishery Development

India has a coastal zone of 7500 km with approximately 25% of its population living along the coast. A large share of this population relies on fisheries for their livelihood. Addressing the efficient use of the ocean resources is hence key for India.

ISRO jointly with ESSO (Earth System Science Organization) developed the techniques for identifying potential fishing zones using ocean colour, sea surface temperature and surface wind vectors, surface currents and GIS along with several other variables. These are indications of food availability and favourable environmental factors for congregation of fish in a specific habitat (Bhatia 2016).

Now INCOIS (Indian National Centre for Ocean Information Services) of ESSO issues three-day forecasts regularly except on cloudy days. These advisories are sent to all major fishing harbours of INCOIS and displayed on electronic boards and disseminated on radio, television and newspapers in the local languages. It is also made available on websites, and about 25,000 fishermen subscribe to receive this information directly.

Evaluation of this exercise indicates that in 80% of the cases the catch per unit effort increases by four times. The search time is reduced by 30% to 70%. Studies have established significant increase in productivity, catch size and reduction in fuel consumption. The total annual net benefit is estimated to be of the order 34,000–

50,000 crores. This is one of the best examples of space-generated information directly benefitting the livelihood of common fishermen. According to Nayak "This is one of the best examples of how satellite-generated information can be provided at the grass-roots level as societal services benefitting the livelihood of common men immensely" (Nayak 2015).

Water Resources
The provision of safe drinking water to hundreds of thousands of villages is a priority for India. Likewise, water is a crucial input required to "enhance agricultural production, as most of the small farmers living in arid and semi-arid regions are deprived of irrigation facilities. With the anticipated global warming and climate change, rainfall is expected to be erratic and the water requirement for crops is likely to increase due to a significant increase in evaporation and transpiration losses. Therefore, concrete efforts are necessary towards sustainable use of all the available water sources, efficient harvesting and storage of rain water, improving the irrigation use efficiency, restoration of reservoir storage capacity etc." (Indian Space Research Organisation 2016a, b, c, d, e, f, g).

The information needs of the water resources sector are diverse, "ranging from mere inventory and monitoring of surface water bodies to more complex ones like irrigation performance, groundwater exploration, snowmelt run-off forecast, flood inundation mapping and forecasting, reservoir sedimentation, etc. Space data has been extensively used in many of these areas operationally" (Nayak 2015). Some of these major areas are described below.

India's remote sensing data, together with ground data, have provided very useful information on the geology, geomorphology, structural patterns and recharge conditions which ultimately define the groundwater regime. Studies have shown that "in the last two decades, significant changes have taken place in India in the use of groundwater for irrigation, and currently about 60% of irrigated agriculture depends on groundwater sources. Depletion of water tables, contamination and over-extraction of groundwater have become critical issues in several regions of India. In addition, close to 90% of rural domestic water supply is from groundwater and due to rapid development and increase in population the demand on groundwater for water supply has grown considerably during the last decade" (Indian Space Research Organisation 2016a, b, c, d, e, f, g).

In this regard, remote sensing satellite data is used to identify the potential groundwater sources and recharge sites in the country, in a phased manner, under the "Rajiv Gandhi National Drinking Water Mission" funded by the Department of Drinking Water Supply. This project addresses the preparation and distribution of groundwater prospect zone maps at 1:50,000 scale. After successful completion of Phase I, II and III of the project covering 20 states in the country, Phase IV activities have been initiated covering the remaining 13 states and 5 union territories. Such work has not only facilitated identifying sources of drinking water for deprived villages; the feedback has shown a more than 90% success rate, when wells were drilled based on groundwater prospect maps generated using remote sensing data, as against 50% when satellite data was not available. These maps have been extensively used for locating prospective groundwater sites in and around villages facing problems (Indian Space Research Organisation 2016a, b, c, d, e, f, g).

Earth observation data have been utilised also for "inventory of irrigated area; cropping pattern and its productivity; monitoring irrigation status through the season; staggering of sowing and transplantation of crops; evaluating system performance, etc. They are also used as a diagnostic tool for poorly performing pockets in the irrigation system. Availability of AWiFS data with five-day repetitiveness has greatly facilitated this activity. Many irrigation systems such as Bhadra and Hirakud have been investigated" (Nayak 2015).

The creation of a database and the implementation of a web-enabled Water Resources Information System in the country, short-named as India-WRIS, have been another important way to address the information needs of the water resources joint project of ISRO and the Central Water Commission and ISRO. It aims to provide a single-window solution for obtaining comprehensive, authoritative and consistent data and information on India's water resources along with other collateral information in a standard GIS framework. Many tools are available for retrieving, visualising and analysing of data and for monitoring, planning and development of water resources. WRIS comprises sub-information systems on base data, surface water, groundwater, hydrometeorology, water quality, snow cover/glacier, inland navigation waterways, inter-basin transfer links, land resources, etc." (ibid).

Bio-Resources and Environment
In India, the increasing population and associated increased demand for resources has led to significant impacts on forest cover and is negatively impacting forest produce as well as various ecosystem services, i.e. hydrology, soil conservation, wildlife support, biodiversity, environmental protection, etc. The loss of biological diversity reduces the ability to adapt to the change compounded by the loss of knowledge of biodiversity especially among people with close relationships with the natural ecosystem. India's focus now is on preserving the overall balance and redefining the short-, medium- and long-term considerations to reflect the benefits of consumption and conservation (Indian Space Research Organisation 2016a, b, c, d, e, f, g). Towards this, a variety of programmes have been undertaken with respect to forest cover and type mapping, wetland inventory and conservation plans, bio-resource characterisation, desertification status mapping, coastal zones studies, snow and glacier studies, water reservoir capacity evaluation, ocean primary productivity assessment and ocean status forecast, to name a few (Ibid).

A periodic inventory of all natural resources under the Natural Resources Census programme is also being conducted on a regular basis. The project uses satellite data at different spatial resolutions to prepare natural resources information layers, namely, land use/land cover, land degradation, wetlands, vegetation, snow and glaciers and geomorphology at 1: 50,000 scale (Indian Space Research Organisation 2016a, b, c, d, e, f, g).

In addition, ISRO has been keeping an eye on climate change and more specifically on phenomena such as desertification or glacier melt, through polar and geostationary satellite constellations. For the past two decades, ISRO has been studying climate change through their ISRO Geosphere Biosphere Programme (ISRO-GBP), addressing "atmospheric aerosols, trace gases, GHGs, paleoclimate, land cover change, atmospheric boundary layer dynamics, energy and mass exchange in the vegetative systems, National Carbon Project (NCP) and Regional Climate Modelling (RCM)" (Indian Space Research Organisation 2016a, b, c, d, e, f, g). In 2012, the National Information System for Climate and Environment Studies (NICES) was established at a Hyderabad remote sensing centre with the aim of bolstering India's climate change research (Indian Space Research Organisation 2016a, b, c, d, e, f, g).

4.3.3   Access to Education

One of the greatest challenges faced by a huge and developing country like India is education. Despite its long-standing efforts, there are still 300 million illiterate people in the country. The task of providing development-oriented education to the lower strata of Indian society involves providing access to sources/media of information and presenting the information in an understandable, acceptable and credible manner. The task poses both a hardware challenge to create system configurations to reach out to these segments and a software challenge to present the information in a manner that will be credible, understandable and acceptable. Above all, it is a managerial challenge to ensure that these systems run efficiently and effectively (Bhatia 2015).

One important feature of the illiterate segments of Indian society is that they are large in number and are spread out in remote areas of these nations where reaching out to them becomes even more difficult (Ibid). Due to its ability to reach people in any region, satellite communications technology has become a powerful tool in development education and subsequent development of areas lacking such resources to better their life quality.

India is considered a pioneer in the use of such space-based applications for development, first taking big strides in this arena with the development of SITE in the mid-1970s (see Chap. 2) and then with the Kheda Communications Project (KCP), which brought important advances in content development programming.

In view of the enthusiastic response by education agencies to use satellite communications, ISRO proposed and launched EDUSAT, a satellite totally dedicated to meet the needs of the education sector programme, in September 2004 (India Space Research Organisation 2016). Before the proposal was prepared, discussions were held with all the national educational agencies, and a presentation was made to the Minister of Human Resource Development in the presence of heads of the country's educational agencies. After the agreement, the proposal was put up to the government, and funding approvals for the space system as well as the ground segment were received and implemented.

India's EDUSAT (or GSAT-3) telecommunication satellite served educational purposes as it provided services such as video and computer conferencing, TV broadcast and Internet-based instructional material. It was the first Indian satellite to be dedicated to such an application that would help add to formal education and fill gaps therein, where gaps were most prevalent – often in remote and impoverished areas of the country (India Space Research Organisation 2016).

The EDUSAT programme was implemented in three phases: pilot, semi-operational and operational phases. Pilot projects were conducted during 2004 in Karnataka, Maharashtra and Madhya Pradesh with 300 terminals. The experiences of pilot projects were adopted in semi-operational and operational phases. During the semi-operational phase, almost all the states and major national agencies were covered under the EDUSAT programme. The networks were expanded in the operational phase with funding from respective state governments/user agencies.

The networks implemented under the EDUSAT programme comprise two types of terminals, namely, Satellite Interactive Terminals (SITs) and Receive Only Terminals (ROTs). A total of 83 networks have been established connecting to about 56,164 schools and colleges (4943 SITs and 51,221 ROTs) covering 26 states and 3 union territories of the country. About 15 million students are benefitting from EDUSAT programmes every year (India Space Research Organisation 2016).

The EDUSAT satellite provided its services until September 2010, supporting Tele-education, Telemedicine and Village Resource Centres (VRC) projects of ISRO. After its decommissioning, the traffic of tele-education networks was migrated to other ISRO satellites. Most of the tele-education networks operating in the Ku-band were migrated from GSAT-3 to INSAT-4CR, and those in the Ext. C-band networks were migrated to INSAT-3A, INSAT-3C and GSAT-12. Migration of remaining few networks is in the pipeline (Ibid).

Today's INSAT system therefore provides development communication. Endeavours such as the Jhabua Development Communications Project (JDCP) and the Training and Development Communication Channel (TDCC) additionally lay the groundwork for GRAMSAT, a satellite communication system focussed on the development of rural areas (Bhatia 2015).

All in all, what started in the 1960s as an experiment in school broadcasting under SITE has now grown into a huge satellite-based operational system for tele-education. It has created a number of institutions for the utilisation of technology (including space technology) for education. With further advancement of technology, all the above agencies are now using offline networks besides the online networks set up earlier. One example of an Internet-based offline use of educational technology is the NPTEL (National Project for Technology Enhanced Learning) portal for engineering education.

4.3.4 Access to Healthcare

As for education, access to primary and secondary healthcare can be problematic, especially for a large part of the 881 million people living in rural India. With 80% of the country's medical professionals located in metropolitan areas, where only

33% of India's population resides, the importance of reaching the other 67% of its population has acted as an important catalyst for the initiation of national telemedicine projects (Bonnefoy 2014). ISRO's telemedicine programme "is an innovative process of synergising the benefits of communication technology and information technology with biomedical engineering and medical sciences to deliver health care services to the remote and underserved regions of the country" (Rao 2015). The programme was heavily promoted through collaboration with private software providers, private hospitals and government hospitals, making Indian telemedicine experience one of the most successful examples of telemedicine in the world.

ISRO launched its first pilot telemedicine project together with the Apollo Hospitals, the largest healthcare provider in Asia. Objectives revolved around the building of a hospital equipped with testing, scanning and operating facilities in Aragonda, a village roughly 400 km from Chennai. The medical centre further received satellite connection, thus allowing consultations over teleconferences with the Apollo Hospitals in Chennai. This was made possible using ISRO's Very Small Aperture Terminal (VSAT). Soon thereafter Narayana Hrudayalaya, a second project, was launched with a focus on delivering cardiac expertise.

In two success stories later, ISRO expanded the idea and created a telemedicine network with the following purposes:

- Connecting remote and specialised hospitals for teleconsultation
- Enabling continuing medical education (CME) for medical personnel through technology and connectivity
- Supporting rural health camps and telemedicine units through technology and connectivity (Rao 2015)

INSAT and GSAT series satellites connect states in the Northeast Jammu and Kashmir states, Nicobar Islands, Lakshadweep and Andaman to the roughly 400-unit nationwide TM network (Rao 2015). A unit in this context could be a mobile unit as well as a remote or specialised hospital as mentioned earlier. ISRO's TM network engaged 384 hospitals, 60 specialised hospitals in connection with 306 remote – or otherwise assistance-receiving – hospitals and 18 mobile units. The latter find application in various medical fields such as cardiology, diabetology, ophthalmology, radiology, women and child care, mammography and general medicine (Indian Space Research Organisation 2016a, b, c, d, e, f, g). Most prevalent are the so-called Hospital on Wheels (HoWs), mobile units part of the Distance Healthcare Advancement (DISHA) Project in collaboration with the Apollo Hospitals, Dhan Foundation, Philips Medical Systems and ISRO. Teleconsultations with specialists are facilitated by equipment on-board HoWs, which carry a team of doctors and paramedics. DISHA's scheme has proven very effective as it has provided 4000 patients with medical consultations over an 18-month pilot timespan. ISRO's TM network furthermore consists of the Disaster Management Support (DMS), continuing medical education and mobile TM. DMS came into application in the aftermath of the 2004 Tsunami (Bonnefoy 2014).

In conclusion, ISRO has established a great range of telemedicine services, which have received a great boost because of the support ISRO provides through connectivity and facilities for utilising the networks. However, as Bhatia notes, India's government health department has never played a large role, as their focus has been on primary healthcare rather than secondary care, under which TM fell. The intent may have been for ISRO not to be actively developing TM software which would then perhaps have become a deciding part of the Indian public health system. Instead, TM spread across private, NGO and government hospitals in remote locations of the country (Bhatia 2016).

4.3.5   Disaster and Environmental Management

India is no stranger to different types of natural disasters – including landslides, droughts, cyclones, floods and earthquakes – that have been causing many deaths each year. Roughly 60% of India's land is subject to the risk of earthquakes, three quarters of the coastline is prone to cyclones, 12% of India's total area is subject to floods, and almost 70% is subject to elevated risks of drought. Tsunamis are cause for concern on parts of the west and east coasts of the nation, along with the Andaman and Nicobar Islands, while forest fires occur in dry areas around the country. Landslides are most common near the Himalayas and the Western Ghats (Indian Space Research Organisation 2016a, b, c, d, e, f, g).

In addressing this plethora of natural disasters, a Disaster Management Support Programme (DMSP) was established in 2003. This programme features an IRS satellite series and Airborne Laser Terrain Mapper (ALTM)/Airborne SAR (ASAR) technology that comprises the observation system, a multilayer database, a secure communication network linking ISRO's NRSC with state governments and the Ministry of Home Affairs and a NRSC decision support centre for early warning services as well as natural disaster monitoring (Navalgund 2015).

Flood
India is one of the single most susceptible countries to flooding on a massive scale – floods occur in virtually every water drainage system in India. 40 million hectares of Indian land and 23 of the 35 Indian states and territories are affected by flooding. Mitigating the devastating results of floods, satellites provide a comprehensive overview of the affected area. Information is relayed from satellite technology to the respective central agency, and hazard zones are established based on past data (Indian Space Research Organisation 2016a, b, c, d, e, f, g).

Cyclone Monitoring
India's vast coastal area experiences a tenth of all global tropical cyclones, with almost three quarters of those cyclones occurring in the ten Indian states of Andhra Pradesh, Goa, Gujarat, Karnataka, Kerala, Maharashtra, Nadu, Orissa, Puducherry, Tamil and West Bengal. An average of up to six cyclones hit the Indian coast annually. Tracking and determining key characteristics of the cyclone are areas where ISRO contributes with satellite technology and SAC-developed methodology that

helps predict many facets of the storm using Oceansat-2 scatterometer data (Indian Space Research Organisation 2016a, b, c, d, e, f, g).

Agricultural Drought

As a large number of Indians rely on agriculture, agricultural drought presents a serious threat. Just over two-thirds of India is generally susceptible to droughts; half of this drought-prone population sees rainfall below 750 mm (chronically drought prone), while another third measures 750–1125 mm (drought prone) (Indian Space Research Organisation 2016a, b, c, d, e, f, g).

Following the drought of 1987–1988, India's remote sensing community started to devise ways of using remote sensing data to combat drought on a watershed basis. In achieving this, it was integral that watersheds were understood in more detail; land coverage, soil, geomorphology and drainage, for instance, were major indicators. This critical information was best found to be provided through remote sensing technology for the purposes of a study conducted in 1992, titled "Integrated Mission for Sustainable Development" (IMSD). 175 involved districts received location-based customised adjustments regarding their agricultural practices, for instance, soil conservation or water harvesting. These adjustments were then applied and observed (Navalgund 2015).

Monitoring drought patterns in such a manner, notably so in the summer and fall months, has worked its way into the Indian political system as the Mahalanobis National Crop Forecast Centre (MNCFC) was created under the Ministry of Agriculture. ISRO introduced this method and continues to seek improvement and development (Indian Space Research Organisation 2016a, b, c, d, e, f, g).

Landslide and Earthquakes

Remote sensing found another valuable application in the establishment of hazard zone maps relating to landslides. Data from space technology can provide great insight into lithology, geological structures or geomorphology, for instance, thus aiding the hazard zone mapping process (Navalgund 2015).

The Decision Support Centre (DSC) gathers the relevant data and swiftly shares it with personnel in governmental departments and other stakeholders, using a VPN and INSAT as a communication channel. The National Database for Emergency Management (NDEM) meanwhile acts as a databank in support of India's disaster management efforts (Navalgund 2015). By addressing the information needs covering all the phases of disaster management such as preparedness, early warning, response, relief, rehabilitation, recovery and mitigation, the value-added products generated using satellite imagery have been leading to better decision-making and intervention.

An overview of ISRO's support to disaster management in 2016 is provided in Fig. 4.10, while Fig. 4.11 illustrates India's support to the International Charter on Space and Major Disasters (Annadurai 2017).

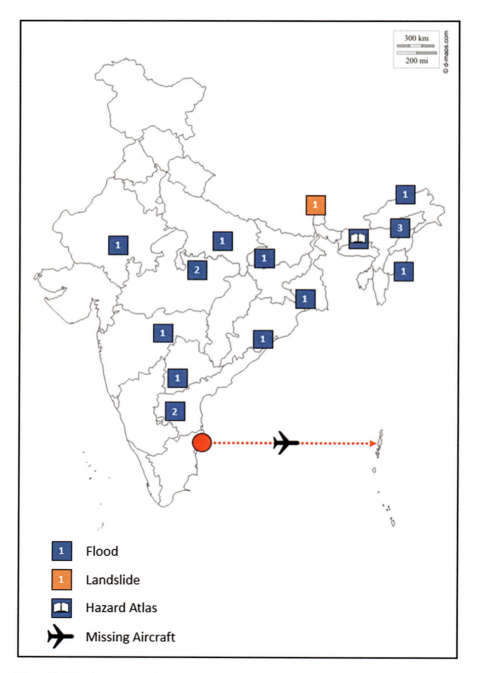

Fig. 4.10 ISRO's support to disaster management in India in 2016 (*Source*: Annadurai 2017)

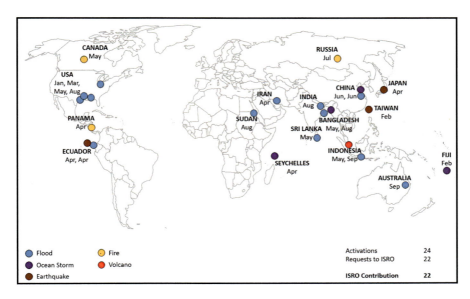

Fig. 4.11   India's support to the International charter in 2016 (*Source*: Annadurai 2017)

4.3.6   Empowering Indian Citizens

It is important to note that space applications have been useful not only for decision-makers at national level but also at local level. A unique feature in which this support has materialised has been through the utilisation of Village Resource Centres (VRCs) (Fig. 4.12).

Established in collaboration with NGOs working in the local area, VRCs integrate satellite communication capabilities with geospatial technologies in order to deliver space-based services at the doorstep of the users. These services include tele-medicine, tele-education, information on natural resources, advisories on agriculture, fisheries, livestock management, vocational training for skill improvement, alternative livelihood education, weather information and even market data. As of 2016, a network of more than 450 VRCs serviced by space systems had been established.

As the VRCs clearly show, space is providing India with a formidable tool to close its "traditional infrastructure" gap, which, as highlighted by the World Bank, is one of the major causes of widespread poverty in the country (Andrés et al. 2013). Equally important, space assets are seen as playing a foundational role towards the establishment of the so-called India Stack, the flagship initiative currently implemented by New Delhi's' government that aims to transform India into a digitally empowered society and knowledge economy by creating a unified software platform through which offering e-governance and services as a core utility to every citizen. A more detailed description of this ambitious initiative is provided in Box 4.1, but the point to consider here is that by eventually providing ubiquitous

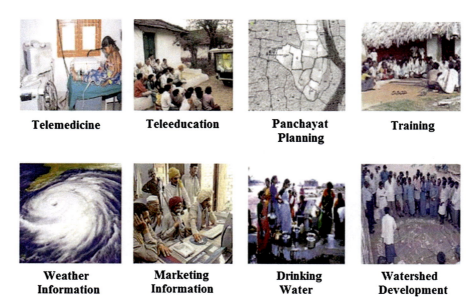

Telemedicine   Teleeducation   Panchayat   Training
                                          Planning

Weather   Marketing   Drinking   Watershed
Information   Information   Water   Development

Fig. 4.12 Village resource centres (*Credit*: ISRO)

Box 4.1: The India Stack

The India Stack is an ambitious project launched by the government in 2009 aiming to create a unified software platform to provide e-services and digitally empower India's citizens. According to the official wording, the India Stack more specifically "is a set of APIs that allows governments, businesses, start-ups and developers to utilize a unique digital Infrastructure to solve India's hard problems towards presence-less, paperless, and cashless service delivery (India Stack 2016)". As the name "stack" suggests, the products and services provided through this digital infrastructure are based on software built on top of the other software.

The first "distinct technology layer" of the India Stack is the *presence-less layer*, which aims to create a unique, universal biometric digital identity that will allow people to participate in any service from anywhere in the country by proving their identity with just their iris or their fingerprint. The Indian government's biometric identity programme, known as Aadhaar (universal ID numbers), enrolled over one billion people in less than five years, making an Indian government programme the fastest platform in the world to enrol a billion users (faster than Facebook or WhatsApp). As explained by IndoGenius, "if identity is something taken for granted in the West, many people in India, especially the marginalized, have not had any forms of such identity (partly because of lost or torn paper forms). This means that there are millions of people

(continued)

Box 4.1 (continued)

invisible to the system. A lack of verifiable identity makes it difficult for both the public and private sectors to interact with these people. A biometric identity on the other hand, in the form of fingerprints and iris scan, cannot be lost. While other countries may collect biometric information, no country has done what India has done, because many other forms of identity were easily available. Aadhaar is only identity, but it can be used for many different things. For example, a bank account can be linked to it; or a government agency can use it to distribute a subsidy, to name a few examples" (IndoGenius 2016b).

The second layer is the *paperless layer*, which enables personal records to be associated with an individual's digital identity, eliminating the need for massive amount of paper collection and storage. Over the past 3 years, more than 150 million eKYCs (electronic Know Your Customer, which enables paperless and rapid verification of address, identity, etc.) and e-Sign (which allow to attach a legally valid electronic signature to a document) have been already used. The third layer is the *cashless layer*, which provides systems like the Unified Payment Interface, a single interface to all national banks and online wallets allowing one-click two-factor authentication, possibly with a biometric smartphone. Over 330 million Aadhaar have been linked to bank accounts in the past 3 years. India aims to demonetise its economy, and it is likely that this initiative will indeed make it the first digital cashless society in the world. Finally, on top of this, is the *consent layer*, which aims to maintain security and control of personal data through a modern privacy data sharing framework.

India Stack is emerging the largest open API in the world and is expected to enable leapfrog scenarios for India's development, particularly in areas like financial inclusion. Indeed, the India Stack "provides frictionless access to financial services through competitive marketplaces bringing millions of Indians into the formal economy and making credit, savings, and investments, and pensions accessible to millions more people" (IndoGenius 2016b). It also "fosters innovation to build products for financial Inclusion, healthcare & educational services at scale. Brings a paradigm shift in the way government services are delivered in a transparent, accountable and leakage free model" (India Stack 2016).

Internet connectivity to any citizen anywhere in the country, space assets enable basic access to the Application Programming Interfaces (APIs) of the India Stack. In addition, they also offer a plethora of citizen-centred applications (e.g. integrated PNT and EO applications) that can be linked to this unique digital infrastructure, thus helping to address some of India's grand challenges in areas such as healthcare, financial inclusion, agriculture, energy, transportation and education.

4.3.7 Space Applications for Development: An Overview

To conclude this section on the societal application of space technologies, the most important uses of space applications are summarised in Table 4.5.

Table 4.5 Space for India's development needs

Area/need	Space applications
Agriculture, food security and water security	Watershed management
	Optimal land use strategy plan
	Control of land degradation
	Recovery of irrigation systems
	Monitoring of crops and cropping systems
	Horticulture development
	Groundwater targeting
	Fisheries forecasting
Access to healthcare and education	Tele-medicine
	Tele-education
	Village Resource Centres
Infrastructure development	Land use mapping
	Road connectivity analysis
	Urban planning
	VSAT communication network
	Support to the digital infrastructure and JAM
Weather and disaster management and response	Regional weather forecast
	Cyclone warning
	Flood damage assessment
	Drought monitoring
	Landslide and earthquake management
Environmental management, natural resource and climate change studies	Vegetation monitoring
	Desertification monitoring
	Ocean productivity and status forecast
	Atmospheric pollution
	Periodic inventory of natural resources
	Land use/land cover, soil, wetland, land degradation, snow and glacier, vegetation
	Biodiversity characterisation
	Characterisation of climate variables
	Methane emission and timberline study
Rural development	National drinking water mission
	Wasteland mapping/updating
	Watershed development and monitoring
	Land records modernisation plan
	Telecommunications for rural audiences

(continued)

Table 4.5 (continued)

Area/need	Space applications
Urban development	Urban sprawl mapping of major cities
	Master/structure plans
	Comprehensive development plans of selected cities/towns
	Base map generation for towns
	National urban information system

Source: PwC (2016)

4.4   Space as a Tool for Macroeconomic Growth: Inspiration and Perspiration

Besides the development of a plethora of applications bringing direct and quality-of-life benefits to large segments of India's society, India's space programme has been aptly leveraged as an enabler of economic expansion also in terms of macroeconomic production functions, particularly inspiration (scientific and technological innovation) and perspiration (industrial upgrading and creation of skilled workforce).

The positive spillover effect of space technologies has been validated in a plethora of studies and through a variety of approaches. One example is offered by the OECD, for instance, which has categorised the diversity of socio-economic impacts derived from space activities in different segments (see Fig. 4.13), namely, new commercial products and services (including "indirect industrial effects" from space industry contracts, meaning new exports or new activities outside the space sector), productivity/efficiency gains in diverse economic sectors (e.g. fisheries, airlines), economic growth regionally and nationally and cost avoidances (e.g. floods) (OECD 2011).

By the early 1990s – when all the components of the Indian space programme had entered the operational stage – also ISRO, together with the Madras School of Economics, started to study the economics of its space programme by analysing cost and time savings, productivity gains and commercial and spin-off benefits during both the construction and exploitation stages (Sankar 2006; Kasturirangan 2011). While providing an economic analysis of the Indian space programme exceeds the objective of this book, this section highlights the contributions Indian policy-makers have intended to harvest from their investments in their space sector in order to foster a process of sustained, dynamic growth through space-related innovation and industrial upgrading. Towards this, a review of India's Science, Technology and Innovation policies is provided first.

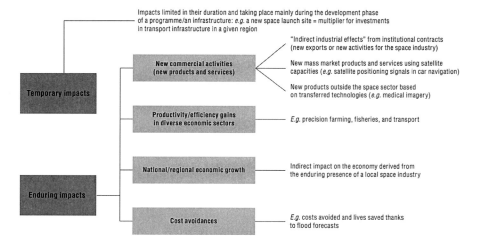

Fig. 4.13 Impacts derived from investments in space programmes (*Credit*: OECD 2011)

4.4.1   Science, Technology and Innovation Policies

Since the turn of the new century, the visibility and prominence of STI topics have progressively increased in India, with its government increasingly striving to move from a factor-driven to innovation-driven economy (World Economic Forum 2016, b).

India's national strategy "Decade of Innovations 2010–20" launched by Prime Minister Manmohan Singh "commits to strengthening science, technology and innovation (STI) capacities, with an objective to increase Gross Expenditure on R&D (GERD) to 2% of the GDP by 2020" (National Innovation Council – Government of India 2011). The commitment to innovation is also reflected in India's 12th Five-Year Plan (2012–2017), which seeks to catalyse growth for faster, inclusive and sustainable development through STI, and in the 2013 STI Policy,[9] which aims to create a robust innovation system and foster a high-technology-led path in India, positioning the country among the top innovative economies.[10]

Over the past 10 years, the visibility and prominence of STI topics have progressively increased, and with the advent of President Modi, a pro-technology, probusiness enthusiast, it has become a national priority, suggesting an aspiration to depart

---

[9] Within this framework, a National Innovation Foundation has been established to support grassroots innovators, and an Inclusive Innovation Fund (IIF) created to mobilise finance to support enterprises developing innovative solutions for the "bottom 500 million".

[10] It is, however, important to stress that only in recent years has innovation become a buzzword in India's policy-making; it suffices to remind that the antecedent 2003 version of the STI Policy was only known as "Science and Technology Policy" and, in contrast to the current and more decentralised scenario, it was essentially within the mandate of the Department of Science and Technology. This in a way reflects that India's understanding of innovation in terms of policy-making had traditionally been focused only on scientific development (Bute 2013).

from India's global image of supplying a competitive IT workforce to becoming a global innovation hub (Bute 2013).

This growing importance is reflected in the plethora of initiatives launched in the first 2 years of President Modi's tenure. To begin with, the government has put forward some important institutional changes in the research policy area. The National Institution for Transforming India (NITI) Aayog, as already noted, replaced the Planning Commission as the country's public policy planning body, while the National Innovation Council – established by Singh's government – was discontinued, with NITI Aayog and various ministries/departments taking over its functions. Under the ambit of NITI Aayog, the government launched the Atal Innovation Mission (AIM) to advance the national innovation ecosystem by providing an innovation promotion platform with related R&D funds (there is currently a budget of €20 million) and promoting a network of world-class innovation hubs connecting federal higher education and research institutions such as the various Indian Institutes of Technology (IITs), the Indian Institutes of Science Education and Research (IISERs), the Indian Institutes of Management (IIMs), the All India Institutes of Medical Sciences (AIIMS), etc. The number of these institutions has also expanded over the last few years: since 2006, six IISERS have been established, and the number of IITs grew from 16 in 2014 to 18 in 2015 and 22 in 2016. In addition, compared to the earlier situation where the Planning Commission together with the Prime Minister's office and the Department of Science and Technology were the sole decision-makers in S&T matters, the current setup has become more decentralised, with various Indian Ministries acquiring more responsibilities over STI topics (Krishna 2015).

Modi's government has also launched a series of National Flagship Programmes/ Schemes having a considerable impact on innovation. These include:

(a) *Digital India*, which focuses on digital empowerment of citizens through better provision of government services building on the success of ICT sector. The India Stack (see Box 4.1) is part of this flagship initiative

(b) *Skill India* (supported by the recently established Ministry of Skill Development and Entrepreneurship), which aims to train 400 million people by 2022 in key industrial sectors

(c) *Make in India* (supported by the Ministry of Communication and Industry), which seeks to position India as a global manufacturing hub by encouraging international firms to manifacture their products in India. The initiative covers 25 industrial sectors, including space and defence

(d) *Start-up India*, which seeks to foster entrepreneurship and promote innovation in a wide range of domains with the ultimate goal of supporting economic growth and employment

(e) *Clean India*, which is a cleanliness drive that commits to enhance the quality of Indian citizens, by inter alia making India free of open defecation by the 150th birth anniversary of Mahatma Gandhi in 2019

(f) *Green India* (supported by the Ministry of Environment and Forests), which aims to respond to climate change by a combination of adaptation and mitigation measures, including the development of renewable energy and electric vehicles

(g) *Smart Cities Mission*, which focuses on urban development, including the building of 100 smart cities in the coming years

In line with the idea of weaker central government and a more local approach, line ministries are responsible for the implementation and management of the programmes. Furthermore, the Modi government aims to deregulate business to create less friction on Indian soil for investors from abroad. In this respect, the government has launched an FDI Policy and a new National IPR Policy in 2015.

All in all, this variety of efforts seems to have already brought important fruits. It suffices to think that in 2016 India's rank in the Global Innovation Index grew considerably, passing from the 81st position in 2015 to the 66th position. Despite still being far from OECD countries in terms of both innovation inputs and outputs, India's standing as the premier economy in Central and Southern Asia along with superb innovation, R&D expertise, academic/scientific excellence, and the global top rank in ICT service exports, speak volumes (Dutta and Lanvin 2016). Weaknesses on the other hand are generally found in innovation inputs, such as education expenses or business creation/environment (OECD 2014). Figure 4.14 provides a nominalised index of performance in India relative to median values in OECD countries.

However, as stressed by Chandrajit Banerjee, Director General of the Confederation of Indian Industry (CII): "The commitment of India to innovation and improved innovation metrics is strong and growing, helping to improve the innovation environment. This trend will help gradually lift India closer to other top-ranked innovation economies" (OECD 2014).

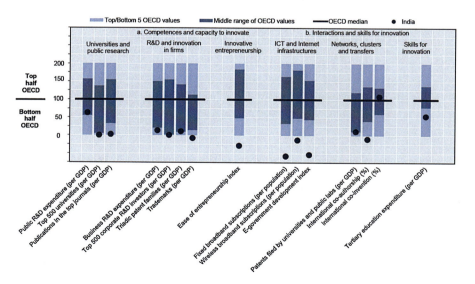

Fig. 4.14 Nominalised index of performance relative to median values in OECD countries (*Source*: OECD 2014)

4.4.2   India's Frugal and Invisible Innovation

Even though innovation inputs lag behind those of the advanced economies of the West, it should not go unnoticed that the country has proved to be a house of so-called grassroots and frugal innovation.

Frugal innovation is an "increasingly fashionable" concept that is having considerable impact on boardrooms around the world, spanning from car companies to engineering courses (The Economist 2010). As the name *frugal* suggests, this kind of innovation essentially implies "achieving more with fewer resources", an idea that is largely inspired by the Indian tradition of *jugaad* or *jugaar*. This colloquial Hindi and Punjabi word, literally meaning a hack, improvisation or mend, is generally used to indicate a context-sensitive solution to a problem through "resourcefulness, given the constraints of any given situation, whether they be financial or simply the materials that are available at the time" (IndoGenius 2016a). For instance, the well-known terracotta clay refrigerator that functions without electricity is a classic example of *jugaad* solution.

Since *jugaad* solutions are generally catalysed by the need to serve markets with low purchasing power, they can be generally intended as *inclusive* innovations, i.e. innovation that aims to serve low-purchasing power market. However, frugal innovation is generally intended as being something more than just making things cheaper. In this respect, authors Navi Radjou and Jaideep Prabhu have in their book *Jugaad Innovation* and their follow-up book *Frugal Innovation* outlined how working with considerable constraints has in fact often resulted in disruptive innovation (Radjou and Prabhu 2010). If it is true that "necessity is the mother of invention, the idea is that lack of resources, or familiarity with thriving under a resource-constrained context, can actually lead to more innovative solutions" (IndoGenius 2016a). In short, India's no-thrill approach has not simply resulted in low-cost solutions, but also in disruptive innovations, as, for instance, demonstrated by the well-known General Electric's ECG Machines and by the Tata Nano.

Another key innovation concept to fully grasp the real weight and potential of India is invisible innovation, a concept introduced by management experts Nirmalya Kumar and Phanish Puranam in their book *India Inside, the Emerging Innovation Challenge to the West* (Kumar and Puranam 2012).

The two authors point out that whereas there are certainly no Indian equivalents to innovative products like Google, iPod or Viagra, there are also no valid reasons to restrict innovation to the products or services visible to the end consumer in the developed world. Kumar and Puranam compare innovation to an iceberg: only a small part is visible to the end consumer. Most of it is invisible (see Fig. 4.15). And, the authors argue, this is where India is really succeeding. Kumar and Puranam highlight four types of innovations originating in India that remain invisible to end consumers.

The first is what they call globally segmented innovation. India is considered a research and development hub for many multinational companies, especially so in the technology industry. According to Zinnov Consulting, 928 MNCs now have

Visible and invisible innovation

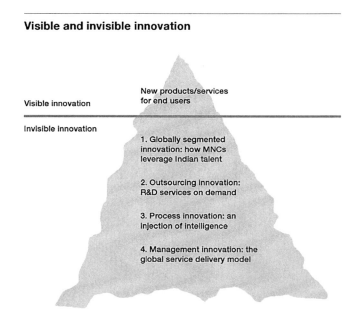

Fig. 4.15 Visible and invisible innovation (*Source*: Kuram and Pruanam 2012)

research and development centres in India, employing more than 300,000 people in product development. In 2015, 40% of all globalised expenditures by MNCs was spent in India. Industry giants such as Microsoft, Intel, General Electric, Hewlett-Packard and IBM found great value in the skill of the Indian workforce, the market, the Indian laws on intellectual property and surely, especially initially, the lower cost associated with outsourcing R&D to India (Zinnov 2015). The involvement of India in the R&D of such huge global firms has put the nation at the forefront of tech innovation.

The second type comprises the outsourcing of research and development services to Indian companies. In this type of invisible innovation, Indian firms offer R&D as a service to client companies. Having grown in double digits in recent years, this R&D outsourcing has become a way for India to be more influential in regard to innovation for major companies yet without them receiving much notoriety for their work. Indian firms thus act as an on-demand service provider, while the concepts are still brought to life outside of the country.

The third kind of invisible innovation encompasses the innovation of the process that creates goods or services, rather than the innovation of these products or services themselves by Indian companies. Prime examples of this can be found in the pharmaceutical industry or the infamous Indian call centres, for instance, where Indian workers have not only contributed to the execution of the outsourced service but have also helped refine the different structural arrangements of delivering such service.

The fourth and final type of invisible innovation is called "management innovation of the global service delivery model", and is a new way of managing globally distributed work "to effectively bring global scale and cost efficiencies to previously locally clustered service processes" (Kumar and Puranam 2012). Here, the end consumer generally has virtually no insight, thus making India's key role as an innovator of management structures and processes most invisible. The concept of breaking single tasks into smaller pieces which have to be performed by them at various locations – whichever place is most suitable for the work to be done – is a structural trend most prevalent among Indian firms in the IT industry in the 1990s. The globe is thus tied together much more closely as work is split and outsourced to what is deemed the most appropriate place and workforce for the task. The Bangalore-based firm Mu Sigma, for instance, has emerged as the world's largest provider of analytics and innovative decision science services for many *Fortune 500* companies. This global delivery model supports the other kinds of invisible innovation and has undoubtedly enabled India to become the global outsourcing force they are today (Kumar and Puranam 2012).

It is important to note that whereas "invisible innovations are all triggered by the availability of highly skilled employees at low prices in India (budget-conserving talent), visible innovations from India such as Nano or GE's new ECG machine are catalysed by the need to serve markets with low purchasing power (budget-constrained customers), though undoubtedly, the budget-conserving talent helps to make these innovations possible" (Kumar and Puranam 2012).

All in all, invisible and frugal innovations from India are bound to pose a formidable challenge to Western leadership in terms of innovation. At the same time, however, much remains to be done to ensure the effectiveness of the various government initiatives; provide more structural, disciplined foundation to frugal innovation; improve innovation inputs; and develop cutting-edge product and processes moving towards even higher value-added activities. And this is where space comes to the fore. Indeed, space activities have been identified as a key domain and tool for achieving these objectives and setting trends for the broader innovation ecosystem.

4.4.3   The Contributions of Space

When looking at the contribution of space, it is not difficult to realise that space activities have proved a hotbed for technological innovation, particularly for the much-heralded frugal innovation. Indeed, ISRO achievements in putting satellites into orbit, and most impressively the Mangalyaan mission to Mars, on a shoestring budget all provide excellent examples of frugal engineering.

Costing $75 million, Mangalyaan's budget was roughly one-tenth of NASA's previous Maven orbiter mission to Mars. While budgetary comparisons among different space exploration missions can be highly misleading, also in light of the different economic conditions in different countries, U. R. Rao, former chairman of ISRO, compared the $75 million spent on the mission to the amount Indians spend on Diwali crackers for 1 day: "For going all the way to Mars, just one-tenth of the money is being spent. So, why are they shouting?" (Patairiya 2013).

However, one should not only consider ISRO as a frugal innovator but also as an inclusive innovator, because many of its applications are intended to benefit common citizens, especially India's rural population (IndoGenius 2016a). As mentioned, ISRO has provided weather forecasting, crop pattern recognition, measurement of groundwater levels and educational and telemedicine services through an end-to-end approach that goes far beyond the normal duties of a space agency. Not only has it provided space-related infrastructure, but it has actually become involved in every step of the user, for data collection, processing, technique development for specific applications, evaluating the effectiveness and providing feedback to both the space segment and the user agency (Bhatia 2016). All these achievements came largely thanks to ISRO's ability to innovate frugally and inclusively.

In addition, it is important to note that India's space programme can also provide more structural foundation to India's *jugaad* innovation. It is indeed widely recognised that excessive dependence on the *jugaad* mind-set may in fact impede the ability to fundamentally transform a situation through disciplined engineering, frugal or otherwise. As Kumar and Puranam note, although frugal innovations from India such as Nano or GE's new ECG machine are catalysed by the need to serve markets with low purchasing power (budget-constrained customers), "there is more to this approach than just making things cheaper". Carlos Ghosn, who heads Renault–Nissan, is credited with coining the term frugal engineering to signify achieving more with fewer resources. Frugal engineering is not mere *jugaad*, that is, making fixes and finding workarounds. Whereas *jugaads* – while undoubtedly creative and inventive – essentially signal resignation to current constraints, frugal engineering can be a systematic approach to making those constraints irrelevant or at least less important (Kumar and Puranam 2012).

It can be argued that ISRO activities can provide a model for making frugal innovation more disciplined. As stressed by ISRO itself, "the Indian space programme has been pursuing a systematic and well-defined policy for transfer of know-how of products and technologies developed by the Indian Space Centres" (Indian Space Research Organisation 2016c, d). The purposes have been multifaceted, namely, "to facilitate greater participation of Indian industries in various space projects, their applications in the commercial domain, and to benefit from 'spin-offs' of such technologies. Implementation of this policy has yielded substantial results in terms of gearing up significant industrial participation in the space programme through provision of various products and services by industry relevant to applications involving space systems, such as satellites, communications, broadcasting, meteorological services and geospatial information services, as well as the technology transfer from ISRO" (Ibid).

The technology transfer mechanism established during the early 1980s enables licensing of know-how from various ISRO centres for commercial exploitation. Over 300 technologies have been transferred to industries in the fields of electronic- and computer-based systems, speciality polymer chemicals and materials, electro-optical instruments, mechanical equipment and ground systems related to satellite communications, broadcasting and meteorology. Industries in the large-, medium- and small-scale sectors have largely been beneficiaries of the technology transfer scheme. ISRO has executed many consultancy projects in high-technology areas to provide support to various industries (Indian Space Research Organisation 2016c, d).

Simultaneously, ISRO has been investing in developing an IPR portfolio, which now consists of around 270 patents, 45 copyrights and 10 trademarks. Even in this area, while ISRO's objective has been to safeguard the technologies developed in ISRO centres, the approach has been tuned towards enabling maximal commercial exploitation of such resources through appropriate technology transfers or licensing schemes (Ibid).

Through the numerous technologies, systems, hardware and services developed by ISRO, India has more broadly sought to boost the growth of its industries, both in terms of factors of production allocations (labour and investments through technology transfer and buybacks) and of high value-added industrial upgrading. This is in turn intended to strengthen the country's overall industrial capabilities, enhance national technical competence and stimulate a knowledge-based economy, which should eventually offer India the ability to compete further in the global economy, even when its competitive advantage in terms of low-cost labour declines.

Despite all these objectives, it is evident that much remains to be done to support the innovation ecosystem in India by improving innovation inputs and stimulating the participation of private industry.[11] As noted by the World Economic Forum's Global Competitiveness Report, "the capacity of a country to be innovative has to be thought of as an ecosystem that not only produces scientific knowledge but also enables all industries – including in the service sector – and society at large to be more flexible, interconnected, and open to new ideas and business models. To be truly innovative, a country should not only file patents and support research and development in science and technology, but should also provide a networked, connected environment that promotes creativity and entrepreneurship, fosters collaboration, and rewards individuals who are open-minded and embrace new ways to perform tasks" (World Economic Forum 2016, b).

The awareness of this largely explains recent initiatives such as "Make in India", the Indian government's campaign to attract foreign companies to produce in India. Interestingly, space has been included as a priority sector in which the above-discussed invisible innovations seen in other sectors could possibly be reproduced.

In addition, such awareness also explains why ISRO's efforts have been not only directed towards the development of technology and infrastructure but have also stretched into developing human resources with a focus on creating thrust for the development of space science and technology in the country and in turn generating highly skilled human capital favouring the rise of a knowledge society and economy. To this end, ISRO is probably the first major space agency to create a flagship

---

[11] More inputs from private sector are essential. This is true also for space activities. In order to stimulate private investment, the government has launched the Make in India initiative to attract foreign companies to invest in India and in which space is identified as a priority sector. Interestingly, the Make in India initiative can also act as a tool to reproduce the invisible innovations in the space sector.

institute (Indian Institute of Space Science and Technology) with undergraduate, postgraduate and doctoral programmes focussed on space science and technology (Prasad 2016a, b, c).

In this respect, it is also noteworthy that ISRO's space activities, particularly its exploration missions, have been intended to serve also as a source of inspiration for stimulating enrolment in technical education in India and pursuing careers in science, technology, engineering and mathematics (STEM) disciplines. This is intended to further enhance the shift towards a knowledge-based economy. This is why creating awareness among the public, especially students, about the benefits to society that have accrued from India's application-driven space programme and the progress made by the country in space science and technology has been given utmost importance by ISRO. Media campaigns on important events, campaigns through social media, webcasting of launches, organisation of exhibitions, educational activities such as lectures, interactive sessions with students, quiz programmes, water rocket making and launching events, publications, video documentaries and SAKAAR – an augmented reality application for Android devices – have helped in not only keeping the public abreast of the latest developments in ISRO's space programme but also evoking interest in them and the nuances of space science and technology (Indian Space Research Organisation 2016c).

Finally, and on a more concrete level, it is important to highlight that space more broadly provides India with a formidable tool to close its digital infrastructure gap and hence achieve the goals encapsulated in the *Digital India* initiative, the flagship campaign that aims to transform India into a digitally empowered society and knowledge-based economy by inter alia improving Internet connectivity in the country and offering a unique software platform (the India Stack) as a core utility to all Indian citizen.

Access to the Internet is indeed generally recognised a "vital enabler of economic growth" in both macroeconomic and microeconomic terms. In the Indian case, some research has even shown that achieving 100% Internet connectivity by 2020 could potentially add an extra $1 trillion to India's GDP (Rajgopalan and Jayasimha 2016). This is not simply because the Internet has always been a catalyser of business but also because in the case of India, it plays a foundational role for the diffusion of a number of empowering cutting-edge technologies (including cloud computing, digital payments, Internet of Things, verifiable digital identity, smart mobility, advanced geographic information systems, etc.) that can enable the country to leapfrog its development path with a new paradigm.[12] Given the inherent advantages that space-based Internet solutions have as compared to terrestrial ones (this is even more true for India, where the sheer scale of the subcontinent and num-

[12] In a recent report, the McKinsey Global Institute has identified a series of 12 enabling technologies that have the potential to radically transform the future of India and enable a strong socio-economic growth through leapfrogging. These include mobile Internet, cloud computing, automation of knowledge work, digital payments, digital identity, Internet of Things, smart mobility, advanced geographic information systems, next-generation genomics, advanced oil and gas exploration and recovery, renewable energy and advanced energy storage (McKinsey Global Institute 2014).

ber of locations involved make space solutions particularly apposite), one clearly needs to acknowledge space as one of the key enablers of India's future growth.

4.5 Towards Commercialisation Efforts and New Space Entrepreneurship

Commercialisation of space-based products and services has been a complementary instrument to better seize the utilitarian benefits stemming from the Indian space programme and enhance the country's performance across various pillars of competitiveness (e.g. business and supply chain sophistication, market efficiency, etc.). Commercialisation of space-related products has not only contributed to generating revenue but also to meeting domestic requirements to progressively boost the growth of a private industrial ecosystem, which is key to further India's economic expansion and productivity gains.

4.5.1 A Growing, Yet Challenging, Commercial Ecosystem

As already noted, the commercial space ecosystem is currently centred on the Antrix Corporation, which is completely owned by the Indian government and serves as the commercial/marketing arm of ISRO (see Fig. 4.16). Leveraging the state-of-the-art capability and infrastructure that ISRO has built through the years, Antrix has been providing commercial services on the global market. Its offerings, as noted in Chap. 3, include selling capacity on INSAT, data from IRS to international customers and launching procurement of foreign satellites, among others. While its commercialisation has in many respects been successful, with an upward trend in terms of revenue, it has also been facing several shortcomings, primarily owing to the very model on which it is based. More recently, however, new space companies building independent business-to-business (B2B) or business-to-customers (B2C) solutions have started to emerge.

Satellite Communication Services
Thanks to the INSAT/GSAT system, which currently comprises 13 operational satellites providing a cumulative capacity of over 250 transponders in C, Ext. C, Ku-band, UHF and S-band frequencies, the provision of satellite communication services is one of the most commercially rewarding offerings by the Antrix Corporation. Although there are some restrictions in areas such as the commercial use of satellite phones due to security reasons (Department of Telecommunications – Government of India 2016), the large majority of INSAT/GSAT transponders has been made available for commercial use for services such as the Direct to Home

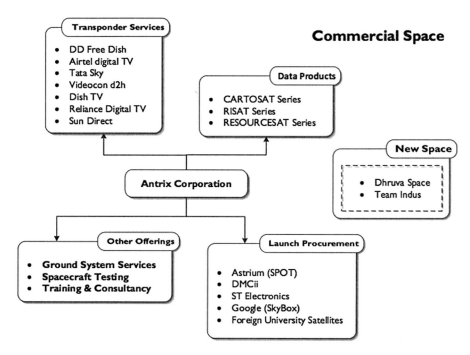

Fig. 4.16 India's commercial ecosystem (*Source*: Prasad 2016a)

(DTH), Digital Satellite News Gathering (DSNG) and Very Small Aperture Terminal (VSATs); the remainder is being used for societal and strategic applications.[13]

With an average annual growth of 13% over the last decade, DTH has separated itself from commercial competition: since the last decade, India has emerged as one of the most competitive and vast markets for DTH operations in a $7.3 billion market as of 2015 (PwC 2016). As of 2016, Antrix has enabled seven DTH players to service the Indian market, which currently has more than 53 million subscribers and still presents huge potential growth (see Table 4.6).

While DHT undeniably is a great commercial success for India, the ever-increasing demand for satellite communication services, coupled with ISRO's inability to meet such demand and India's SATCOM policy inhibiting private sector

---

[13] As reported by Antrix, the company "enables news channels to provide live coverage from the source and pitch-in the TV sets of nearly 160 million households in the country through Digital Satellite News Gathering (DSNG) Services. The service is operational in C and Ku bands. Antrix enables VSAT services by public and private sector for variety of operations including Banking and Financial Services. Antrix is also serving the needs of various users for meeting telephony and backhaul requirements. Some of the private TV channels are uplinking on the INSAT/GSAT fleet using their own teleports and ground stations set-up across the country. Antrix has also enabled commercial teleport operators, who make use of space segment capacity as well as fiber optics technology, thereby providing total solution for channels to 'play-out' the contents from their studios located anywhere in India" (Antrix Corporation 2016).

Table 4.6 Private DTH operators in India

DHT operator (as of January 2016)	Total channels	HD channels
Dish TV	431	50
Tata Sky	474	51
Airtel Digital TV	469	41
Videocon d2h	553	45
Reliance Digital	292	12
Sun Direct	230	15
DD Direct Plus	75	0
Total	2524	214

investments in the establishment of satellite systems, has created a transponder supply gap and forced India to adopt a policy of leasing out foreign transponders via the Antrix Corporation as a gap filler (The Hindu Business Line 2014; PwC 2013, 2016).

It is estimated that only 25% of India's transponder capacity is being served by Indian satellites, while the rest is leased from foreign satellites via 3-year deals made by Antrix, which subsequently sublets foreign transponders to Indian DTH operators through back-to-back agreements (PwC 2016). While envisaged as a short-term measure to ensure that the service could be brought back to the INSAT system when Indian satellite capacity is available, this modus operandi has created several drawbacks (PwC 2013). For one thing, it delays the whole process significantly while also limiting Indian operators' ability to procure favourable commercial terms through direct negotiations (Ibid). In addition, it creates uncertainty for foreign satellite operators, which logically pass the cost of such uncertainty to Indian DHT service providers.

Furthermore, the absence of independent regulation between governmental entities working alongside each other to deliver services constitutes a key issue (Prasad 2016a, b, c; Ashok and D'Souza 2017). Indeed, although officially adopting an open-sky policy in satellite operations, "Antrix/ISRO plays overlapping and conflicting roles as a supplier, an intermediary, policy formulator and an arbitrator. This results in a direct conflict of interest scenario,[14] which generates further negative fallouts" (PwC 2016). The Antrix–Devas fiasco offers a clear example of these regulatory fallouts and has more importantly shed serious doubts on the effectiveness of the commercial efforts put in place by the Antrix Corporation (see Box 4.2: The Antrix–Devas Imbroglio).

While ISRO is working hard to replace foreign-leased transponders and move towards self-reliance – also by progressively encouraging private investments in the establishment of communication satellites systems – it is manifestly unable to keep up with the pace of demand: suffice it to mention that during the 12th Five-Year Plan (2012–2017), ISRO has planned to increase its operational transponder capacity from INSAT/GSAT satellites from 198 to 398 against a projected demand of 794 transponders. The latest reports show that this supply gap is only going to increase

---

[14] It is against the regulatory best practice adopted in other sectors.

Box 4.2: The Antrix–Devas Imbroglio
In January 2005, an agreement was signed between Antrix and Devas Multimedia, a telecommunication company headquartered in Bangalore and established by US venture capitalists together with former ISRO employees. Under the agreement, Antrix would lease S-band transponders on two ISRO satellites (GSAT-6 and GSAT-6A) to Devas, which would use them to provide Internet services in India. Between 2005 and 2010, Devas started to build its terrestrial network and to conduct field trials while making high profit from the selling of its shares, including 17% of its stake to the Deutsche Telekom (Prasad 2016d).

In 2010, however, an audit report from the Comptroller and Audit General of India pointed out several irregularities in the agreement, including non-compliance of the standard operating procedures and financial mismanagement, particularly in the financial load and risks to be borne by DOS. The report highlighted that Antrix and ISRO "virtually gifted a valuable and potentially high-profit earning band to Devas", neither conducting a proper bidding process nor engaging in interdepartmental consultations and informing the Cabinet that the two satellites would be built specifically for Devas (Comptroller and Auditor General of India 2012). The ensuing investigations found some former ISRO employees guilty of "acts of commission" and "acts of omission" and eventually resulted in the decision to terminate the deal. The official reason, however, did not refer to corruption or conflict-of-interest issues, but to the impossibility for Antrix to obtain orbital slots in S-band for commercial activity because of the Cabinet Committee on Security (CCS) decision to use the spectrum to meet the country's strategic requirements, and hence because of a "force majeure event". In response, Devas and Deutsche Telekom, respectively, demanded US $2 billion and US $1 billion in damages and started arbitration proceedings against Antrix.

The first arbitration by the International Chamber of Commerce ruled against India, ordering it to pay $672 million to Devas for unilaterally terminating the contract with Devas. The ICC determined that Antrix was wrong to argue that the termination of the contact was a force majeure event stemming from an independent decision of the Indian government (Deloitte 2010). The second arbitration, which was filed by the company's investors Columbia Capital and Telecom Ventures, took place at the Permanent Court of Arbitration (PCA) at The Hague and produced similar results but with more dramatic consequences for India. The PCA ordered the Indian government to pay a minimum financial compensation of $1 billion for having expropriated the investment made by foreign shareholders of Devas through the contract annulment (Prasad 2016d). The third arbitration, filed by the Deutsche Telekom under the Germany-India Bilateral Investment Treaty, is currently experiencing a battle over jurisdiction but may produce similar results.

in the future, particularly in view of the above-mentioned initiatives such as *Digital India* and India Stack that require additional transponder capacity to connect thousands of villages to enable e-governance services (Rao et al. 2015; Alawadhi 2016).

With the aim of addressing the transponder supply gap and reducing dependence on foreign satellite operators, in May 2016, the Telecom Regulatory Authority of India (TRAI) proposed consolidating all satellite television broadcasts by sharing satellite capacity on fewer INSAT satellites (Telecom Regulatory Authority of India 2016). The proposal, published in the form of a pre-consultation paper, was based on the assumption that "there is scope for better utilization of available infrastructure, considering that multiple DTH providers are broadcasting the same channels even as they compete with each other for subscribers" (and pointed out the benefits to be reaped if broadcasters banded together on one or two spacecraft, instead of beaming the same programs on different satellites. The proposal, however, inevitably received a lukewarm response by satellite operators, which considered it as wrong-headed and counterproductive in view of the "inherent issues in migration of satellite capacity such as substantial migration expenses for the DTH service provider and inconvenience to the millions of customers in reorienting their TV dish antennas" (de Selding 2016). Not surprisingly, it is doomed to be withdrawn.

It appears evident to many that if the government wants to reduce dependence on satellite operators and most effectively seize this market opportunity, it must put in place a sound growth strategy that supports the participation of the private sector and "move ahead with new technology and service offerings in the satellite communications segment".[15] To be sure, the government is very keen to encourage investments in the establishment of satellite systems and Satcom services, as demonstrated by the recent increase to 100% of the FDI cap on establishing satellite systems, as part of its *Make in India* initiative. However, ease of doing business requires not only support in terms of the cap on investments but also regulatory transparency in terms of planning and allocation of resources and more broadly "novel institutional frameworks (such as establishing an Office of Space Services/Commerce) that shall gather all stakeholders in creating a transparent, time-bound policy/regulatory mechanism to support rapid growth" (Prasad 2016a, b, c).

As also pointed out by legal experts Ashok G.V. and Riddhi Souza, "as it stands now, the Department of Space, the ISRO and Antrix Corporation represent the administrative, executive and commercial wing of India's space programme. These three agencies share a symbiotic relationship and therefore India's space regulator, despite a history of integrity and transparency, has nevertheless failed to win the confidence of the private sector. It is perhaps for this reason that despite the 100-percent FDI in the SATCOM sector, the investments flowing into the country for the SATCOM field is only a fraction of its full potential. In principle, there are two ways of addressing this problem:

---

[15] These include electric propulsion, high-capacity satellite Ka-band links, large-class Ku-satellites, high-throughput satellite (HTS), machine to machine (M2M) and ultra-HD technologies (Rao et al. 2015).

(a) Creating an independent body similar to the approach of the Commercial Space Launch Act, 1984 which vested regulatory functions and powers to the US Department of Transport.

(b) Reducing the role of ISRO in operations by transferring the same to the private sector and retaining only research and development functions with ISRO" (Ashok and D'Souza 2017).

Whereas the latter avenue is the one currently under implementation by ISRO, it would remain nevertheless essential to introduce more transparent regulations that clearly define the role of the various public stakeholders.

Remote Sensing Services

While the generation and provision of geospatial data within the country are predominantly driven by ISRO through its National Remote Sensing Centre (NRSC), the export of data products and downlink services from the IRS constellation on the international market is managed by the Antrix Corporation (Antrix Corporation Limited 2016a, b). The IRS data and services offered to international customers are more specifically those from RISAT-1, RESOURCESAT-2, CARTOSAT-1 and OCEANSAT-2 satellites. As part of this offering, Antrix can establish the IRS Ground Systems (IGS) for customers outside India, offering turnkey solutions for the direct reception and processing of IRS data with mission-specific hardware and software. Currently operational IRS Ground Systems are located in Germany for receiving Resourcesat-2 and Oceansat-2 data, in Svalbard/Tromso, Norway, for RISAT-1, and in Iran and Algeria for Cartosat-1 (Antrix Corporation Limited 2016a, b).

Although the EO products offered by Antrix are gaining increased commercial success, India remains far from fully seizing the opportunities in the global EO satellite market, primarily because of the lack of commercial exploitation of various Geospatial Information Systems (GIS) solutions within the country.[16] Admittedly, as reported by Prateep Basu, the Indian market is theoretically well primed for satellite EO services – as demonstrated by the signature of more than 170 MoUs between ISRO and various public agencies for a variety of EO applications, such as infrastructure monitoring, crop insurance, watershed development, asset management and mapping – and the government is aggressively taking steps to integrate satellites into the digital economy (Basu 2016).

---

[16] A study conducted by the National Institute of Advanced Studies (NIAS) has shed some light on the most critical issues in the application of remote sensing data for terrestrial applications. The study has identified the non-availability of regularly updated geographical information (GI) content for the nation; lack of a coordinated, aligned and professional effort at furthering the national goals of GI generation/usage; and the lack of a holistic national GI policy that shall lay forward a roadmap for all elements of GI in aiding to make GI usage all-pervasive and easily possible. Remarkably, these aspects are inherently related to the lack of commercial exploitation in various Geospatial Information Systems (GIS) solutions with B2B/B2C delivery models within the country. The downstream applications are, in fact, predominantly government-driven societal applications such as disaster management, meteorology and resource management (Rao and Murthi 2012).

However, all the services are offered through government-to-government (G2G) service platforms such as Bhuvan, rather than through B2B/B2C delivery models. As Narayan Prasad aptly notes, "the current alignment of the space program leaves very little traction for the establishment of commercial remote sensing avenues in India. Unlike the United States, where the private sector was incentivized to exploit remote sensing as a commercial venture, India's approach is more rooted in the collection and use of space data for the benefit of its society. This also translates into a lack of any interest in financing the private sector for ventures that are themed around space commercial activities seen in the West [… and] limits the involvement of the private sector in the utilization of geospatial data (data processing, and tasks of data conversion and enrichment for extended use) to an outsourcing business model" (Prasad 2016a, b, c).

Consequently, "India also does not have any established practice of the private sector utilizing the geo-location of the country to possibly provide uplink-downlink solutions as a service in international markets" (Prasad 2016a, b, c). While North American and European satellite EO industries are further expanding their business also by shifting their focus from products to integrated services combining image, weather, location and other datasets with Big Data analytics, "India remains oblivious to this growth story" (Basu 2016). Even though the country has significant expertise in ICT, there are only a handful of firms capitalising on such expertise to deliver B2B or B2C services in the EO market. Examples include CropIn, Cyient and ESRI India. But these examples do not hide the fact that the downstream commercialisation of remote sensing data in India is far from reaching its full potential and that there is a visible gap in scaling up commercial applications by the private sector (Ibid).

The biggest challenge, though, is arguably not a technical one. As in the case of satellite communication services, it is mainly a policy and regulatory challenge. As argued by many, India's Remote Sensing Data Policy does not provide a consistent framework that can stimulate the emergence of a robust commercial ecosystem for satellite EO industry in India, capable of seizing the increasing opportunities in the global EO market (Rao and Murthi 2012). Towards this, in April 2016, India's Ministry of Home Affairs proposed a draft law, the Geospatial Information Regulation Bill 2016, which intends to provide an updated policy framework to regulate the use of geospatial technologies within the country (Ministry of Home Affairs – Government of India 2016). The draft, however, has been criticised for not being fully empowering of private sector undertakings. It hence remains to be seen whether the draft will be further upgraded to allow the growth of a robust and internationally competitive commercial EO industry.

Satellite Platforms, Subsystems and Mission Support Services
Leveraging on ISRO's state-of-the-art capability in satellite manufacturing, the commercial offerings of Antrix also extend into the upstream sector, with the delivery of satellites and satellite subsystems for different applications. In terms of satellite systems, Antrix offers to customers the various satellite platforms developed by ISRO over the years; these include the highly proven INSAT/GSAT platforms,

namely, I-1K (1000 kg class), I-2K (2000 kg class) and I-3K (up to 3500 kg class), with varying payload and power configurations, as well as the versatile IRS satellites IMS-2 (350 kg class) platform for EO missions. Antrix in addition offers satellite subsystems such as Structures, Thermals Systems, Mechanisms, Power Systems, Attitude Control, Digital and RF Systems as well as payload Subsystems for a variety of applications. In this context, Antrix can also undertake spacecraft integration and environmental testing for customers by using the infrastructure available at ISRO centres.[17]

To date, however, the presence of Antrix in the commercial satellite manufacturing market has been rather marginal. The company, in alliance with EADS Astrium (presently Airbus Defence and Space), has delivered only two satellite platforms to international customers, namely, "W2M" based on the I-3K satellite platform for Eutelsat (32 Ku transponders, 3460 kg mass) launched on 20 December 2008 and "Hylas" based on the I-2K satellite platform for Avanti Plc, UK (2 Ku, 6 Ka transponders, 2100 kg mass), launched on 26 November 2010 (Antrix Corporation Limited 2016a, b). Moreover, in the immediate future, it might be even more difficult for Antrix to gain a larger share in the international market, due to the technological changes in satellite platforms (such as all-electric systems) to which the INSAT/GSAT buses have to be upgraded as well as to India's own domestic requirements, which already beset ISRO's ability to produce satellites in volume within current facilities (Prasad 2016a, b, c).

Commercial Launch Services

Although commercialisation of launch services has not been a priority for India, being its launcher vehicles production intended to satisfy domestic demand, since 2002 Antrix has been offering excess capacity to international customers with a view to progressively increasing the country's presence on the worldwide commercial launch market. Given the relatively low GTO performance of the GLSV launcher and the limited number of flight opportunities due to the need to meet India's institutional market demand, launch services for international customers have been limited to the LEO market.

The most prevalent form of commercial LEO launches has been piggyback missions where smallsats are launched with domestic payloads on the PLSV[18] and in the form of full-on ISRO cooperation in the joint development of a payload.[19] The

---

[17] The facility offering includes checkout facility for the test protocols of large high-power satellites, thermo-vacuum chamber (CATVAC) for thermo-vacuum performance qualification of the spacecraft, vibration shaker (CATVIB) for dynamic tests, physical parameter measurement facilities and the Comprehensive Antenna Test Facility (CATF) (Antrix Corporation Limited 2016a, b, c).

[18] Up until the end of 2016, ISRO had launched 38 foreign piggyback payloads on PSLV, among which 16 European institutional payloads, namely, DLR-TUBSAT, BIRD and AISat 1 for the DLR, Agile for the ASI, the ESA PROBA and RUBIN-8 payloads and 10 university microsatellites.

[19] For instance, two joint CNES/ISRO payloads Megha-Tropiques (launched in October 2011) and SARAL (launched in February 2013)

first dedicated PSLV launched for foreign institutional customers (ASI and Israel's MoD) was carried out in 2007 for ASI and in 2008 for Israel's Defence Ministry.[20] The list of all foreign satellites launched by India is provided in Annex 3.

It has now become customary for at least one annual PSLV launch to include small satellites. Even though until today Antrix has not seen the launch of cubesats and nanosats as a revenue source but has provided these launch services to institutional customers on a cost-basis, there is now a clear path towards nanosatellite launches on a commercial basis.

Beginning in 2008, India also started to adopt a more aggressive commercial stance "by offering launch service prices below the level of its Western competitors and partially even below the price levels of converted Russian ICBMs or Chinese launch vehicles. This strategy has appeared to be attractive for small payloads left with a limited number of launch alternatives incurring increasing prices and recurring delays (e.g. Dnepr, Cosmos and Rockot)". Besides this, in a push to boost global competitiveness, India has focused on strengthening partnerships with the United States and Asian nations such as South Korea and Kazakhstan.[21] PSLV has become an attractive tool for India and stands as a cornerstone of ISRO's commercial launch services for US satellites, for instance, which has largely helped create access to the Western market. As a first step, a Technology Safeguards Agreement (TSA) between the United States and India was signed in 2009 to facilitate the launch of non-commercial satellites built in the United States or including US components on Indian vehicles. Collaborative work between Europe and India in this area was also enhanced through an agreement that allows for the marketing of EO satellites launched aboard PSLV.[22]

In the meantime, an increase in PSLV production rates (aided by the completion of a second launch pad at Sriharikota and the construction of a second vehicle assembly building) has allowed India to offer one annual launch dedicated solely to commercial customers and in turn led to increased marketing of PSLV launch opportunities: the first two launches with a commercial main payload were carried out in 2012 and 2014 for the Spot Image.

The focus on long-term commercial activity moved ahead smoothly as PSLV launch service contracts were signed in 2013 and 2014, setting aside one PSLV launch per year for a purely commercial launch service for foreign payloads. As of 2016, 74 satellites from international customers had been launched on-board the

---

[20] Agile (350 kg, ASI, launched on 23 April 2007) and TechSAR (300 kg, Israel MoD, launched on 21 January 2008).

[21] The placing of India among the groups of closest US allies for export control opens a possibility to accelerate ITAR-related procedures. Cooperation involving civilian nuclear technology has also become possible under the "United States-India Nuclear Cooperation Approval and Nonproliferation Enhancement Act" of 2010.

[22] The agreement was signed in 2008 between the ISRO/Antrix, Secretary of DOS and Astrium for marketing Earth Observation Satellites launched by PSLV. India also signed a framework agreement on cooperation with France in the same month for the launch of Megha-Tropiques and SARAL (see Sect. 6.3).

PSLV rocket. Appendix 2 provides a detailed list of the foreign satellites launched by the ISRO/Antrix Corporation.

With the exponential increase of small satellites and with the PSLV hitting a sweet spot in the international LEO market that is currently underserved, there is certainly great potential for India to seize a larger share of the commercial launch market. However, the success of India's commercialisation efforts is directly related to the ability to increase launch frequencies by enhancing production rates.

To achieve this objective, ISRO has planned to progressively outsource PSLV engineering design services as well as its assembly, integration and testing (AIT) processes to a consortium of private firms[23] by promoting the creation of a joint venture between its suppliers. Such privatisation should allow not only to increase the rate of launches up to 18 per year but also enable the adaptation to advanced manufacturing methodologies while benefiting from synergy effects with other sectors (Prasad 2016a, b, c; Laxman 2016).

Towards this, in April 2016, ISRO set up a Steering Committee to outline the transition of Indian industries from mere vendors to integrators of the launch vehicle. As detailed during a Parliament questioning: "the terms of reference of the Steering Committee for stepping up the Launch Capacity include – (i) establishing the production profile of the launch vehicles; (ii) assessing the gaps in meeting the production requirement and (iii) arriving at a strategy for production including industry linkages, infrastructure build-up, technology sharing methods, quality assurance support" (Lok Sabha – Ministry of Space 2016).

While the setting up of an industrial consortium or a joint venture appears a necessary step towards creating economies of scale able to meet increasing launch frequency requirements, there are also legal and regulatory issues that must be overcome. The privatisation of the PSLV requires a well-developed domestic law developed in consonance with India's international obligations (Abhijeet 2016). On the international front, there is still no clear resolution to the regulatory barriers put in place by the United States. However, as Prasad noted, "with the outsourcing AIT strategy paving the way for systematic increased participation of the industry, there is a foundation for the emergence of transparency in pricing due to private sector participation. This can allay concerns raised by COMSTAC that India is subsidizing access to space, hindering the competitiveness of U.S. companies" (Prasad 2016a, b, c).

The PPP is scheduled to be operational in 2020 and is for the moment intended to cover the PSLV rocket only. While GSLV is theoretically open to commercial customers, Antrix has chosen not to market its launch services due to its low GTO payload capacity and low production rates. So far Antrix has not signed any commercial launch service agreements for GSLV launches. This situation could gradually change once GSLV Mark III becomes available, which will allow medium to

---

[23] As already explained in Chap. 3, although a number of SMEs in India already deliver subsystems and components for the PSLV, the final AIT is done by ISRO with the support of the Satish Dhawan Space Centre (SHAR), Liquid Propulsion Systems Centre (VSSC) and ISRO Propulsion Complex (IPRC). Sensitive elements such as propellants are likely to remain excluded from the outsourcing and should remain institutional.

heavy GTO payload launches. In addition, the mandate of the PPP could possibly be extended to cover also the AIT of the GSLV launch vehicle for both Indian and foreign customers, so as to enable ISRO to focus its efforts on new launchers' technologies, including the reusable launch vehicle (RLV).[24]

The GSLV Mark III rocket is currently on track to reach full operational flight status. A test launch of the GLSV Mark III rocket (with a passive cryogenic third stage and a 4-tonne payload capacity) was undertaken in 2014, while the first mission to deliver a communication satellite to GEO was undertaken in June 2017. With the new semi-cryogenic engine under development, GSLV Mark III may also expand its capacity to launch payloads up to 6 tonnes to GEO. It has been noted that GSLV Mark III with this expanded payload capacity of 6 t to GEO will be produced at approximately $30 million (Prasad 2016a, b, c). Interestingly, this highly competitive cost – which is directly linked to the substantially lower operating costs in India for infrastructure and manpower – would translate into about the same costs estimated for the reusable rockets of SpaceX (Ibid).[25]

Be this as it may, it remains unlikely that GSLV Mark III will seize a significant market share in the near term since full operational reliability and sustained production capability are still some way off. All in all, whereas India may in the coming years position itself as a stronger competitor in the international market for LEO launches thanks to the PSLV privatisation, the potential for its competitiveness in the GEO launch market does not look feasible in the short-term but only in a medium- to long-term timeframe.

4.5.2 New Space India

Whereas India's commercialisation efforts in the worldwide space markets are essentially centred on the offerings of the Antrix Corporation, over the past few years, India has also witnessed the emergence of a New Space phenomenon.[26] Partly

---

[24] This would be an important development considering that India is still at a very fundamental stage in the development of reusable rockets. It has been estimated that in light of the development and test cycles involved in achieving this feat, India might reach this milestone only by 2030.

[25] According to SpaceX, reusing the first stage will likely entail 30% cost savings, which in turn may reduce the Falcon 9's price to $42.8 million from today's $61.2 million. The projection of reduced cost to access to space for a reusable rocket such as Falcon 9 translates into ~$5000/kg (de Selding 2016).

[26] Although no broadly accepted definition of New Space exists today, in a recent research, ESPI has pointed out that the New Space can be intended as a "disruptive sectorial dynamic featuring various end-to-end efficiency-driven concepts driving the space sector towards a more business- and service-oriented approach. Within this new dynamic, six major interrelated trends can be isolated, namely: (a) new entrants in the space sector including large Information and Communications Technology (ICT) firms, start-ups and new business ventures; (b) innovative industrial approaches with announcements and initial developments of ambitious projects based on new processes; (c) disruptive market solutions offering, for example, integrated services, lower prices, reduced lead times, lower complexity or higher performance among other value proposition features; (d) sub-

inspired by the undertakings of US companies such as Space X, Planet and Blue Origin and partly propelled by the high dynamism of Indian start-up culture, several Indian entrepreneurs have seen the opportunity from companies doing businesses based on space products and services.

The emergence of a New Space entrepreneurship in India is at a very nascent stage and hence understandably centred on start-up companies trying to build offerings in either the upstream or downstream segments of the value chain. To be sure, as noted in Chap. 3, India can already boast an industrial base of more than 500 small, medium and large enterprises that are seizing commercial opportunities within the current value chain. However, unlike these traditional private companies currently working as vendor/suppliers of Tier-2, Tier-3 and Tier-4 level work for ISRO, India's New Space companies – like most of the New Space companies worldwide – focus on providing end-to-end space upstream and downstream systems or services independently from governmental procurement. In other words, they try to seize commercial opportunities by building independent B2B or B2C solutions, with a particular focus on positioning their offerings in the international market and integrating in the global space industry ecosystem (Prasad 2017a, b).

An overview of notable Indian space start-ups and their offerings is provided in Table 4.7.

Recent analyses have opined that there is a large scope for these New Space companies to provide independent services and achieve economies of scale (Prasad 2017a, b; Prasad and Basu 2015; Rao et al. 2016), particularly if considering such factors as India's burgeoning demand for space products and services, its huge and growing market size as well as its favourable market conditions (in terms, for instance, of low labour costs, talent pool, innovation competences and space technology know-how).

While the total number of New Space companies is certainly set to become higher in light of these factors, when comparing the amount of India's New Space companies to the landscape of space ventures in other spacefaring nations, one will clearly notice how still embryonic this ecosystem really is.[27] Especially if compared to the US and European private space sectors, it becomes perceptible that India's space start-up ecosystem not only lags behind in terms of entrepreneurs but also in terms of venture capitalists, mentors, accelerators and incubators. Despite various

---

stantial private investment from different sources and involving different funding mechanisms; (e) new industry verticals and space markets targeting the provision of new space applications; (f) innovative public procurement and support schemes involving new R&D funding mechanisms and costs/risks sharing arrangements between public and private partners" (Vernile 2017).

[27] According to the statistics provided by the New Space Ventures, a discovery platform for New Space industry around the world [5], the United States stands out as the unparalleled world leader in the private space industry with 535 American New Space enterprises and organisations, followed by the United Kingdom 82 and Germany 61. India ranks 12th in the world with 16 New Space companies. Ahead of India are advanced Western European and North American economies as well as Australia, Poland and Japan; however both global space powers Russia (11) and China (9) have, for obvious reasons, shown less private start-up activity than India (New Space Ventures 2017).

Table 4.7   Major new space Indian companies

Name	Incorporation	Segment	Mission/offering
Aeolus Aero Tech	2012, Chennai	Education	Development of smallsat hardware and training kits
Aniara Space	n.d, Bangalore	Engineering	Satellite services (VSAT connectivity, capacity leasing, IoT, etc.)
Astrome Technologies	2015, Bangalore	Internet	Satellite constellation-based Internet connectivity
Bellatrix Aerospace	2015, Bangalore	Propulsion	R&D company; launch vehicle and electric propulsion systems
Dhruva Space	2012, Bangalore	Satellites	Development of small satellite platforms with a focus on AIT
Earth2Orbit	2009, Mumbai	Engineering	Provision of remote sensing value-added products, technology consulting
Naika Statlogic	2015, Bangalore	Consulting	Space-based statistical data analysis
SatSure	2014, Bangalore	GIS	Crop yield risk assessment for Indian farmers
SPACE India	n.d.	Education	Hands-on astronomy education and space tourism services
Team Indus	2012, Hyderabad	Lander	Indian contender at the Google Lunar X Prize competition
Xovian	2014, Kanpur	Satellites	Low-cost, sustainable satellite fabrication solutions, sounding rockets and HAPS

Source: Aliberti et al. (2017)

success stories are progressively emerging, the overall enterprise readiness level of India's New Space companies remains rather low, with an extremely small number of companies having reached technological and operational maturity.

A number of elements can be identified as the major causes for this situation. For one thing, the involvement and capabilities of private industries in the Indian space programme remain rather marginal in comparison to the United States and Europe, because of the quasi-monopolistic role played by the ISRO/Antrix in both the upstream and downstream segments of the space value chain. In addition, the private investment base to support the growth of Indian start-ups has thus far been very small, not to say inexistent. Most of them have been self-funded or sponsored by grants from high net-worth individuals like Tata, Mohandas Pai, and Rajan Anandan. Interestingly, this is in stark contrast to India's broader start-up ecosystem (particularly in areas as e-commerce and consumer services companies), which has a well-organised community of venture capitalists, mentors and business angels (see Box 4.3).

What is more, unlike New Space companies in the United States or Europe, Indian astropreneurs cannot count on any dedicated start-up support programme such as the Small Business Innovation Research fund offered by NASA or the Business Incubation Programme of the European Space Agency (ESA). There is no real case of institutional capital investment for these companies and no dedicated support structure like incubators or accelerators. More broadly, the level of institu-

Box 4.3: India's Start-Up Ecosystem

With over 4000 start-ups that are expected to nearly quadruple by 2020, India is the third largest start-up ecosystem (behind the United States and the United Kingdom) with the second fastest growth rate in the world. In addition to the sheer number of start-ups, the ecosystem itself is likewise surging: India now has an expanding community of venture capitalists, mentors, incubators, accelerators, institutions and, perhaps most importantly of all, a culture among its young people who are interested in technology and in starting up (NASSCOM 2016). In identifying the ecosystem's components and characterising the different players, Microsoft Ventures – in cooperation with the Venture Intelligence – set the scope of start-ups considered for their analysis at total funding reaching up to $ 20 million.

Twelve unicorns (i.e. start-up companies that are valued above $ 1 billion) have been identified in the Indian market, accounting for 10% of the unicorns' club worldwide. India is best known for its e-commerce start-ups, with rapid e-commerce growth in India projected to become a $220 billion market by 2030. E-commerce has spread through India, projected to evolve into a $220 billion market by 2030. As observed elsewhere in the world as well, India's shift to mobile from a web-based structure has fundamentally changed basic processes such as payment. Indian online shopping firm Myntra, for example, completely scraped their website and replaced it with a mobile platform instead. The shift to mobility makes sense in India especially given the prevalence and dominance of mobile devices. Moreover, a boom in on-demand consumer services has added $ 2 billion in funding to those start-up enterprises in 2014–2015 alone. Besides e-commerce, there are other sectors that are actively pushing new and innovative solutions. These include banking, manufacturing, retail and supply chain, healthcare, media and EdTech.

Three vital needs are identified as India's start-up ecosystem moves forward. Firstly, a cultural and ideological change must occur on the side of investors. The idea that US start-ups reign supreme acts as a serious limitation to the potential of Indian start-ups. Secondly, a shift also manifested in culture and ideology needs to be sought by the consumers. The Indian market generally prioritizes cost over value, thus making it hard for start-ups providing high value at a relatively high cost to gain customers in India. Until Indian consumers are ready to pay for value rather than seek the cheapest option, many Indian start-ups will struggle in the domestic market and be forced to seek foreign consumers. Thirdly, and finally, the importance of a favourable regulatory environment for the success of start-ups cannot be overstated. The proliferation of successful Indian start-ups is directly tied to the government's legal limitations enforced upon these newfound companies (Microsoft Venture 2016).

tional sponsorship, which has proved to be a key in the emergence of New Space companies in the United States, has been thus far negligible. Support for private industries has been limited to a policy of "transfer and buy back", in which ISRO has been procuring products manufactured by industry as a result of a technology transfer enabled by ISRO itself, and to the execution of some consultancy projects in high-technology areas to provide support to various industries.

Whereas Indian astropreneurs today see increasing opportunities to create synergies and cross-fertilisation with other industrial sectors like IT as well as the broader start-up ecosystem,[28] those externalities will not suffice without a firm support on the institutional side. To be sure, to India's policy-makers, it is increasingly becoming clear that enabling the New Space to flourish will not only have a direct benefit on these businesses but more broadly generate positive spillover effects on the country's industrial base and overall footprint in the global space economy. And this, in turn, will be key to contribute boosting an innovation-based macroeconomic growth. It is perhaps an indication of this that the Indian government has included space among the key areas of the *Make in India* initiative, which aims to attract foreign companies to produce in India by, inter alia, introducing a significant relaxation of the foreign investment threshold (Government of India 2016a, b).

However, if the move of raising the cap on FDI to 100% contemplated in the *Make in India* initiative can certainly provide foreign companies with incentives to invest in or create partnerships (e.g. through merger and acquisitions) with Indian space start-ups, to fully encourage investments, there must also be transparent and comprehensive regulatory mechanisms that are supportive of private space business. A sound policy, legal and procedural framework is quintessential for any industry in any geographic location. However, in the case of India's space sector, governmental policies and practices have been unsupportive, not to say obstructive, of private sector undertakings. Indeed, while such policies acknowledge the existence of commercial activities, they are not by any measure comparable to the probusiness legislations earmarked by the United States or Luxembourg. In other words, they are still far from fully ensuring predictability, transparency, responsiveness and efficiency, in short what the private sectors demand. It suffices to remind the cumbersome procedures for establishing satellite systems or for accessing and using space-based information and data from Indian satellites. From an overall perspective, these policies not only hinder potential foreign investment but also prevent "the acceleration of new technology applications by indigenous innovations and start-up enterprises to assist in the timely achievement of national economic development goals" (Kaul 2017).

---

[28] This is a clear possibility considering that the New Space phenomenon is typically characterised by a shift of innovation towards downstream exploitation of space infrastructures and in particular data processing and analytics, an area where India is really thriving. In this domain, India has a very impressive community of venture capitalists and institutions that could be ideally leveraged by Indian astropreneurs to access critical capital investment and, more importantly, to improve the overall conditions of the space industry ecosystem.

All in all, whereas the emergence of a New Space dynamic in India is per se a remarkable development, its future growth trajectory will be essentially determined by the level of support public institutions will offer. Given India's inherent assets, should the government start to back the private sector through a series of institutional, financial and legal tools along the line of the US government, New Space India has the potential to become a very active private industry, exercising a great impact on Indian and the global space economies alike.

4.5.3   India's Commercial Ecosystem: An Assessment

As emerges from the above sections, India's current commercial efforts have several achievements to look back on. Whereas such efforts are still led by the public sector, Antrix Corporation's performance during the last few years has been remarkable, with a clear upward trend (see Fig. 4.17). In 2014–2015, the company's revenue was ₹1860.71 crores (approximately $200 million), an increase of 17% compared to the previous year, and future prospects appear bright (Antrix Corporation Limited 2015).

However, even with a projected growth in revenues, it is undeniable that if compared to the global space markets, "India's commercial engagement as a provider to the outside world has been in tiny pockets, with a couple of dedicated rockets flown for outside customers with most of the payloads being piggyback. Similarly, on the spacecraft side, India has built two spacecraft for foreign customers, while there may be more potential to exploit" (Prasad 2016a, b, c). To put things in perspective, the annual revenues reported by Antrix currently represent less than the 0.5% of the global satellite market, which is estimated to be worth $300 billion (Al-Ekabi 2015; Satellite Industry Association 2016).

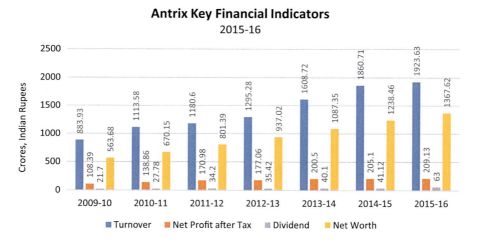

Fig. 4.17 Antrix performance (*Source*: Antrix Corporation)

Considering the inherent advantages India has at its disposal (e.g. cost-competitive skilled workforce, innovation skills, low cost of operations, etc.), it is hence quite evident that global commercial possibilities are not sufficiently seized by India. The bottom line, however, is that Antrix/ISRO proves unable to scale up its offerings on the commercial market owning to the large backlog in satellite and launch vehicles for the country's own requirements (Prasad 2015; Matheswaran 2015).

To overcome the current gap in the timely provision of space-based services for meeting domestic needs while enabling more substantial efforts on the commercial side, there is a clear need to improve the country's capacities in the supply of space assets and services, by more specifically moving away from the government's monopoly in the sector and enabling participation of private industry.

In this respect, as noted before, the Indian government has already expressed interest in further enabling the already present space industry to move from vendor to turnkey solution provider and in turn foster active engagement of the Indian space industry in both upstream and downstream commercialisation efforts, nationally and globally. Clearly, this move is not limited to the two industry consortia under implementation for the production of launchers and satellite systems. It also extends to the broader industrial base and hence to "traditional" space industries and New Space companies alike.

Another key issue closely related to the need to enhance industry participation will be the ability of India to leverage New Space initiatives by supporting the emergence of a suitable ecosystem (Prasad 2017a, b). India's government is arguably becoming aware that supporting the emergence of a mature New Space dynamic in India will have a positive impact not only on the business of these start-ups; it will more broadly enable the country to meet its domestic requirements while also boosting India's presence on the global commercial market.

The Indian government has already made moves to incentivise investment, evident, for instance, in the full opening for FDI of upstream and downstream applications. At the same time, it is clear that instilling confidence in private sector investments also requires a comprehensive and transparent regulatory system. As aptly noted by Prasad, since the "marketing of space technology applications can involve several issues – such as allocation of frequencies for satellite operations, data distribution policies, and Intellectual Property Rights (IPR) – lack of transparency and clarity in steps and processes in allotment of these resources hinders private industry's resolve to participate in a comprehensive manner. Ultimately, these translate into reduced risk appetite by venture capitalists to entrepreneurial ideas. A comprehensive national space legislation that shall create transparent regimes of liability and procedural aspects of engagement in outer space activities is the need of the hour for greater engagement of the private sector" (Prasad 2016a, b, c).

Instilling confidence in investors is key; this is why comprehensive policy framework could help unlock investment potential by creating a fruitful environment for such activity to take place independently (Prasad 2016a, b, c).

In conclusion, whereas India's commercialisation efforts still present a large untapped potential, in order to fully exploit such potential, several requirements must be addressed. Figure 4.18 summarises the necessary steps the Indian government

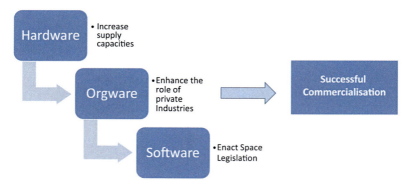

Fig. 4.18 India's commercialisation requirements

would need to undertake in order to achieve successful commercialisation of its space products and services in the global market and create a stronger India brand.

As the figure shows, the potential for successful commercialisation is first linked to an expansion of the *hardware*, i.e. an expansion of the country's space infrastructure and supply capacities. This possibility, in turn, proves to be directly linked to changes in the *orgware* of space activities (i.e. the organisational structures set up to develop and run the hardware) and more specifically to the need to enhance private industry participation through privatisation operations and creation of specific support structures for the emergence of the New Space. These changes are, in fine, linked to an evolution of India's *space software* (i.e. the norms and rules applied to run the programme) that can enhance the confidence and the level of private investments in space. Hence, the bottom line of India's commercialisation efforts lies in the ability to introduce a comprehensive and probusiness legislation that can enable the future expansion of the Indian space programme.

It should go without saying that such regulatory mechanisms can only be put in place in an optimal manner if there is an overarching policy spelled out at the high political level, rather than being entrust to specific organisations like ISRO. To date, however, a comprehensive policy document driving India's posture in space remains wanting.

References

Abhijeet, K. (2016). Privatisation of the PSLV: What the law of outer space demands? In *67th international Astronautical Congress* (pp. IAC-16,E7,5,8,x32950). Guadalajara: IAF.

Agraval, N. (2016, October 4). *Inequality in India: What's the real story?* Retrieved November 2, 2016, from World Economic Forum. https://www.weforum.org/agenda/2016/10/inequality-in-india-oxfam-explainer/

Alawadhi, N. (2016, June 6). *Government must use satellite technology for digital India initiative: TV Ramachandran.* Retrieved November 15, 2016, from The Economic Times. http://economictimes.indiatimes.com/tech/ites/government-must-use-satellite-technology-for-digital-india-initiative-tv-ramachandran/articleshow/52615913.cms

Al-Ekabi, C. (2015). *Space policies, issues and trends in 2015–16*. Vienna: European Space Policy Institute.

Aliberti, M., Nader, D., & Vernile, A. (2017). Understanding India's new space potential: Implications and prospects for Europe. In *68th international Astronautical Congress (IAC)*. Adelaide: International Astronautical Federation.

Andrés, L., Biller, D., & Dappe, M. (2013, December). *Reducing poverty by closing South Asia's infrastructure Gap*. Retrieved October 30, 2016, from The World Bank. http://documents. worldbank.org/curated/en/881321468170977482/pdf/864320WP0Reduc0Box385179B000P UBLIC0.pdf

Annadurai, M. (2017). Recent Indian space missions. In *54th session of the scientific and technical subcommittee of United Nations Committee on peaceful uses of outer space*. Vienna: UNOOSA.

Antrix Corporation Limited. (2015). *Antrix Annual Report 2014–15*. Bangalore: Antrix.

Antrix Corporation Limited. (2016a). *Remote sensing services*. Retrieved November 30, 2016, from Antrix Corporation Limited. http://www.antrix.gov.in/business/remote-sensing-services

Antrix Corporation Limited. (2016b). *Satellite platforms*. Retrieved December 9, 2016, from Antrix Corporation Limited. http://www.antrix.gov.in/business/satellite-platforms

Ashok, G. V., & D'Souza, R. (2017). SATCOM policy: Bridging the present and the future. In R. P. Rajagopalan & N. Prasad (Eds.), *Space India 2.0 commerce, policy, security and governance perspectives* (pp. 119–139). New Delhi: Observer Research Foundation.

Basu, P. (2016, December 1). *New horizon for satellite Earth observation*. Retrieved December 2, 2016, from Observer Research Foundation. http://www.orfonline.org/expert-speaks/ new-horizons-satellite-earth-observation/

Bhatia, B. S. (2015). Satcom for development Education. The Indian experience. In P. M. Rao (Ed.), *From fishing hamlet to red planet: India's space journey*. New Delhi: Harper Collins.

Bhatia, B. (2016). Space applications for development – The Indian approach. In *ESPI 10th autumn conference*. Vienna: European Space Policy Institute.

Bonnefoy, A. (2014). *Humanitarian telemedicine*. European Space Policy Institute. Vienna: ESPI.

Bute, S. (2013, March 7). *Innovation: The new mantra for science and technology policies in India, Pakistan and China*. Retrieved November 2, 2016, from Institute for Defence Studies and Analysis. http://www.idsa.in/backgrounder/ScienceandTechnologyPoliciesinIndiaPakistanandChina

Corrigan, G., & Di Battista, A. (2015, November 5). *19 Charts that explain India's economic challenge*. Retrieved October 2, 2016, from World Economic Forum. https://www.weforum.org/ agenda/2015/11/19-charts-that-explain-indias-economic-challenge/

Credit Suisse. (2015). *Global wealth databook 2015*. Zurich: Credit Suisse.

Comptroller and Auditor General of India. (2012). *Report of the Comptroller and Auditor General of India on hybrid satellite digital multimedia broadcasting service agreement with Devas*. New Delhi: Department of Space.

de Selding, P. (2016, June 8). Satellite operators give negative reviews of Indian regulator's satellite-TV proposal. *Space News*.

Deloitte. (2010). *Overview of Indian Space Sector*. Bangalore: Deloitte.

Department of Telecommunications – Government of India. (2016, February 2). *Restrictions on the use of satellite phone*. Retrieved November 10, 2016, from Department of Telecommunications. http://www.dot.gov.in/restrictions-use-satellite-phone

Dutta, S., & Lanvin, B. (2016). *The Global Innovation Index Report 2016*. Retrieved September 30, 2016, from Global Innovation Index. https://www.globalinnovationindex.org/gii-2016-report

Government of India. (2016a). *Economic survey 2015–2016* (Vol. 1). New Delhi: OUP India.

Government of India. (2016b). *SPACE – Make in India*. Retrieved November 13, 2016, from Make in India. http://www.makeinindia.com/sector/space

Haub, C. (2012, March). *2011 Census shows how 1.3 billion people in India live*. Retrieved from Population Reference Bureau. http://www.prb.org/Publications/Articles/2012/india-2011-census.aspx

India Space Research Organisation. (2016). *Tele-Education*. Retrieved October 3, 2016, from ISRO. http://www.isro.gov.in/applications/tele-education

India Stack. (2016). *About*. Retrieved December 2, 2016, from India Stack. http://indiastack.org/about/

Indian Space Research Organisation. (2016a). *Climate & environment*. Retrieved October 26, 2016, from ISRO. http://www.isro.gov.in/applications/climate-environment-0

Indian Space Research Organisation. (2016b). *Disaster management support programme*. Retrieved October 20, 2016, from ISRO. http://www.isro.gov.in/applications/disaster-management-support-programme

Indian Space Research Organisation. (2016c). *ISRO Annual Report 2015–2016*. Bangalore: ISRO.

Indian Space Research Organisation. (2016d). *ISRO technology transfer*. Retrieved September 18, 2016, from ISRO. http://www.isro.gov.in/isro-technology-transfer

Indian Space Research Organisation. (2016e). *Natural resource census*. Retrieved October 27, 2016, from ISRO. http://www.isro.gov.in/applications/natural-resources-census

Indian Space Research Organisation. (2016f). *Tele-medicine*. Retrieved October 11, 2016, from ISRO. http://www.isro.gov.in/applications/tele-medicine

Indian Space Research Organisation. (2016g). *Towards ensuring water security*. Retrieved October 1, 2016, from Indian Space Research Organisation. http://www.isro.gov.in/applications/towards-ensuring-water-security

IndoGenius. (2016a, August). Five types of India innovation. *The importance of India*. IndoGenius (Online Lecture).

IndoGenius. (2016b, August). India's Stack and JAM Trinity. *The importance of India*. IndoGenius (Online Lecture).

IndoGenius. (2016c, August). Rural vs. urban India. *The importance of India*. IndoGenius (Online Lecture).

Kasturirangan, K. (2011). *Relating science to society*. Mysore: University of Misore.

Kaul, R. (2017). A review of India's geospatial policy. In R. P. Rajagopalan & N. Prasad (Eds.), *Space India 2.0 commerce, policy, security and governance perspectives* (pp. 141–150). New Delhi: Observer Research Foundation.

Kochhar, R. (2015a, July 8). *A global middle class is more promise than reality*. Retrieved October 21, 2016, from Pew Research Center. http://www.pewglobal.org/2015/07/08/despite-poverty-plunge-middle-class-status-remains-out-of-reach-for-many/#poverty-retreats-in-india-but-the-middle-class-barely-expands

Kochhar, R. (2015b, July 15). *China's middle class surges, while India's lags behind*. Retrieved from Pew Research Center. http://www.pewresearch.org/fact-tank/2015/07/15/china-india-middle-class/

Krishna, V. V. (2015). *Science, technology and innovation policy in India. Some recent changes*. INDIGO Policy.

Kumar, N., & Puranam, P. (2012). *India inside. The emerging global challenge to the West*. Boston: Hardward Business Review.

Laxman, S. (2016, February 15). *Plan to largely privatize PSLV operations by 2020: ISRO chief*. Retrieved November 10, 2016, from The Times of India. http://timesofindia.indiatimes.com/india/Plan-to-largely-privatize-PSLV-operations-by-2020-Isro-chief/articleshow/50990145.cms

Lele, A. (2013). *Asian space race: Rhetoric of reality?* New Delhi: Springer.

Lok Sabha – Ministry of Space. (2016). *"Questions: Lok Sabha,"* Government of India. Retrieved September 2, 2016, from Lok Sabha. http://164.100.47.192/Loksabha/Questions/QResult15.aspx?qref=35204&lsno=16

Mallisena. (1933). Syādvādamanjari, 19:75–77. *Dhruva, A.B*, 23–25.

Matheswaran, M. (2015, May 9). *Hot on Mars, but short everywhere else*. Retrieved December 12, 2016, from Business Line. http://www.thehindubusinessline.com/opinion/hot-on-mars-but-short-everywhere-else/article7224259.ece

McKinsey Global Institute. (2014). *India's technology opportunity: Transforming work, empowering people*. New Delhi: McKinsey.

Microsoft Venture. (2016, November 10). *Enterprise Readiness of Indian startup ecosystem.* Retrieved November 15, 2016, from Microsoft Accelerator: https://www.microsoftaccelerator. com/blog/entry/EnterpriseReadinessofIndianStartupEcosystemReport%7C6205

Ministry of Home Affairs – Government of India. (2011). *2011 Census data.* Retrieved October 23, 2016, from Office of the Registrar General & Census Commissioner. http://www.censusindia.gov.in/2011-Common/CensusData2011.html

Ministry of Home Affairs – Government of India. (2016, April). *National Geospatial Policy [NGP 2016].* Retrieved December 9, 2016, from http://www.dst.gov.in/sites/default/files/Draft-NGP-Ver%201%20ammended_05May2016.pdf

National Innovation Council – Government of India. (2011). *Decade of innovations: 2010–2020 Roadmap.* Retrieved November 3, 2016, from National Innovation Council. http://innovation-councilarchive.nic.in/index.php?option=com_content&view=article&id=36:decade-of-innova tion&catid=7:presentation&Itemid=8

Navalgund, R. (2015). Remote sensing applications. In P. M. Rao (Ed.), *From fishing hamlet to red planet: India's space journey.* New Delhi: Harper Collins.

NASSCOM. (2016). *Start-up report - Momentous rise of the Indian start-up ecosystem.* Retrieved from NASSCOM: http://www.nasscom.in/knowledge-center/publications/ start-report-momentous-rise-indian-startecosystem

Nayak, A. (2015). Potential fishing zones. Science to service. In P. M. Rao (Ed.), *From fishing hamlet to red planet: India's space journey.* New Delhi: Harper Collins.

New Space Ventures. (2017, September 5). *Analytics.* Retrieved September 5, 2017, from New Space Ventures. https://mission-control.space/newspace_ventures

OECD. (2011). *The space economy at a Glance 2011.* Paris: OECD Publishing.

OECD. (2014). *India – Addressing economic and social challenges through innovation.* Paris: OECD.

Office of the Register General and Census Commissioner. (2011). *Percentage of households to total households by amenities and assets.* Retrieved October 1, 2016, from Census of India. http://www.censusindia.gov.in/2011census/hlo/Houselisting-housing-PCA.html

Patairiya, M. K. (2013, November 22). Why India is going to Mars? *The New York Times.*

People Research on Indian Consumer Economy. (2014). *Snapshots – ICE360° Survey.* Retrieved November 20, 2016, from People Research on Indian Consumer Economy. http://www.ice360. in/en/projects/data-and-publications/snapshots-ice360-survey-2014

Pew Research Center. (2015, July 8). *Global population by income.* Retrieved October 10, 2016, from Per Research Center. http://www.pewglobal.org/interactives/global-population-by-income/

Prasad, N. (2015, July 6). Is India turning a blind eye to space commerce? *The Space Review.* Retrieved December 12, 2016, from The Space Review. http://www.thespacereview.com/ article/2782/1

Prasad, N. (2016a, March). Demystifying space business in India and issues for the development of a globally competitive private space industry. *Space Policy, 36,* 1–11.

Prasad, N. (2016b). Diversification of the Indian space programme in the past decade: Perspectives on implications and challenges. *Space Policy, 36,* 38–45.

Prasad, N. (2016c). Industry participation in India's space programme: Current trends & perspectives for the future. *Astropolitics: The International Journal of Space Politics and Policy, 14*(2–3), 234–255.

Prasad, N. (2016, July 2016). ISRO/Antrix-Devas arbitration asco explained. *New Space* India.

Prasad, N. (2017a). Space 2.0 India – Leapfrogging Indian Space Commerce. In N. Prasad & R. P. Rajagopalan (Eds.), *Space India 2.0 commerce, policy, security, and governance perspectives.* New Delhi: Observer Research Foundation.

Prasad, N. (2017b). Traditional space and new space in industry in India: Current outlook and perspectives for the future. In N. Prasad & R. P. Rajagopalan (Eds.), *Space India 2.0 commerce, policy, security, and governance perspectives.* New Delhi: Observer Research Foundation.

Prasad, N., & Basu, P. (2015). *Space 2.0: Shaping India's leap into the final frontier.* New Delhi: Observer Research Foundation.

PwC. (2013). *Easing India's capacity crunch. An assessment of demand and supply for television satellite transponders.* Hong Kong: Casbaa.

PwC. (2016). *Capacity crunch continues assessment of satellite transponders' capacity for the Indian broadcast and broadband market.* Hong Kong: Casbaa.

Radjou, N., & Prabhu, J. (2010). *Jugaad innovation.* New Delhi: Jossey-Bass.

Rajgopalan, R., & Jayasimha, S. (2016, December 6). *Non-geostationary satellite constellation for Digital India.* Retrieved from Observer Research Foundation. http://www.orfonline.org/expert-speaks/digital-india-non-geostationary-satellite-constellation/

Rao, P. (2015). Applications of communication satellites. In M. Rao (Ed.), *From fishing hamlet to red planet: India's space journey.* New Delhi: Harper Collins.

Rao, M., & Murthi, S. (2012, September). *Perspectives for a national GI policy – Including a national GI policy draft.* Retrieved November 5, 2016, from National Institute of Advanced Studies. http://ficci.in/sector/report/20034/Perspectives-for-National-GI-Policy-NIAS.pdf

Rao, M. K., Murthi, S., & Raj, B. (2015). *Indian space – Towards a "National Eco-System" for future space activities.* Jerusalem: 66th International Astronautical Federation.

Rao, M. K., Murthi, S. K., & RaJ, B. (2016). Indian Space – Towards a "National Ecosystem" for future space activities. *New Space, 4*(4), 228–236.

Sankar, U. (2006). *The economics of the Indian space programme. An exploratory analysis.* New Delhi: Oxford University Press.

Satellite Industry Association. (2016, September). *State of the satellite industry report.* Retrieved December 20, 2016, from SIA. http://www.sia.org/wp-content/uploads/2014/05/SIA_2014_SSIR.pdf

Staley, S. (2006, June). The rise and fall of indian socialism. Why India embraced economic reforms. *Reason.*

Telecom Regulatory Authority of India. (2016, May 26). *Pre-consultation paper on infrastructure sharing in broadcasting TV distribution sector.* Retrieved November 15, 2016, from Telecom Regulatory Authority of India. http://www.trai.gov.in/Content/ConDis/20774_0.aspx

The Economist. (2010, April 15). First break all the rules. *The Economist.*

The Hindu. (2012, March 14). Half of India's homes have cell phones, but not toilets. *The Hindu.*

The Hindu Business Line. (2014, November 28). *DTH satellite services: Dept of space lost out to foreign players.* Retrieved November 10, 2016, from The Hindu Business Line. http://www.thehindubusinessline.com/economy/dth-satellite-services-dept-of-space-lost-out-to-foreign-players/article6644047.ece

The World Bank. (2016a). *Internet users.* Retrieved October 11, 2016, from The World Bank. http://data.worldbank.org/indicator/IT.NET.USER.P2?locations=IN

The World Bank. (2016b). *India.* Retrieved October 2, 2016, from The World Bank. https://data.worldbank.org/country/india

The World Bank. (2016c, October 1). *India – Country snapshot.* Retrieved November 3, 2016, from The World Bank. http://documents.worldbank.org/curated/en/341851476782719063/pdf/109249-WP-IndiaCountrySnapshots-highres-PUBLIC.pdf

The World Bank. (2016d). *Urban population – Data.* Retrieved November 25, 2016, from The World Bank. http://data.worldbank.org/indicator/SP.URB.TOTL?locations=IN

United Nations. (2014). *World urbanisation prospects: The 2014 revision, highlights (ST/ESA/SER.A/352).* Department of Economic and Social Affairs. New York: United Nations.

United Nations. (2015). *UNDP human development reports.* Retrieved November 2, 2016, from India Human Development Indicators. http://hdr.undp.org/en/countries/profiles/IND

Vernile, A. (2017). *The rise of the private actor in the space sector.* Vienna: European Space Policy Institute.

World Economic Forum. (2016). *The Global Competitiveness Report 2015–2016.* Davos: WEF.

Zinnov. (2015). *The Indian promise.* Retrieved Octoober 12, 2016, from Zinnov.com: http://zinnov.com/the-indianpromise/

Chapter 5
Understanding India's Uses of Space: Space in India's Changing Geopolitics

Although primarily driven by the need to serve developmental aims, India's space programme is not limited to the sole pursuit of utilitarian, socio-economic benefits. Space has also been a tool in India's geopolitical calculus, a dimension that has progressively strengthened since the past decade. Thanks to exploration missions such as Chandrayaan-1 and Mangalyaan, space has increasingly become an asset for boosting national prestige, a prestige directed as much towards its citizens as the international community. Similarly, space has also become a tool of Indian foreign policy and diplomacy, as well as an element in its military modernisation. It is important to see the context of a growing geopolitical significance of space in light of India's own growing role on the world stage. Accordingly, the next section will shed light on the most recent developments in India's foreign policy agenda. The chapter will subsequently elaborate on India's emerging uses of space for both diplomatic objectives and security needs. Further, considering that space is not only a *tool* for achieving broader policy objectives but also an *arena* of strategic interaction, India's approach vis-a-vis diplomacy and the security of space will also be analysed.

5.1 Disentangling India's New Grand Strategy

Following independence, India's first Prime Minister, Jawaharlal Nehru, devised the doctrine of non-alignment to guide India's foreign relations. This tradition has loomed large over Indian foreign policy in subsequent decades, making the country reluctant to declare formal alliances (IndoGenius 2016b). Even though it originated in the early Cold War period, the doctrine of non-alignment "had an almost talismanic quality for India's foreign policy establishment" and continued to be India's official policy also during the 1990s (Ganguly 2016).

However, with the turn of the century and even more so following the election of Prime Minister Narendra Modi, India has started assuming a much larger role both

© Springer International Publishing AG 2018   157
M. Aliberti, *India in Space: Between Utility and Geopolitics*, Studies in Space Policy 14, https://doi.org/10.1007/978-3-319-71652-7_5

in regional and international fora. This new assertiveness has become evident on multiple occasions, for instance, when upgrading the Look East policy towards ASEAN countries and East Asia to an Act East policy, when expanding its economic, energy, resource and security cooperation in Africa and the Middle East, when deciding to join the Shanghai Cooperation Organisation (SCO) in June 2017, when voicing the interests of the developing world at the 2016 climate change negotiations in Paris and when extending the country's naval outreach through the Red Sea into the Eastern Mediterranean Sea. In parallel to this, it did not go unnoticed that India PM Modi decided to skip the 2016 "Non-Aligned Summit" held in Margarita, Venezuela, for the first time, thus further feeding the impression that India may, in fact, take a more strategic and "aligned" posture on the international stage.

From an overall perspective, all these developments indeed suggest a clear departure from India's traditional foreign policy posture. Hence, the key questions are: What is this new foreign policy aiming at? Where is it heading? And more importantly, what grand strategy is behind it? The latter is a hard question to answer, because the most striking feature of India's grand strategy is the apparent lack thereof.

India has rarely articulated its grand strategy or national doctrine to guide its foreign policy. As highlighted by Deepa M. Ollapally and R. Rajagopalan:

> India's rise has not been accompanied by White Papers, Prime Ministerial doctrines, or any other clear and open statements by the government about what its objectives are for India's global role. This is not surprising – official India rarely spells out its long-term vision, preferring discrete steps to be taken to achieve its goals. Without considering whether this is due to a lack of purposeful thinking or whether it is a clever attempt to maintain flexibility, the outcome is either way a certain amount of ambiguity. Thus, it would be fair to term India "an ambiguous rising power (Ollapally and Rajagopalan 2013).[1]

The one factor that perhaps contributes most to the lack of long-term planning is given by the very way India's foreign policy establishment operates, and in particular by the structure of the Indian Foreign Service (IFS), which produces a decision-making process that is highly individualistic.

All three major governmental bodies responsible for making India's foreign policy – namely, the Prime Minister Office, the National Security Office and the Ministry of External Affairs – are filled in their top positions by officers of the Indian Foreign Service (IFS). As Chatterjee Miller contends:

> since foreign service officers are considered the crème de la crème of India and undergo extensive training, they are each seen as capable of assuming vast authority. What is more, the service's exclusive admissions policies mean that a tiny cadre of officers must take large portfolios of responsibility. In addition to their advisory role, they have significant leeway in crafting policy. This autonomy, in turn, means that New Delhi does very little collective thinking about its long-term foreign policy goals, since most of the strategic planning that takes place within the government, happens on an individual level (Chatterjee Miller 2013).

---

[1] Interestingly, also during the "non-alignment period", India showed a high dose of ambiguity, with regard for instance to non-proliferation issues.

Because most policy choices within the government are seldom the result of institutional/bureaucratic politics, the role of key personalities in foreign policy cannot be overestimated (Ollapally and Rajagopalan 2013).

As a former Indian ambassador poignantly commented in an interview:

I was completely autonomous as ambassador. There is little to no instruction from the [Prime Minister's Office], even in cases of major countries. I had to take decisions based on a hunch. I sometimes got very, very broad directives. But I violated virtually all of them." Another Indian ambassador to several European countries confirmed: "I could never find any direction or any paper from the foreign office to tell me what India's long-term attitude should be toward country X. Positions are the prerogative of the individual ambassador (Chatterjee Miller 2013).

Unlike many world capitals in which policy planning is an integral part of the foreign policy-making processes, in India "there is very little institutional capacity to provide intellectual guidance and forward-looking strategies" (Ollapally and Rajagopalan 2013). The two departments within the Foreign Ministry allegedly in charge of long-term strategy, namely, the Policy, Planning, Research Division and the Public Diplomacy Division are "virtually defunct" and largely unutilised.

The lack of a comprehensive, long-term assessment of India's grand strategy and associated means is further exacerbated by the policy-making insularity and organisational culture of Indian bureaucracy. In stark contrast to countries such as the United States and even China, whose foreign policy-makers constantly turn to a plethora of organisations to supplement the long-term planning taking place within the government, the intellectual input by Indian think tanks remains very marginal. Foreign policy officers in India rarely seek inputs and analysis from non-governmental institutions, which are generally seen as lacking clout rather than useful sources of strategic advice. And this is true even for India's most renowned think tanks, such as the semi-official National Security Advisory Board (NSAB), which was created in 1998 by the government of Atal Vajpayee to provide non-governmental expertise to decision-makers and whose members have complained of not being taken into serious account by the government. Thus, the influence of opinion-makers outside the government is not particularly strong. If non-governmental opinion matters, "it appears to matter negatively as a constraint on or as a veto against particular choices, rather than positively in promoting policy initiatives such as through policy lobbying" (Ollapally and Rajagopalan 2013).

All in all, due to this policy-making insularity and the fact that most policy choices by the government are rarely the output of unified institutional vision, the contending foreign policy perspectives promoted by domestic élites prove a key determinant in the making of Indian foreign policy.

Compared to structural determinants (i.e. the policy choices imposed by the *power structure* of the international system), *domestic*-level variables exercise an equal – if not greater – weight in the crafting of India's foreign policy.[2] Indeed, since

---

[2] This is so not only because India's international position is progressively making structural determinants less potent but also because the democratic nature of India's political processes, its size and the extreme heterogeneity in terms of ethnicity, languages, religions and caste diversity and

much of India's foreign policy ambiguity appears to stem from the lack of the strong internal consensus that India enjoyed after independence, navigating the ocean of India's foreign policy requires analysing both domestic and structural determinants.

5.1.1 Structural Determinants

Just like any other country, India's foreign policy is strongly shaped by the international condition in which New Delhi finds itself, that is, by structural determinants and particularly the power distribution within the regional and international order, which in line with neorealist views of international relations represents the most significant determinant in the foreign policy of a state. India has demonstrated a clear responsiveness to the logic of "power distribution", as evidenced by its foreign policy practice at key juncture points (e.g. when using a power balance approach vis-à-vis China after 1962, or when it sought rapprochement with the United States after the collapse of the USSR as a result of the changed structure of global power in the post-Cold War era). But alongside the power logic, India's international condition proves to be essentially shaped by structural factors related to its geographic, historical, and cultural legacy.

There is a wide consensus among India's foreign policy analysts that "India is possessed of geographical logic" (Kaplan 2013, p. 229; Singh 2009; Pillai 2016; Tanham 1992) – the great impact exercised by geography on its strategic thinking and posture. Insulated as it is by the Himalayan range in the north and the Indian Ocean in the three other directions, the Indian subcontinent has been typically described as a "natural geographic unity", a very distinct geopolitical zone, "possessed of geographical logic". As articulated by George Tanham in his well-known *Indian Strategic Thought* (Tanham 1992):

> Geography has imparted a view of the Indian subcontinent as a strategic entity, with various topographical features contributing to an insular perspective and a tradition of localism and particularism. India's unique culture reinforced this unity and imparted, first, a tendency towards diversity and accommodation to existing realities and, second, a highly developed capacity to absorb dissimilar concepts and theories. This tolerance was strengthened by the caste system, which also helped to maintain an extraordinarily durable system and ethic for social relations.

With respect to the external world, India's geographical location has proved one of the most important determining factors for its essentially defensive strategic orientation, with no expansionist military tradition. For the past 2000 years, the major strategic problem confronting India has been – as Stephen Cohen elegantly put it – "how to achieve strategic unity of the subcontinent and protect it from the incursion

---

fractured and democratic nature of its polity create several disagreements or even active sociocultural and political cleavages that have competing or conflicting claims (Stepan et al. 2010; Ollapally and Rajagopalan 2013).

of outside powers" (Cohen 2001, p. 10). Indeed, India's strategic location at the centre of the Asian arc has always made it an extremely important pivot in the geopolitics of Eurasia, despite its insularity. This can be clearly observed not only in the context of the Mughal invasion and the British domination but also during the World War II with the so-called tripartition of Eurasia as well as Japan's Greater East Asia Co-Prosperity Sphere. Again, during the Cold War, India was an important crux in the global strategies of both the United States and the USSR.

Today, India's strategic geography offers it a key role in regional – and by reflex – international dynamics due to its position on the main trade route between Europe and Asia and on the world's key energy checkpoints. It has been noted that its locational relationship to Europe and East Asia is so crucial that under certain conditions India could assume a decisive role in the strategy of world control (Karan 1953). At the same time, it must also be recognised that in spite of its strategic location, India is confronted with what Robert Kaplan calls a crucial "geographical dilemma" (Kaplan 2013). As he illustrates "while as a subcontinent India makes eminent geographical sense, its natural boundaries are, nevertheless quite weak in places. (…) In fact, the present Indian state still does not conform to the natural boundaries of the subcontinent, and that is the heart of its geographical dilemma: for Pakistan, Bangladesh, and to a lesser extent Nepal also lie within the subcontinent", and deprive India of its so-called strategic unity. To a large extent, this dilemma has been the inevitable by-product of the dynamics surrounding the end of the British Raj and the very birth of Independent India. The borders with Pakistan came into place after the partition of British India into two separate states defined on the basis of religious boundaries, with Indian territory separating Pakistan in two noncontiguous parts and creating the condition for the birth of a third independent state, Bangladesh. Similarly, the borders with Afghanistan and China were also defined on the basis of the artificial boundaries drawn by the British Empire in the nineteenth century, thus setting the stage for successive disputes and even conflicts. As also explained by Henry Kissinger:

> India's role in world order is complicated by structural factors related to its founding. Among the most complex (are) its relations with its closest neighbours, particularly Pakistan, Bangladesh, Afghanistan, and China. Their ambivalent ties and antagonism reflects a legacy of a millennium of competing invasions and migrations into the subcontinent, of Britain's forays on the fringes of its Indian realm, and of the rapid end of British colonial rule in the immediate aftermath of World War II. No successor state has accepted the boundaries of the 1947 partition of the subcontinent in full. Treated as provisional by one party or another, the disputed borders have ever since been the cause of sporadic communal violence, military clashes, and terrorist infiltration. (Kissinger 2014).

These structural geographical factors are not only thought to have deprived India of "vital political energy" for its geopolitical projection across much of Eurasia but also to pose significant threats to its security. Whereas none of India's small neighbouring countries such as Bangladesh, Nepal and Bhutan are seen to pose a direct military threat, they are nonetheless viewed as sources of vulnerability, particularly because of the nondemocratic governments controlling them, their ethnic and tribal fragmentation as well as their systematic recourse to foreign support to counterbalance India's preponderance in the region (Tanham 1992). Conversely, India's large neighbours China and Pakistan, with whom it has fought four wars since independence, are seen as posing direct threats to its survival, with their political posture inevitably triggering a self-help dynamic for India.

Islamabad's menace to India's security and core interests manifests in multiple ways. For one thing, its very existence disrupts the unity of the subcontinent, drawing "Indian power away from the defence of the subcontinent as a whole". In addition, Pakistan "sits astride the main land invasion routes to India" (Tanham 1992), and its navy could potentially disrupt India's access to its energy markets in the Middle East. Needless to say, Pakistan's nuclear capabilities are also a source of great concern among Indian strategists. Further, Islamabad has allowed – not to say openly sponsored – violent extremist and separatist movements, including terrorism, inside India and amplified the risk of Islamic fundamentalism within the country. Equally worrisome have been Islamabad's close relations with antagonistic foreign powers, including several Islamic states, the United States and more importantly China – which is considered India's major rival.

Although India and the Central Kingdom historically have no real track record of major conflicts, with the exception of the 1962 war, India views China as the major external threat, primarily because of its sheer size, population and military might, including its nuclear arsenal. China poses both a *direct* challenge to India along their 3500 km-long borders and *indirect* challenge through its quasi-military alliance with Pakistan and its expansion in South and Central Asia. Border disputes remain a serious problem for India, as do China's nuclear capabilities. As opined by strategist Chad M. Pillai, it was the development of China's nuclear weapons in the 1950s that, together with the distrust of US relations with Pakistan and China, pushed India to power balance with the USSR during the Cold War and develop its own nuclear capabilities, which in fact were primarily "designed as deterrent to China, not Pakistan as many claimed" (Pillai 2016). Since the 1962 war – the greatest humiliation of post-Independence India – relations have remained mistrustful and invariably headed towards competition. What in addition drives their more recent competitive relations is the progressive rise (or more properly) resurgence of both Beijing and New Delhi as major powerhouses on the international stage. As Robyn Meredith observes in her *The Elephant and the Dragon*, the fact is that "after a century-long hiatus, India and China are moving back toward their historical equilibrium in the global economy and that is producing tectonic shifts in economics as well as in geopolitics" (Meredith 2008).

More broadly, China plays a key role in shaping India's strategic posture also because of its systemic impact on the very configuration of the global order. Indeed, from a systemic perspective, the two most salient features of the current international system that India is confronted with are the progressive weakening of the unipolar configuration and the parallel renaissance of China as an international superpower. Driven by its sense of historical greatness, China has indeed rediscovered its geopolitical centrality and joined the United States as a vigilant *producer* of global governance and *shaper* of the international system, starting to mark the end of the American era.

There appears to be common understanding among analysts from the two sides of the Pacific that the *systemic* and *strategic* interaction between these two powers will ultimately craft the upcoming global order and exercise the greatest impact on the foreign policy directions of all other powers, including India. It is also clear, however, that India's directions are, in turn, also bound to have important implications. As succinctly put by Robert Kaplan, "as the United States and China become great power rivals, the direction in which India tilts could determine the course of geopolitics in Eurasia in the twenty-first century. India, in other words, looms as the ultimate pivot state" (Kaplan 2013). Indian policy-makers seem to be fully aware of this and, as explained in the next sections, have started to shape their foreign policy accordingly.

5.1.2 Domestic Determinants

Alongside structural determinants, India's foreign policy is also clearly shaped by so-called domestic determinants, i.e. by the contending perspectives promoted by domestic groups in the making of Indian foreign policy choices.[3] The very existence of these perspectives is not surprising in view of the multinational nature of India's democratic polity.[4] Delineating the various perspectives on India's foreign policy is, however, not an easy task because they do not represent a well-defined spectrum from the left to the right, but rather fall within a transversal range, cross-cutting the major parties.[5] In addition, there are overlapping views, with plenty of commonality across different perspectives and difference across the same groupings. Generally

---

[3] This section is mainly based on the perspectives offered by Indian international relations' scholars Ollapally and Rajagopalan (2013).

[4] Because of its extreme heterogeneity in terms of ethnicity, languages, religions and caste, Indian polity indeed presents a significantly robust multinational dimension that does not fit well into the classic French-style nation-state model based on a "we-feeling" resulting from an existing or forged homogeneity. According to Stepan, Linz and Yadav, India should be more properly analysed under the lenses of the state-nations' model. The term indicates "a political-institutional approach that respects and protects multiple but complementary socio-cultural identities. State-nations policies recognise the legitimate public and even political expression of active socio-cultural cleavages, and they include mechanisms to accommodate competing or conflicting claims without imposing or privileging in a discriminatory way any one claim. State-nation policies involve creating a sense of belonging (or "we feeling") with respect to the state-wide political community, while simultaneously creating institutional safeguards for respecting and protecting politically salient sociocultural diversities" (Stepan et al. 2010, p. 54).

[5] It is also important to note that Western understandings of left- and right-wing politics are not easily applicable to India's political life. As explained by Nick Booker, India parties' ideologies fall at different places on the liberal–conservative spectrum, socially and economically. For instance, India's longest ruling party, the Congress, is considered both socially and economically centrist, while the Communist Party of India, of Marxist origins, is liberal from a social perspective and conservative from an economic perspective. The current ruling party at the national level, the Bharatiya Janata Party (BJP), is the exact opposite: it is conservative from a social perspective but liberal from an economic perspective (IndoGenius, India's Election Calendar 2016).

Table 5.1 India's foreign policy school of thought

Major schools of thought	Goals and attitude	Roots
Nationalists		
Standard nationalists (1947–)	Aim for developed country status Pursue balanced growth	Nehruvianism
Neo-nationalists (Post-1991)	No to idea of great power Domestic consolidation first South=South solidarity	Soft Nehruvianism Gandhian Indian civilisation Socialist theory
Hyper-nationalists (Post-1998)	Achieve global power India first Tight internal security	Kautilya Selective realist theory Hindu nationalism
Great Power Realists (Post-1998)	Become global player	Kautilya Realist theory
Liberal Globalists (Post-1991)	Aim for global economic power	Economic theory
Leftists (Post-2004)	Third world coalition against united S hegemony	Marxist theory

Source: Ollapally and Rajagopalan (2013)

speaking, according to political scientists Deepa Ollapally and Rajesh Rajagopalan, four separate schools among these domestic groups can be distinguished, namely, the Nationalists, the Great Power Realists, the Liberal Globalists and the Leftists. The Nationalist school can be further refined into three sub-groups: Standard Nationalists, Neo-Nationalists and Hyper-Nationalists. The different goals, attitudes and roots of these contending schools are summarised in Table 5.1.

Each of these schools has advanced specific viewpoints concerning India's actions in the international arena. Their views on the major issue areas of India's foreign policy are summarised in Table 5.2.

As evident from the Tables, the variations among these groups concerning India's strategic goals and positioning vis-à-vis the major issue areas of its foreign policy action are not extremely dramatic. Even though some perspectives are very distinct, the majority fall within a rather narrow spectrum. The most important area of disagreement is in relations with the United States, the other differences in viewpoints (e.g. Arms Control Policy and Iran) arguably being subsumed as fallouts of this key issue area. Vice versa, there are several areas – e.g. relations with Pakistan, China and Myanmar – in which the differences between these perspectives are not too substantial and potentially convergent.

Overall, the existence of all these contending domestic viewpoints helps to explain the apparently high degree of ambiguity that characterises India's strategic posture. Indeed, all these different viewpoints translate into a balanced foreign policy, extremely cautious in making a public "grand strategic statement", and in which the dominant opinion constantly needs to converge to an ambiguous middle ground. This is evident in a number of issue areas. In the posture vis-à-vis the United States, for instance, the support towards closer working ties with Washington manifested by the

Table 5.2 Domestic positions in key foreign policy areas

	Nationalists		Hyper-nationalists	Great power realists	Liberal globalists	Leftists
	Standard nationalists	Neo-nationalists				
Military power	Need more	Adequate	Much more	Need more	Adequate	Need less
Use of force	Oppose	Oppose	Support	May support	Oppose	Oppose
Nuclear weapons	Adequate	Adequate	Much larger arsenal	Larger arsenal	Adequate	Oppose
US nuclear deal	Support/mixed	Opposed	Opposed	Supported	Supported	Opposed
Strategic ties/alliances	Possible	Suspicious	Dangerous	Possible	Necessary	Oppose
Globalisation	Cost and benefit	High cost	High cost	Cost and benefit	Benefits	Oppose
US relations	Support/mixed	Suspicious	Suspicious	Support	Support	Oppose
China relations	Suspicious	Seek rapprochement	Suspicious	Suspicious	Seek rapprochement	Strong support
Iran relations	Mixed	Favourable	Favourable	Mixed	Mixed	Very favourable
Pakistan relations	Suspicious; reluctance to compromise	Rapprochement; sentimentalists; for compromise	Military response; no compromises	Resolve to facilitate global role for India	Peace through dialogue, trade; for compromise	Favour peace even with big compromise

Source: Ollapally and Rajagopalan (2013)

Indian government since the end of the Cold War has been tempered by opposition from the Leftists as well as Neo-Nationalists and Hyper-Nationalists, leading the Indian government to deny any intention of creating a military bloc with the United States. Similarly, in relations with China, the difference in position among opinion-makers has translated into a two-pronged, cautious posture in which efforts towards rapprochement combine with a degree of suspicion and confrontational stances.

5.1.3   Implications: Old and New Axes in India's Foreign Policy

Despite all the multifaceted domestic and structural variables, it is possible to distil the major tenets and axes in India's grand strategy on the international scene.

From an overall perspective, it can be argued that India's self-restraint and defensive posture in regional and global politics has progressively weakened, replaced by an emerging aspiration to achieve great power status (Ghosh 2009). To be sure, New Delhi remains rather cautious in trumpeting its geopolitical upswing on the world scene and raising international expectations because of fears that "a growing India might have to take on responsibilities commensurate with its power" (Chatterjee Miller 2013). It more specifically fears that "the West, and particularly the U.S., might pressure India to step up its global commitments (…) to abandon its status as a developing country" by inter alia making concessions on environmental issues and on trade (Ibid, p. 18).

India Engages

However, even if India does not yet want to assume an international leadership role, it is also manifest that its footprint in global politics has been rapidly growing, with a continual extension of its diplomatic and military projection beyond its *immediate neighbourhood* of South Asia and the Indian Ocean Region (IOR). Indeed, since the turn of the new century – thanks to the momentum created by the 1998 nuclear tests and the parallel upsurge of its economy – the Indian government has started to adopt a strategy of "total diplomacy" aimed at developing an increased international profile and deeper interactions across the world. As a sign of this new assertiveness, New Delhi has forged strategic partnerships both with global powers such the United States and tier-two powers such as the EU, France, Israel and Japan. It has also participated in the construction of diplomatic groupings such as Brazil–Russia–India–China (BRIC) and India–Brazil–South Africa (IBSA) and broadened its political and security horizons to an *extended neighbourhood* that encompasses the quasi-totality of Asia. Today India pursues a foreign policy that– in Henry Kissinger's words – is:

> In many ways similar to the quest of the former British Raj as it seeks to base a regional order on a balance of power in an arc stretching halfway across the world, from the Middle East to Singapore, and then North to Afghanistan. Its relations with China, Japan and Southeast Asia follow a pattern akin to the nineteenth-century European equilibrium. Like

China, it does not hesitate to use distant 'barbarians' like the United States to achieve its regional aims (... And just) like the nineteenth-century British who were driven to deepen their global involvement to protect strategic routes to India, over the course of the twenty-first century India has felt obliged to play a growing strategic role in Asia and the Muslim world to prevent these regions' domination by countries or ideologies it considers hostile (Kissinger 2014).

In a similar line of reasoning, Indian scholars and analysts opine that in light of the inherent security dilemma triggered by India's neighbours China and Pakistan, it is imperative for India to tailor its foreign and security policy calculus by behaving in a similar way to its rivals, that is, by striving for a hegemonic status in the region as a tactic for continued survival and growth (Rajagopalan and Sahni 2008; Pardesi 2005). As more specifically argued by Pardesi, "As India becomes wealthy, it will work towards maximising its political and military power and to attain regional hegemony in accordance with the core features of offensive realism" (Pardesi 2005). At the same time, the ever-present veneer of morality and a postcolonial identity are likely to continue deterring Indian governments from adopting a solely militaristic confrontation with China or Pakistan as a permanent principle of its foreign policy.

While the former argument about the pursuit of regional leadership can be clearly detected in India's recent initiatives in South and Southeast Asia, including the increased military presence in IOR and the upgrading of its Look at East policy with an Act East policy, the latter element is visible in New Delhi's uncertainty about the proposed creation of a Democratic Security Diamond (Global Times 2014) to geopolitically constrain China's power projection.[6] Indeed, although New Delhi is willing to increase its cooperative relations with Tokyo as well as to gain strategic advantage vis-à-vis Beijing, Modi's government has remained somehow reluctant to take part in initiatives that would limit the freedom of manoeuvre of its diplomacy and generate undesirable rifts in its relationship with Beijing.

Looking East, West and at the Global South
Irrespective of this geopolitical dilemma facing New Delhi, what these dynamics suggest is that India has now become able to punch with increased weight in the areas of its *extended neighbourhood* and has emerged as a key pillar in the economic and security architecture of the Asia-Pacific, with ever-growing economic, political and even military links with the United States, Japan, South Korea and Australia.

What is also remarkable to highlight is that during Manmohan Singh's decade (2004–2014), such an eastward shift in India's foreign policy was progressively

---

[6]This diplomatic and strategic initiative was first launched during Shinzo Abe's first tenure as prime minister of Japan (September 2006–September 2007) and named the Quadrilateral Security Dialogue (Japan, the United States, India and Australia) as a solution to the maritime disputes involving China. The new strategy was launched by Prime Minister Abe in 2014 based on "three pillars: (1) reinvigorating the U.S.-Japan alliance; (2) a reintroduction of the UK and France to Asia's international security realm; and (3) bolstering international cooperation between key democracies in the Indo-Pacific, such as India and Australia" (Miller 2013).

complemented by an equally important westward shift, with the adoption of a "Look West" strategy. As succinctly pointed out by David Scott, this emerging "Look West" strategy "is evident with regard to India's strategic partnership with Iran, where constraining Pakistan and energy motifs feature uncomfortably alongside links with the USA, and with regard to the Gulf, where energy access entwines with emerging military-naval links with actors like Oman and Qatar" (Scott 2012).

Furthermore, building upon the recognition that also the dynamics in the Middle East–North Africa (MENA) region greatly affect India's security and vital interests, over the past years "India's sense of westwards *extended neighbourhood* has (…) been extended still further beyond the Gulf into the further reaches of the Middle East and North Africa" (Kumaraswamy 2012). Indicative of this expansion are India's naval outreach through the Red Sea into the Eastern Mediterranean Sea, the normalisation of its relationship with Israel and the subsequent launch of a security–military partnership (which nonetheless still proves rather ambivalent in view of New Delhi's close relations with Iran) and its nascent links with Turkey.

Together with its increasing involvement in its *extended neighbourhood* of Southeast, Central and West Asia, another emerging cornerstone of India's "total diplomacy" is offered by its intense engagement in the dynamics of the Global South and particularly of Africa. To be sure, a look at Indo-African relations would reveal that India's presence in the subcontinent is not new. However, since the launch of the India–Africa Forum Summit (IAFS) in April 2008, such relations have received a new impetus, and India has become more assertive in economic, energy, resource and security cooperation, joining Japan and China in what scholars have dubbed a "new scramble for Africa" by Asian powers (Naidu 2009).

There are also several indications that New Delhi has started to propose itself as a new source of leadership for African, and more broadly developing, countries. In a 2010 speech on the eve of President Obama's visit to India, the then economic advisor Larry Summers outlined what he called the *Mumbai Consensus*, suggesting that more and more countries could adopt this model. As he highlighted, it is possible that over time:

> The discussion will be less about the Washington Consensus or the Beijing Consensus, than about the Mumbai Consensus – a third way not based on ideas of laissez-faire capitalism that have proven obsolete or ideas of authoritarian capitalism that ultimately will prove not to be enduringly successful. Instead, a Mumbai Consensus based on the idea of a democratic developmental state, driven not by a mercantilist emphasis on exports, but a people-centred emphasis on growing levels of consumptions and a widening middle class.[7]

India's attentiveness to developing countries stems not only from its particular model of economic development and gradual approach in decentralisation, privatisation and entrepreneurship but also from its key values and institutions (such as democracy, federalism, secularism and official recognition of different languages), which "hold great promise for managing multi-ethnic societies, especially in the developing world" (Paul 2014). What is noticeable here is that all these features have been ably used by Indian diplomacy to enlarge the toolbox of New Delhi's soft

---

[7] Larry Summers. Quoted from (Bajaj 2010).

power assets and complement its equally swelling hard power in exercising greater influence in the global arena. As US analyst John Lee already pointed out in 2010 "the overall result is a strategic boon for India: political and strategic elites increasingly see India as a muscular, but also predictable, stabilising, co-operative and attractive rising power. The notable lack of apprehension of India's re-emergence is demonstrated by the remarkable speed with which regional militaries are conducting extensive exercises with the Indian navy" (Lee 2010).

A final and closely related point to highlight is India's role on global governance issues, which, in fact, is progressively becoming more assertive and of greater impact. The Indian government and MEA has been actively using diplomacy with a greater sense of purpose and pragmatism in global institutional forums such as the UN – where New Delhi's larger role as "provider of peace-keeping forces goes hand in hand with its quest to achieve a permanent seat on the UN Security Council" (Scott 2012) – the WTO, the G20 and associations like BRICS and IBSA, as well as with regard to global issue areas such as economic governance, climate change, terrorism, nuclear proliferation and space (Mohan 2017). India's evolving geopolitical calculus is fully reflected in this latter arena, which has become both a *source* of hard power projection and a *tool* for achieving diplomatic objectives. The rest of this chapter will look at the emerging uses India is making of its increasingly ambitious space programme.

5.2 Space and Indian Diplomacy: The Emerging Nexus

Although the case for India's utilisation of outer space has always been on harnessing space technology for national development, in recent years, space activities have gradually emerged as a central element in India's foreign policy and diplomacy (Prasad 2016; Reddy 2017; Moltz, Asia's Space Race. National Motivations, Regional Rivalries, and International Risks, 2012). Three dimensions are discussed in this section: India's soft power projection through national space efforts, the utilisation of space activities as a tool for pursuing India's foreign policy objectives and India's diplomacy of space.

5.2.1 Space as a Soft Power Tool for India

Soft power has often proved to be an ambiguous concept. For the sake of clarity, soft power is here intended as a specific matrix of power which enhances the ability of a state to influence other without the use of coercive measures, but on the basis of the reputation and attractiveness it has built through national achievements in specific fields (Nye 2004). India's space programme could be viewed as having started to make deliberate efforts in this soft power projection game, as evidenced by its recent opening to space exploration and human spaceflight. After all, complex space endeavours like planetary exploration have always been powerful symbols of a country's technological prowess, economic strength as well as future aspirations.

India's Grandeur Through Space

Although since the inception of space activities Indian leaders had typically rejected any type of effort not directly linked to a clear development rationale, with the successful mastery of space technologies, India has found itself in a "position to exploit its success and extend its influence towards geopolitical pursuit" (Lele 2016). Consistently, since the turn of the century, Indian space efforts have started to be driven by the need to boost the nation's reputation and attractiveness on the international scene. The Chandrayaan-1 and Mangalyaan missions are excellent case in point, as they have made it clear that India no longer views space as only enhancing the living conditions of its citizens but also as a measure of global standing (Gopalaswamy 2013).

The success of these missions has earned India much international prestige, underpinning the idea that India, once a recipient of space technology from advanced countries, had now become an advanced space power. In addition, thanks to the successful deployment of the Mars Orbiter in 2014, not only did India became the first country to reach Mars on a first attempt, it also became the first Asian country to do so, thus signalling "India's ascension in the global space order where it can claim a status at par with the advanced space faring nations" (Reddy 2017).

By the same token, the recent overture towards more daring and high-profile missions such as human spaceflight underpins the view that prestige-related considerations are likely to grow ever stronger in India's political calculus. As Asif Sidiqi observes "from a purely practical perspective, the manned space programme seems unnecessary to ISRO's original mandate; it is clear that the manned programme is not about the pursuit of scientific or technological knowledge or about alleviating poverty – it is first and foremost about prestige" (Brown 2010b). Prestige, however, can in turn provide geopolitical and diplomatic leverage to the nation, and this is an ever-important driver for "rising India".

In addition to these undertakings, India's soft power has also seen a progression thanks to its increasingly successful participation in the global space economy. ISRO's commercial arm, the Antrix Corporation, has progressively made forays into the global space market for launch services – where it has built a reputation of high reliability and low cost among various customers including those from the United States, Germany, Singapore, the United Kingdom and Canada, among others (see Sect. 3 in Annex for the list of foreign satellites launched by India) – and has developed an international market for the sale of its IRS satellites' imageries. India's increasing influence on the global economy is indicative of the fact that its stature is changing from a developing state to a developed one. As noted by Lele, "this provides India with opportunities to locate itself strategically as a focal point for soft power projection on the various activities taking place in space" (Lele 2016).

Soft Power Narratives

The soft power dimension of the Indian space programme has been constructed not only upon the pursuit of prestige-oriented undertakings but also upon the deliberate creation of specific narratives that have served the purpose of elevating the space programme's profile and, in turn, the international image of India.

One of the most interesting efforts in this respect has been the successful projection of the country's technological self-reliance in the space arena. India has taken particular pride in this and has continuously and emphatically presented its technological feats (particularly in the rocket and space development programmes) as having been achieved entirely through India's own efforts. Although the degree of genuine self-development could in part be questioned, given the continuing reliance on external partners for the development of key technologies (see Chap. 2), this narrative has certainly contributed to boosting the image of a technologically powerful nation. Indeed, the international space community sees India's capacities as a dramatic achievement for a country that at the end of the Cold War was still one of the poorest in the world. Consistently, India has been seen to have reached "the second-tier of space-faring states alongside countries such as France, China and Japan, with only the United States and Russia possessing more advanced capabilities". This in turn means that "India can cooperate with other states and organisations on a peer-to-peer basis". As Michael Sheehan notes, "International cooperation has always been a feature of the Indian space programme, and is itself a mechanism for developing India's influence and demonstrating its technological sophistication by working with the leading states in the developed world" (Sheehan 2007).

Alongside the issue of self-reliance, an equally powerful narrative that India has thoroughly used to elevate its international reputation has been the successful projection of the space programme's cost-effectiveness. The country's space leaders have missed no occasion to convey the message that the country's space achievements are the most cost-effective and successful programmes in the world. The MOM is an excellent case in point. Following the successful deployment of its orbiter, the mission has been ably presented to both domestic and international audiences as a successful example of India's ability to innovate frugally and keep the cost down thanks to a clever use of technologies and ISRO's low-cost – yet highly talented – labour.

These efforts have proven to be successful, as evidenced by descriptions made by the international media on Indian space efforts, which has frequently hailed ISRO's prolific space programme and success despite a shoestring budget. Buzzwords like *jugaad innovation* and *frugal engineering* have nowadays entered in the common lexicon also thanks to the space programme's achievements. What is more important, however, is that these efforts have arguably accrued relevant benefits, in terms, for instance, of boosting investments in the county by a growing number of multinational enterprises looking for low-cost innovation (see Chap. 4).

Closely associated with this fruitful effort is the equally powerful narrative about the "virtuous purposes" of India's space efforts. Thanks to its impressive achievements, the Indian space programme has been presented, in the country and elsewhere, as an outstanding model for the successful development of technology and its utilisation in the achievement of societal goals such as access to education and healthcare, crop monitoring and natural resource management and rural and urban development, among others. In this process, the country's space leaders have not only succeeded in changing the perception from "why should poor India have a space programme" to "India should have a space programme exactly because it is poor"; they have more importantly showcased their programme as a helpful guide for developing countries making forays into the space arena. The message conveyed here is simple and compelling: the experience of India could be relevant for a developing country wanting to realise a cost-effective and socially relevant programme.

It should go without saying that this aspect is inherently linked to India's broader goal of advancing its development model on the international arena – what several scholars have called the *Mumbai Consensus*. As India prepares itself to become one of the pillars in the future geo-economic triad (together with China and the United States), it is natural that it wants to perorate a development model alternative to the *Washington Consensus*, based on laissez-faire capitalism, and to the *Beijing Consensus*, based on state-led, export-powered authoritarian capitalism. In the words of Larry Summers, the Indian model is to the contrary based on a democratic, developmental state driven not by mercantilists' emphasis on exports but a people-centred emphasis on growing levels of consumption and a widening middle class (Tellis 2015).

India's space programme clearly plays a role in the advancement of the *Mumbai Consensus*, and, as clarified in the next section, a more proactive diplomacy has been progressively built around this objective.

5.2.2   Space for India's Diplomacy

In a similar way to what Asian powers China and Japan have been doing, in recent years India has begun to explore the utilisation of its space programme to serve its broader foreign policy objectives. As Prasad notes, Indian diplomatic channels have increasingly "turned towards using the success of the space programme to leverage benefits in cooperative agreements with other states in an effort of extending India's relationship with them" (Prasad 2016).

In this context, it is important to highlight that whereas international cooperation is not a new issue for the Indian space programme, most of the past cooperation experiences have been primarily driven by programmatic needs, rather than foreign policy goals. Over the last 10 years, however, the progressive expansion of India's technical competences in the field has increasingly enabled the country to deliberately leverage the programme for achieving diplomatic goals.

From an overall perspective, the space programme has been mainly used to achieve two core foreign policy objectives of India: strengthening its dominant position in South Asia and claiming political leadership of the Global South.

Regional Capacity-Building Initiatives
As emphasised in a previous section, given that a peaceful and prosperous neighbourhood is a prerequisite for India's national security and economic expansion, it is natural that New Delhi has been according high priority to its development.

Within the SAARC framework, a wide range of regional projects for both physical and data connectivity have been proposed, and India has consistently kick-started efforts to augment these projects using its space infrastructure.

The first initiatives were taken during the tenure of Indian Prime Minister Manmohan Singh, who offered to make Indian satellite resources data available to Southeast Asian countries for managing natural disasters. He also offered Indian assistance in launching small satellites built by them (Lele 2013). While this latter initiative did not materialise, in 2014 Prime Minister Modi proposed the development of a communication satellite as a "gift" to SAARC members. Accordingly, the ISRO designed "a satellite hosting 12 transponders expected to optimise DTH broadcasting, tele-medicine, tele-education, disaster management, and a host of communications services in the region. The costs associated with building and launching of the satellite [were] borne by India while respective countries contribute for their ground stations" (Reddy 2017). Because Pakistan did not join the project, the satellite (GSAT-9) was officially named South Asia Satellite.

Following its successful launch, on 5 May 2017, Prime Minister Narendra Modi said, "With this launch we have started a journey to build the most advanced frontier of our partnership. With its position high in the sky, this symbol of South Asian cooperation would meet the aspirations of economic progress of more than 1.5 billion people in our region and extend our close links into outer space" (Annadurai 2017). The positive spillover effect of the satellite's launch on India's "neighbourhood first" diplomacy was well demonstrated by the warm responses given by all the leaders of South Asian countries. Maldives President Abdulla Yameen Abdul Gayoom, for instance, stated that the satellite "underlined India's neighbourhood first foreign policy and showed its commitment to the development of the region", while Bhutan Prime Minister Tshering Tobgay praised the launch as an "impressive milestone in the history of the world with one Country launching a satellite for the free use of its neighbours" (Annadurai 2017).

Along the same line of this cooperation, India has decided that the IRNSS will be shared with the SAARC nations, augmenting regional terrestrial and marine navigation, disaster management, vehicle tracking and other activities. India's diplomatic reach to neighbouring countries also includes bilateral initiatives. In Afghanistan, for instance, India included remote sensing satellite transmitters for acquiring space-based data in a $1.2 billion aid package to the Kabul government (Moltz 2012). More recently, India has initiated the construction of a ground station in Vietnam, which will have the ability to downlink data directly from India's remote sensing satellites, thus helping the Hanoi government in its own development efforts (Reuters 2016).

Although not to the same extent as China or Japan, India has also sought to position itself as a focal point for space-related capacity building in the region with the underlying goal of further strengthening its position in South Asia. For instance, DOS has established the Indian Institute of Space Science and Technology (IIST) to offer Indian and foreign students from the region graduate education in the space field. In addition, India has started to launch several initiatives through the ISRO-hosted Centre for Space Science and Technology Education in Asia and the Pacific

(CSSTEAP), which was established in 1995 in response to a resolution of the UN General Assembly recommending the creation of centres for space education in developing countries.[8]

The CSSTEAP, which works under the auspices of the United Nations Economic and Social Commission for Asia and the Pacific (UNESCAP), promotes educational activities in space science and technology, targeting in particular emerging space nations in South Asia (Aliberti 2015).[9] However, it should be also stressed that the centre has limited scope and less ambitious targets than the Chinese-led APSCO or the Japanese-led APRSAF. It is thus unlikely that it will play a major role in giving political shape to cooperative undertakings in the region.

India's Inter-Asia and India–Africa Telemedicine Projects

In addition to the abovementioned construction of a dedicated South Asian satellite and capacity-building activities, another initiative having positive spillover effects for India's foreign policy objectives is offered by its Inter-Asia and India–Africa Telemedicine Projects. Whereas medical aid in developing countries predominantly originates from Western nations, India has affirmed itself as an un-discussed leader in terms of healthcare provision through satellite-based telemedicine. Hosting the biggest telemedicine network in South Asia, India has expanded not only to other Asian countries but also to the African continent. It has been involved in different telemedicine projects with its regional neighbours, both at the governmental and non-governmental levels (Bonnefoy 2014).

The first efforts were initiated by the Indian government in 2007, with a pilot healthcare project connecting hospitals in SAARC countries with specialty hospitals in India. Two years later, the project was officially inaugurated and named the SAARC e-Network Telemedicine project. As detailed by Bonnefoy, "through an ISRO's INSAT satellite, the network currently connects hospitals in Thimphu (Bhutan), Kathmandu (Nepal) and Kabul (Afghanistan), with two specialty hospitals in the north of India (one in Lucknow and the other in Chandigarh. In total, since 2009 there have been 148 tele-education sessions, 110 Continuing Medical Education sessions, and 15 teleconsultations; which include 19 distinct specialisations, ranging from paediatrics to neurosurgery" (Bonnefoy 2014).

Other relevant projects in South Asia have been launched by the Indian not-for-profit organisation Apollo Telemedicine Networking Foundation (ATNF). The Foundation has been involved in the SAARC e-Network Telemedicine project, and it has also managed its own projects. The foundation has telemedicine services with dedicated centres in countries such as Iraq, Yemen, Kazakhstan and Myanmar. The doctors in the foreign hospital are connected to specialists from the Apollo Hospitals in India for "referrals, consultation, second opinion, reviews, [and] post treatment follow-ups besides facilitating tele-continuing medical education" (Dutta 2012).

---

[8] UN General Assembly Resolution 45/72 of 11 December 1990.

[9] The main task of the Centre is to develop the skills and knowledge of university educators, environmental research scientists and project personnel in the design, development and application of space science and technology for subsequent application in national and regional development and environment management.

Furthermore, India has not only taken a keen interest in assisting South Asian countries but has also extended its reach up to the African continent. The Government of India, its Ministry of External Affairs and its former president, Abdul Kalam, set up the Pan African e-Network in 2007. The project has been allocated a budget of over USD 120 million by the Indian government. Its first phase, which involved 11 countries, including Burkina Faso, Ghana and Mauritius, was officially launched in 2009 (Pan-African E-Network 2016). Through satellite connectivity (VSATs) and a fibre-optic network, the project connects 17 super-specialty hospitals in India to 8 regional specialty hospitals, 45 learning centres, 40 patient-end hospitals and 37 VVIP nodes (video conference) in Africa. The network allows for tele-education, telemedicine and conferencing (Bonnefoy 2014). The focus seems to be on tele-education, with over 1100 continuing medical education sessions having taken place between hospitals located in India and in several African countries since 2009 (Pan-African E-Network 2016).

The activities undertaken by India in South Asia and in Africa demonstrate its will to assume a larger role in the promotion of international development and hence in global governance issues. As noted by Reddy, As India charts:

> its path towards becoming a leading power in the world, it cannot afford to ignore the international responsibilities arising from this aspiration. The distribution of global public goods for the benefit of developing and underdeveloped countries, especially in the immediate neighbourhood, is one such responsibility. Space services have emerged as the new global public goods. [...] India's willingness to distribute space services to the countries in South Asia, Africa and South East Asia is a hallmark of its status as a leading power (Reddy 2017).

5.2.3 India's Diplomacy of Space

Since space can be regarded as both a *tool* for diplomacy and a *domain* in which state diplomacy interacts, it appears heuristically useful to make a distinction between space *for* diplomacy and the diplomacy *of* space. While the former indicates the utilisation of space technologies to achieve diplomatic objectives, the latter refers to the diplomatic interactions revolving around space issues. This hence includes a country's posture in bilateral or multilateral fora such as UNCOPUOS, the Conference on Disarmament (CD), the IADC, ISEF, CEOS, etc.

India's diplomacy of space is very indicative of its changing international stature, geopolitical concerns and world views. For the first several decades after 1959 – when UNCOPUOS was established – India maintained that outer space was a domain that must be promoted as a realm of peace and cooperation. In one of the earlier statements in 1964, India's representative at the UN, Krishna Rao, said that "outer space was a new field and there were no vested interests to prevent the international community from embarking upon a regime of co-operation than conflict. The problems of outer space were fortunately not those of modifying an existing regime but of fashioning a new pattern of international behaviour".[10] In extending

---

[10] Rao, Krishna. Quoted from (Rajagopalan 2017a).

this view and articulating India's commitment to international cooperation in space explorations, the then Prime Minister Indira Gandhi too gave a clear message to the UN secretary general in 1968:

> The peaceful uses of outer space, particularly in the fields of telecommunications and meteorology, promise to confer great benefits to developing nations. India looks forward to expanding areas of international collaboration and would take initiatives as she has at the United Nations sponsored International Rocket Launching Station in Trivandrum and at the Experimental Satellite Communication Earth Station.[11]

Since that year, when Vikram Sarabhai also became the vice president and scientific chairman of the first UN Conference on Peaceful Uses of Outer Space, India consistently sought to utilise this forum to project its views of using space for national development and on the importance of international cooperation.[12]

The logic behind this diplomatic posture was clear: India was still at the very infant stage in the development of its space capabilities and wanted to maintain cooperative relations with the rest of the world, including the two superpowers, to create a favourable environment for its growth (in space and elsewhere). This policy posture was more also a reflection of India's broader foreign policy of non-alignment based on a "moralpolitik", which focuses on the importance of morality in foreign affairs and is critical of a military or power-hungry focus in governance – policies reminiscent of colonialism (Mohan 2013). According to political scholar Manjeet Pardesi during the Cold War, India viewed international relations through the lenses of "moral realism", defined as "a drive towards power maximisation [...] under the veneer of morality. In other words, since independence, weak India "has used morality as a tool of realpolitik to wield more influence than its actual capabilities afforded it" (Pardesi 2005).

This attitude is particularly visible in India's proactive diplomacy against US–USSR space competition, including their space arms race with ASAT weapons' testing, and, later, in its effort towards the reorganisation of the modus operandi principles of the ITU and UNCOPUOS. Another example is offered by India's ardent opposition to the US Strategic Defense Initiative (SDI) during the 1980s. The then Indian Minister of External Affairs, PV Narasimha Rao, strongly attacked such efforts by affirming "Extension of arms build-up to outer space would mean a permanent goodbye to disarmament and peace and [will] plunge mankind into a perpetual nightmare".[13]

---

[11] Gandhi, Indira. Quoted from (Jayaraj 2004).

[12] Over the years, these diplomatic efforts would be successful. In 1982, the UN recognised the potential of space technologies for these purposes, noting example states such as India are "leapfrogging over obsolete technologies and getting away from percolation and trickle down models of development for which developing countries do not have the time". Remarkably, Yash Pal was the secretary general for the second UN Space Conference held in August 1982 called UNISPACE-82, and U.R. Rao was the chairman the UNISPACE-III held in July 1999.

[13] Rao, Narasimha. Quoted from (Lele 2017c).

India's concerns about a possible weaponisation of space urged it to promote a ban on space weapons in all available multilateral fora, particularly the CD (Lele 2017a, b, c). As Rajagopalan observes:

India's pro-activism at that time also led then Prime Minister Rajiv Gandhi to sponsor in January 1985 'a declaration of six nonaligned countries opposing an arms race in outer space and nuclear testing'. The Indian debate, seeking a total ban in all global commons, was also being increasingly pursued from a morality and sovereignty angle, which did not find too many takers except for the non-aligned community (Rajagopalan 2017a, b).

It is also important to note, however, that during that period India had also used much rhetoric about its resolve to promote the peaceful uses of outer space and to ensure that space remains the common heritage of humankind by sharing the benefits deriving from its utilisation. For instance, like the vast majority of spacefaring nations, India did not ratify the 1979 Moon Treaty, which calls for the dissemination of benefits deriving from the exploitation of lunar resources. Hence, a gap remains between India's diplomatic rhetoric and its actual behaviour. This, to a large extent, corroborates the views of Manjeet Pardesi about India's "moral realism" during the Cold War (Pardesi 2005).

Be this as it may, India's diplomatic stances towards the promotion of the peaceful and sustainable uses of outer space as well as the application of space technologies for the benefit of developing countries continued even after the end of the Cold War throughout the 1990s, the major difference being the progressive growth of its bargaining power in pursuing its objectives. A clear case in point is the first restructuring of the bureaux of UNCOPUOS in 1997. As recollected by M.Y.S. Prasad:

India worked with many developing countries to change the old and almost permanent leadership of the UN Committee and its Subcommittees, and a new formula of equitable distribution of positions was worked out in 1996 and 1997. India played a crucial role in raising the subject in the context of the changing world order and in carefully creating a new structure without disturbing the useful activities of the committee. After restructuring of the Committee, all countries agreed on the first chairmanship. Recognising the key role played by India, U.R. Rao was unanimously elected as Chairman of the UNCOPUOS for a three-year term from 1997 (Prasad 2015).

The UNISPACE-III, organised in Vienna, was also chaired by U.R. Rao.

As India's political, economic and military might progressively grew, by the early 2000s, its diplomatic posture regarding the peaceful and sustainable uses of outer space began to evolve accordingly, despite the official rhetoric. An interesting case in point is the veiled objections it raised vis-à-vis the EU Code of Conduct for Space Activities. While India was in principle in favour of such code, the European perception was that New Delhi opposed the proposal because it considered it as a means of protecting some vested Western interests, making it too intrusive and potentially restrictive for other space operations. What irritated India the most, however, was that the EU formulated the code without consulting it. The gaffe of EU diplomacy in advancing the code is well-known. What is interesting to note is that India did not want to become party to an instrument that would not give New Delhi "ownership", even if it probably was in India's interest to have such a code in place (Rajagopalan 2011).

Another telling example of India's evolving diplomacy of space is India's threatened rejection of the UN Debris Mitigation Guidelines during their negotiations. As articulated by Moltz based on his interviews with participants in the debris mitigation discussions during 2006–2007:

> India initially appeared to be playing the role of potential spoiler. In light of past practices by the Soviet Union and the United States, which had put the bulk of the then-extant orbital debris into orbit, Indian representatives argued that unfairly forcing India and other developing countries to abide by strict debris mitigation guidelines now amounted to 'cultural imperialism'. Indian diplomats initially threatened to block agreement over the proposed guidelines. In the end, after receiving promises of technology from the United States and other space powers, India altered its policy and allowed the guidelines to be approved by acclamation at the United Nations in December 2007 (Moltz 2012).

All in all, India's diplomacy of space is evolving, and, consistent with its vision of a larger international role for itself in the twenty-first-century world, it is striving to achieve a greater voice in space governance issues. Yet, it is also undeniable that India has thus far failed to offer alternative proposals for international space security and cooperation, let alone specific public goods (such as the GPS) that can make it a true leader in space and on Earth.

5.3 India's Military Space: The New Dimension

Whereas remarkably the Indian space programme has had a strong civilian focus throughout most of its history,[14] since the last decade, the country has dramatically increased its interest in the area of security- and defence-related space activities and has started to bring its programme in line with the rationales typical of the other major space powers (Sheehan 2007; Paracha 2013). There are also indications that India has started to make a transition from dual-use satellite technology to dedicated defence satellites and that active military uses of space are also under consideration by the political leadership.

From an overall perspective, the policy shift in India's policy posture has entailed three broad dimensions: the use of space assets for passive military applications, the possible development and testing of offensive space capabilities and a change in the institutional architecture to better address India's space-related security needs.

---

[14] Once again, this is not because the country's space leaders did not recognise the military value of space activities, but they made a deliberate effort to harness space technology for socio-economic utility, hence avoiding possible confrontational stances in the space arena and benefiting instead from cooperation with both the United States and the Soviet Union.

5.3.1   Space Assets for Indian Security: India's Progressive Militarisation of Space

One of the first drivers that account for this unfolding policy shift in India's posture is the 1999 Kargil War. Indeed, the failure to detect Pakistani intrusion in the Kashmir region and alert the military in 1999 revealed serious gaps in the capabilities of the Indian space programme and served as a wake-up call to expand India's priorities in the utilisation of space. In 2001, ISRO launched the Technology Experiment Satellite (TES), which was the first Indian satellite capable of military surveillance with imagery close to one-metre resolution. The then ISRO chairman described it as a "satellite for civilian use consistent with the state's security concern" (Lele 2017a, b, c). TES was used to transmit high-quality images of US operations during Operation Enduring Freedom in Afghanistan (Paracha 2013). Within the next few years, the first two satellites of the new Cartosat series were launched for topographic mapping. Cartosat-1 and Cartosat-2 were high-resolution satellites, with 2.5-m and 1-metre resolution, respectively, and had a significant military utility, particularly in terms of surveillance and reconnaissance. During the same period, ISRO also laid the groundwork for its indigenous navigation satellite system, the IRNSS, and in 2002 the Indian Defence Committee in the Lok Sabha (the lower house of the Parliament) issued a recommendation for the establishment of an aerospace command within the Indian Air Force as part of military modernisation. The issue was discussed for several years, but no action was taken.

The real policy shift in India's posture came only after China's ASAT test in 2007. The test was both a surprise and a shock for India, urging top level Indian attention on the need to commit more dedicated resources and efforts in this domain (Rajagopalan 2017a, b). Because of India's rising anxieties about China's strategic modernisation, Beijing's continuing assistance to Islamabad, and the fear of losing strategic ground vis-à-vis the technological development of other major powers, the Indian Air Force, the Indian Navy and the Indian Army all began to express their resolve in acquiring dedicated satellites and advancing their expertise and capabilities in terms of space assets. Since then, a greater sense of purpose in India's military uses of space assets has clearly emerged.

In order to expand India's still limited capabilities, in 2010, the Headquarters Integrated Defence Staff (IDS), Ministry of Defence, issued the Technology Perspective and Capability Roadmap (TPCR) to define the requirements of the armed forces, which included space-based requirements. The roadmap highlights the need to develop satellites for meeting reconnaissance, surveillance, meteorological, navigation and communication requirements, and the need for strategic forces to be facilitated by real-time information in areas of interest which includes satellites that produce sub metric resolution imagery, aid in communications, all-day-all-weather capability and signal intelligence among others (Headquarters Integrated Defence Staff – Ministry of Defence 2013).

Consistent with the TPCR, India's approach to utilisation of outer space for security-related purposes has been focusing on securing assets for intelligence gathering – including Signal Intelligence (SIGINT), Geospatial Intelligence (GEOINT), Imagery Intelligence (IMINT), Cyber Intelligence (CYINT) and Measurement and Signature Intelligence (MASINT) – as well as for navigation and communications (Prasad 2016).

Remote Sensing for Surveillance and Reconnaissance

In the field of remote sensing, over the last decade, ISRO has developed and launched two series of dual-use remote sensing missions with high-resolution optical and SAR capability, namely the Cartosat and RISAT series. Both have been used by the military to address some of the required intelligence gathering requirements.

The first dual-used satellite was Cartosat-2A, a 0.8-metre high-resolution optical satellite, launched in April 2008. It was officially said to be for civilian applications, such as urban and rural development mapping, but it is also known to have significant military utility due to its technological specifications. A second optical imaging satellite, Cartosat-2B, was launched in July 2010 for providing scene-specific spot imagery with a similar resolution, and after a gap of 6 years (during which relevant progress was made in sensor technology), ISRO launched Cartosat-2C in 2016, with a resolution of a few centimetres. As highlighted by Lele, an important feature of Cartosat-2C was the use of adaptive optics and acoustic–optical devices. This satellite has micro electromechanical systems and adoptive optics that offer better visibility of objects on the ground. Here, the optical system adapts to compensate for optical effects introduced by the medium between the object and its image, while acoustic–optical devices enable interaction between sound waves and light waves. Together, the Cartosat satellites offer the Indian security establishment 24/7 capability to monitor various sensitive areas. These optical satellites, however, do not perform correctly when the sky is overcast (Lele 2017a, b, c).

In order to overcome these limitations in optical and sensor-based surveillance, since 2009 the Cartosat series has been complemented by the RISAT (Radar Imaging Satellite) series of reconnaissance satellites. These are India's first Earth observation satellites using Synthetic Aperture Radar (SAR) technology and providing all-weather and day-and-night visibility. As highlighted by Rajagopalan:

> this is of huge strategic importance to India as SAR technologies provide for better reconnaissance, surveillance and location targeting for guidance and navigation. India's requirement for such missions grew significantly after the Mumbai terrorist attacks in November 2008. Therefore, even though India was developing RISAT-1 indigenously, security needs following the terrorist attacks pushed India to partner with the Israel Aerospace Industries (IAI) to expedite the development and launch of RISAT-2 (Rajagopalan 2017a, b).

The first satellite in this series (RISAT-2) was launched in April 2009, while the in-house-made RISAT-1 went into orbit in 2012. RISAT-1 has a resolution of 1 m and carries a C-band SAR payload, operating in a multi-polarisation and multi-resolution mode to provide images with coarse, fine and high spatial resolutions. Both these satellites have day-and-night viewing capacity and are not blinded by cloud cover/bad weather. They have the capacity for continuous surveillance and are hence considered a "force multiplier" in the defence context. The RISAT series has been put to great use for security-related functions: border surveillance, detection of insurgent infiltration and facilitation of counterterrorist operations.

Military Communications

Together with remote sensing, military communications have also emerged as key focus areas in India's military space applications. For a long time the Indian Air Force, the Indian Navy and the Indian Army had been expressing interest in acquiring dedicated communication satellites, but had had to rely on foreign partners (in particular Inmarsat, a British commercial satellite communication provider) for many of their satellite-based communications and data services' needs.

On 30 August 2013, however, ISRO launched GSAT-7, the first dedicated defence communication satellite made available for the Indian armed forces. Specifically built by ISRO for the use of the Indian Navy, the launch of this satellite was an important advance in India's uses of space for strategic and national security-related operations, potentially marking a transition from dual use to dedicated defence satellites.

GSAT-7 carries relay capacity in UHF, S-band, C-band and Ku-band, which provides high-density data transmission facility, both for voice and video. In addition, the satellite has been provided with additional power to communicate with smaller and mobile (not necessarily land-based) terminals. As detailed by Lele:

> this dedicated satellite provides the Indian Navy a 3,500-4,000-km footprint over the Indian Ocean region and enables real-time networking of all its operational assets in water (and on land). It also helps the Navy to operate in a network-centric atmosphere. The Indian peninsula is an extremely tricky region for operations because of its geographic location. One of the deadliest terrorist operations on Indian soil, the 2008 Mumbai attack, was launched using the Arabian Sea route. GSAT-7 is useful for gathering communications and electronic intelligence in respect to moving platforms in the sea, particularly through its UHF facility. GSAT-7 also helps the Navy monitor activities over both the Arabian Sea and the Bay of Bengal regions. Broadly, India's strategic area of interest extends from the Persian Gulf to the Malacca Strait, and now a significant portion of this region is covered by this satellite (Lele 2017a, b, c).

The second dedicated communication satellite for India's military, GSAT-6, was launched by ISRO in August 2015. GSAT-6 is a two-tonne class satellite mainly developed for the Army and is considered of utmost strategic importance, given that Indian soldiers operating in diverse terrain and topographic conditions (from the peninsular region to deserts to snow-clad mountains) have on many occasions encountered breaks in communications. GSAT-6 provides quality and secure communication. Equally important is that the introduction of this satellite has also freed Indian soldiers from carrying bulky communication equipment since very small handheld devices can be now used.

Navigation

In the field of satellite navigation, India's armed forces have been dependent on GPS/GLONASS. Independent navigation-related functions have, however, become critical in India's military space utilities. Whereas India has never been denied access to the two systems, following the 2003 Iraq War, it felt the need to put in place an indigenous system to "avoid being exposed to the possibility of it being denied in the future, especially during a crisis or a war" (Rajagopalan 2017a, b).

Accordingly, the IRNSS or NAVIC constellation, the deployment of which was completed in April 2016, was designed to provide an encrypted restricted service (RS) for special authorised users, especially government agencies and the three armed forces. While IRNSS's coverage is still limited, it provides absolute position accuracy throughout India and within the 1500–2000-km region around it – which is India's strategic area of interest.

Overall, over the past 10 years, India has made consistent efforts towards using space assets for strategic purposes. It remains, however, far from fully exploiting the full-spectrum capabilities offered by space technologies in this domain.

5.3.2 Security for Indian Space Assets? India's Deterrence Considerations

Whereas the progressive militarisation of India's space activities has become manifest in recent years, the country's stance vis-à-vis weaponisation issues has remained much more ambiguous. Traditionally, India's leaders (from Indira Gandhi to Manmohan Singh) have been proactive champions of the cause of non-weaponisation and peaceful uses of space both in the domestic and international arena. For the first several decades after the beginning of the space age, India maintained the view that outer space was to be a realm of peace and cooperation. During the 1980s, India's concerns about the US–Soviet space competition, including their anti-satellite tests, also pushed the country's diplomacy to seek a ban on space weapons in all the multilateral platforms such as the UN and the Conference on Disarmament (CD) (Paracha 2013; Rajagopalan 2017a, b). In January 1985, Rajiv Gandhi's government also sponsored a declaration of six non-aligned countries opposing an arms race in outer space and nuclear testing. Shaped by India's broader "moralpolitik", this policy posture continued also after the end of the Cold War throughout the 1990s.

However, with the recent proliferation of technological demonstrations of offensive capabilities at both regional and international level, the Indian government has started to review its position and to consider developing deterrence capabilities to protect its space assets. The instigator of this policy shift was once again China's ASAT weapons test of 2007.

Even though few weeks after that test, during the visit of President Putin to India, Prime Minister Manmohan Singh released a joint statement calling for a "weapons free outer space", subsequent statements by senior Indian officials revealed that India's approach towards space security was, in fact, changing. Following China's 2007 test, the then chief of army staff of the Indian Army, General Deepak Kapoor, was reported by *Times of India* as saying that China's space programme was expanding at an "exponentially rapid" pace in both offensive and defensive capabilities and that space was becoming the "ultimate military high ground" to dominate in the wars of the future (Pandit 2008).

Since then, this trend has been progressively reinforced. In 2008, the then head of the Integrated Defence Staff (IDS), Lieutenant General H.S. Lidder, affirmed "there is every possibility that we might get sucked into a military contest either to protect our assets in space or to launch an offensive" (Samson and Bharath, Introduction, 2011). More importantly, in January 2010, the DRDO Director General V.K. Saraswat openly recognised Indian efforts to develop ASAT capabilities, claiming "we are also working on how to deny the enemy access to its space assets" (Brown 2010a, b). This was corroborated by the *Technology Perspective and Capability Roadmap* issued on the same year by the Ministry of Defence's Integrated Defence Staff, which called for the development of both defensive measures, like electromagnetic pulse hardening as well as ASAT capabilities "for electronic or physical destruction of satellites", both in LEO and GEO (Bagchi 2011; Paracha 2013).

India's possible development of offensive space capabilities eventually came to the scrutiny of the international space community in 2011 when DRDO's V.K. Saraswat announced that India had all the building blocks in place to integrate an anti-satellite weapon to neutralise hostile satellites in low Earth and polar orbits (Unnithan 2012). In the same interview, Saraswat suggested that India's Anti-Ballistic Missile (ABM) defence programme could be utilised as (or incorporated in) an anti-satellite weapon, along with its Agni series of missiles.

Although India has publicly claimed that it has brought together all the basic technologies needed to create a kinetic-kill fully fledged ASAT weapon based on Agni and the ballistic missile interceptor, it remains doubtful whether India has, in fact, acquired such capability. As Michael Listner pointed out "integrating the necessary technologies may give India an ASAT capacity, but does not necessarily give India a proven ASAT capability. The only way for India to demonstrate that it has a proven ASAT capability is to perform a test on a target satellite" (Listner 2011).

Apparently, a large part of India's scientific, military and academic communities has been open to showcase this capability, on condition, however, that such test is done in a responsible manner without creating a huge amount of long-lasting debris that could damage existing satellites (Vasani 2016). According to Narayan Prasad and Rajeswari Pillai Rajagopalan:

> A possible template to showcase technological capabilities may lie in following a strategic engagement of a low- flying asset that may not sustain any in-orbit debris but will be completely destroyed in entry and upper atmosphere. The U.S. performed such a mission as recently as 2008, launching a single Standard Missile-3 (SM-3) and destroying a 5,000-pound satellite with nearly 100 per cent of the debris safely burned-up during re-entry within 48 hours and the remainder safely re-entering within the next few days. One has to note that India does not have any space asset at such a low altitude; and if such a capability has to be demonstrated, one of the dying satellites will have to be lowered to perform such a test (Prasad and Rajagopalan 2017).

Similar views have been expressed by a number of analysts and commentators of the Indian space programme (Chandrashekar 2016; Bommakanti 2017; Bagchi 2011). The key question, however, is whether New Delhi's policy-makers will support such a test.

The Case for and Against an ASAT Weapon Demonstration

For India's civilian leadership, the situation is particularly difficult, as there are both pros and cons for demonstrating India's ASAT weapons capability. One of the major reasons that could encourage India to kinetically test an ASAT weapon is clearly to mature deterrence tools and cover all possible contingencies for future conflicts (Rajagopalan 2017a, b). Equally powerful is the perceived necessity for India not to fall behind its political regional rival, China. India sees China's space capabilities as their main space threat and points to the 2007 and 2010 Chinese ASAT and hit-to-kill missile defence tests as examples of the pressing need for reciprocal capability. As, for instance, contended by Kartik Bommakanti:

> for deterrence and compellence to have any credibility in the India-China space dyad, New Delhi will require a potent retaliatory capability [...] The risks are substantial in the absence of Indian ASAT capability. False optimism could lead India to underprepare, thereby easing a Chinese attack against Indian space-based assets and the ground nodes. Logical and justifiable optimism could also lead China to attack, if India unduly or unilaterally subjects itself to self-imposed constraints (Bommakanti 2017).

Another reason behind India's possible interest in showcasing ASAT capabilities is the perceived need to establish a precedent in case of future space arms control talks and hence avoid a Non-Proliferation Treaty (NPT)-like situation. As aptly commented by Victoria Samson about India's past disappointment with being left out by the 1968 NPT as a nuclear weapon state:

> There are lessons learned from previous arms control debates that have probably affected India's decision to seek a missile defence/ASAT capability. One strong one is that Indians remember well that the 1968 Nuclear Non-Proliferation Treaty (NPT) made a concrete division between the nuclear haves and the have-nots. This partition was largely based on who had held a nuclear test prior to the treaty's creation. India missed becoming an official nuclear weapon state by six years by having its first nuclear test—or, as India termed it, a "peaceful nuclear explosion"—in 1974. There are some within India who have taken that lesson to heart and want India to develop an ASAT capability so that India would be grandfathered in, should any future treaty or international agreement ban ASATs. This is probably to gain the prestige of being one of a select few states and the wish to avoid being hemmed in, should future Indian military officials decide that an ASAT capability is needed for their national security needs (Samson 2010).

Along those same lines, it can be argued that India fears to be seen as a "paper tiger" by the arms control and intelligence community if it does not conduct a test and demonstrate its ASAT capability explicitly (Listner 2011). In other words, "even if New Delhi does have an anti-satellite weapons capability, it will only be acknowledged if it comes out in the open with a successful test" (Vasani 2016). It is, however, clear that such a demonstration could also generate undesirable consequences for India.

Along with potentially causing the creation of a large amount of debris posing a hazard to other spacecraft, it is possible that such a test would be viewed as an aggressive military action, particularly if India performs such a test unilaterally without consulting the international community. But even if India fulfilled its obligations under Article IX of the OST – which requires prior international consultations if a state believes that a planned activity might be harmful for others – it is likely that the test may lead to serious international repercussions and could jeopardise India–US space cooperation or, even more importantly, lead to unwelcome sanctions. Although China avoided tangible international responses from its 2007 ASAT test, according to several observers "it is unlikely that India would enjoy similar immunity and could find itself at the centre of a serious political and diplomatic tempest, a fact that India's officials are likely aware of" (Samson 2010; Listner 2011; Vasani 2016).

India's space leaders also appear conscious that such a test would be inconsistent with its declared alignment with the OST's provisions about the peaceful use of outer space and that it may in addition impact its credibility in international fora that deal with orbital debris and, more broadly, space sustainability issues. As a member of the IADC that significantly contributed to crafting its mitigation guidelines, India knows that a successful test of an ASAT and the ensuing creation of orbital debris could seriously hamper its credibility in that forum.

Even more worrisome, however, is the possibility that an ASAT test by India may inadvertently create the exact thing that India is trying to avoid: a space arms race with China or, even worse, a conflict in or about space. In fact, in New Delhi's eyes there is a chance that China might consider such a test as a provocative action and decide to respond with its own military one-upmanship, thus leaving India further behind (Moltz 2012). Finally, as Listner interestingly pointed out:

> India also has to consider the possibility that a test could fail, and such a failure might not go unnoticed. Even though India may have the technology to produce an ASAT capacity it does not guarantee that it will work the first time out. […] A failure would not only be a blow to the technical and scientific community of India, but it could also affect India's national security as it would provide China a level of certainty that India does not have an effective ASAT capability (Listner 2011).

For many, it is the uncertainty surrounding India's ASAT capability that serves India better than holding an actual test and removing all doubt as to whether it actually has that capability. In fact, such uncertainty can be its most powerful tool for ensuring deterrence and for protecting its space assets. It is well possible that Indian officials have consistently opted not to cross the line so as to retain the country's best weapon of choice. For other commentators, however, a strategy based on exploiting the risks inherent in uncertainty, while necessary, is not sufficient to insure deterrence (Bommakanti 2017). In short, the dilemma of whether to test or not a kinetic ASAT capability continues to dawdle.

Should an ASAT test be undertaken in the future, it will be primarily because new geopolitical developments will force India to do so (Gopalaswamy 2013). In the final analysis, however, the question of whether India will or will not perform an ASAT test boils down to a purely political choice that will be probably taken at the highest political level.

5.3.3 Institutional Transformations: Towards an Aerospace Command

With India's armed forces increasingly seeking to expand their capabilities in areas such as space-based intelligence gathering (ELINT, COMINT, SIGINT, IMINT), early warning, navigational aids, communications and possibly – as just discussed – active space defence, the Indian government is increasingly confronted with some challenging tasks. Alongside technological and resource-related issues, one the main challenges for the government is clearly institutional in nature and lies in the creation of an effective organisational architecture for space activities that can simultaneously allow ISRO to continue focusing on the delivery of civilian applications while giving enough room for the military to systematically integrate space capabilities into their operations (Prasad 2016; Rajagopalan 2017a, b).

Some steps in this regard have already been taken in an incremental – though rather slow – fashion. The first such step was taken in June 2008, when the Defence Minister A. K. Antony announced the establishment of an Integrated Space Cell (ISC) under the HQ Integrated Defence Staff of the Ministry of Defence.[15] The minister articulated the logic behind the ISC with the need to create greater coordination and coherence of purposes in India's military uses of space. Consistently, the ISC has been operating as an integrating window between the armed forces, as well as ISRO, the Department of Space and the Ministry of Defence (Prasad and Rajagopalan 2017). It is also important to stress that since the beginning, the establishment of the ICS was intended as functional building block towards the creation of an operational tripartite aerospace command, an institution that would eventually integrate all the existing capabilities and functions that pertain to the military uses of space.

The possible creation of such an institution, however, has been in the making for almost 20 years now (as of 2017) with no concrete results. Indeed, the need for an aerospace command was first articulated as early as 1998 by the then Chief of Air Staff, Marshal S.K. Sareen. In 2002, the Indian Defence Committee in the *Lok Sabha* recommended that the Indian Air Force establish an aerospace command to address the needs of India's military modernisation. The Committee also recommended that the Air Force have a "space interface for real-time situational awareness, for ballistic missile defence, for jamming enemy satellites, and for real-time intelligence about enemy radars and missile" (Paracha 2013). However, no decisive actions were taken. The same proposal was reiterated in 2003, and 3 years later, in 2006, the Indian Air Force eventually established a Directorate of Aerospace in Thiruvananthapuram, Kerala, which "can be referred to as the initial avatar of the Indian aerospace command" (Rajagopalan 2013). The idea was to have a separate

---

[15] In explaining the drivers behind its establishment, the minister affirmed: "Although we want to utilize space for peaceful purposes and remain committed to our policy of non-weaponisation of space, offensive counter space systems like anti-satellite weaponry, new classes of heavy-lift and small boosters and an improved array of Military Space Systems have emerged in our neighbourhood" (Rajagopalan 2013).

command with communications, navigation and surveillance as major functionalities. However, the government refused once again to take actionable measures over the additional expenditure involved, particularly for fear that it might be accused of militarising outer space (Rajagopalan 2013).

The debate on the possible creation of an aerospace command received a new boost in the aftermath of China's 2007 ASAT test. Soon after the test, the then Indian Air Force Chief Air Marshal S. P. Tyagi publicly articulated the logic behind an aerospace command by stating: "As the reach of the Indian Air Force is expanding it has become extremely important that we exploit space and for it you need space assets. We are an aerospace power having trans-oceanic reach. We have started training a core group of people for the 'aerospace command'" (Space Daily 2007). By the same token, also the Indian Navy has started to introduce a certain upgrading in its institutional architecture, by creating a new office for Communications Space and Network Centric Operations to manage the Navy's military space capabilities.

While the logic of and need for a single aerospace command in India is today widely acknowledged, there still appears to be some political, strategic and bureaucratic hesitance in its eventual creation, particularly because such move would eventually sanction the official militarisation – or, from another angle, normalisation – of the Indian space programme. Additional intermediate steps might hence be required for its acceptance. As documented by Prasad and Rajagopalan, "following recent turn of events, there has been a movement towards opening-up India's space security by a possible creation of a Defence Space Agency (DSA) as an interim arrangement until a full-fledged dedicated Aerospace Command is in place. This seems a case for the expansion of ISC for a more active role in utilisation of outer space within the armed forces" (Prasad and Rajagopalan 2017).

To conclude, even though in the last decade India has dramatically increased its interest in the area of security-related space activities and has achieved important steps forward, its approach to military space issues has remained quite piecemeal, as it still lacks both an appropriate institutional architecture and an overarching policy articulation by the country's political leadership, having greater sense of purpose and directions for the future.

5.4   India's Evolving Space Doctrine; An Assessment

As described throughout Chaps. 4 and 5, the Indian space programme is an increasingly complex and multilayered endeavour that, over the years, has come to entail a wide range of dimensions at the nexus of societal utility, commerce and geopolitics. India's use of space technologies is now driven by a combination of factors that simultaneously span from societal development and economic growth to military security and diplomatic goals. In the process, India has succeeded in joining the club of major spacefaring nations and has established itself as fully fledged space power with key independent technological capabilities.

Notably, despite a 40-yearlong involvement in space activities, the Indian government has not yet issued an overarching and integrated policy document that "provides the *raison-d'etre* of national space activities and explicates the philosophy behind the development and use of space assets" (Sachdeva 2016). Admittedly, there has been a debate on its adoption for more than a decade now. But thus far, the outcome seems to be that the pressing need to get a national space policy is not equally shared among decision-makers. Whether this seems to be related more to India's resolve to maintain strategic flexibility than to the lack of grand political vision,[16] either way the result has been a certain degree of ambiguity, at least from the perspective of foreign observers.

India's space officials have certainly missed no occasion to stress that the overall objective of India's involvement in space is to assist the all-round development of the nation and that there is no ambiguity of purpose. But concerns have been often raised over the "alleged contradiction between the development goals that Indian governments historically used to justify the programme, and the obvious military rationale that lies behind the country's development of both launchers and satellite technology" (Sheehan 2007; Thomas 1986). As a widespread line of argument has gone since the latter part of the Cold War "whether originally intended or not [...] space programmes serve dual purposes" (Thomas 1986), and a developing country like India has been investing in such programmes despite their huge cost exactly because of their military applications. Similarly, many have pointed to India's clash between its declared interest in developing international norms regulating behaviour in space and its underlying suspicion of a more stringent regime that might limit its freedom of action in the field of military space.

However, this line of reasoning proves particularly limiting, if not misleading. For one, the historical evolution of the Indian space programme makes it clear how, for a long time, the military dimension has been strictly firewalled because of political considerations. In addition, India's current uses of space prove to have been brought in line with the rationales typical of all other spacefaring nations, the difference being that the military dimension of the programme has arisen well after the societal one. Hence, it would be more appropriate to look at the progressive expansion of the Indian space programme and in particular its forays into the realm of military space as a "normalisation" rather than a "militarisation".

Even more importantly, viewing India's rationales for engaging in space through the traditional development/military dichotomy is not particularly useful to guide understanding of its space policy. Rather than in opposition to each other, these rationales should be more properly seen as complementary and embedded within an overarching *security driver*, which, from the Indian perspective, encompasses "both

---

[16] As also contended by Ajey Lele, it would be erroneous to equal the absence of a space policy to the absence of vision. Various specific policy documents, articulations of space agenda by top ISRO leadership on various occasions, documents such as Vision 2020, etc. elucidate India's expectation for the future (Othman 2017).

the traditional military dimension and extended categories of economic, environmental, societal and human security". As ISRO official Narayana Murthy explained:

> the concept of security applied to space has been broadened "from its traditional conception
> in military terms, to encompass other threats (including those emanating from poverty, lack
> of education, health hazards, environmental degradation, and natural disasters). Seen in this
> light, the space programme clearly shows no ambiguity of purposes and, as Michael
> Sheehan also opines, "is distinct and coherent in the way that it simultaneously addresses
> the requirements of the Indian people and the state in all these [security] dimensions
> (Sheehan 2007).

The Growing Need for a Space Policy

Be this as it may, the very existence of international speculation on the nature and
directions of the Indian space programme reveals one incontestable fact: that the
lack of a declared and comprehensive space policy may be detrimental to India.
Although the lack of an articulated policy posture certainly continues to be a useful
means to provide flexibility, the benefits of defining policies in the open can no
longer be overlooked. National space policies have indeed been demonstrated as a
very useful tool to optimise the benefits of space by ensuring clarity in goals and
intentions as well as a degree of stability. And they do so with respect to both the
domestic and foreign contexts, reducing the risks of miscommunication on the
international stage. A declared space policy is, in fact, not only a tool of communication to inform domestic stakeholders and generate national support for space
activities but also – and perhaps more importantly – for foreign ones. It sends them
a message on the *principles*, *boundaries* and *aspirations* upon which the programme
operates, as well as on the opportunities for engagement it wants to open.[17] Without
such a clear message, these audiences can be left with wrong impressions about
India's underlying goals, impressions that can inevitably fuel speculation or even
fears. To the contrary, an open policy articulation can do a lot to lessen such uncer-
tainties and support regional and international stability. As ORF Senior Fellow
Rajeswari Rajagopalan also stressed:

> Open policy statements and declared policies have remained the best means to assuage
> fears, build confidence and avoid ambiguities. These are important measures for building
> transparency and reducing tensions in regional and global contexts. Since the Asian context
> is characterized by growing competition and rivalry and the potential for conflict, even rela-
> tive openness and transparency will go a long way in diluting the levels of regional insecuri-
> ties .... There are apprehensions particularly in the immediate neighbourhood because of
> India's dominant presence in South Asia. India's growing capabilities in the space arena as

---

[17]As highlighted by Mazlan Othman, "these principles can be used to reaffirm or demonstrate a
state's adherence to international agreements and treaties, and to outline national principles that
have a historical, cultural, or ideological basis. The principles in a national space policy can also
form the foundation for lower-level government policies in specific sectors. As for International
cooperation, it is "rarely pursued haphazardly, but is instead often part of larger policy and strategic considerations. International cooperation is often considered both a mechanism and a goal, so
it may feature in policy documents. As a mechanism, space cooperation enables actors to leverage
the expertise, investments, and resources of others in the development of programs, whether
through the direct acquisition of hardware or the joint development of technical capacity".

in several other areas has the potential to heighten insecurities among these smaller neighbours. It will benefit India in the longer run if it were to showcase its interests and benefits in the regional context. Moreover, India should be able to cultivate regional interests that are akin to its own interests. … Articulating an open policy would also add to the credibility of India as a major spacefaring nation. It would be a major transparency and confidence building measure, which is particularly important in the Asian context (Rajagopalan 2017a, b).

In short, having a declared and comprehensive space policy vision is becoming key for India to inform other actors about its own interests and concerns over certain behaviour or activities, besides enabling the country to better advance those interests at negotiating tables and unlock the full potential of G2G dialogue and cooperation in areas of mutual interest.[18] The principles in a national space policy can also form the foundation for lower-level government policies in specific sectors such as military or commercial space.

In parallel to these considerations related to India's international posture, the need for a declared space is becoming more pressing also to bring much-needed internal clarity and support development of the entire domestic space ecosystem in India or what can be termed – with a nod to the famous India's Stack – as India's Space Stack (see Fig. 5.1). As already stressed in Chap. 4, at the first layer the success of the Indian space ecosystem is linked to an expansion of its *hardware* (i.e. to a scaling-up of activities, enhanced funding and maturation of supply capacity to meet the country's demand for increased launch frequency, transponder capacity,

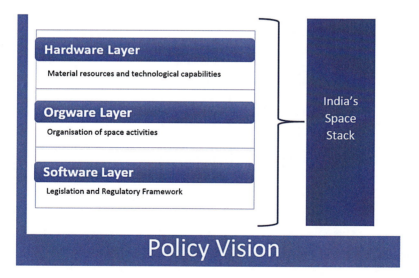

Fig. 5.1 India's space stack

---

[18] Knowing India's interests, needs and priorities for the future can help foreign partners enlarge the scope of potential cooperation with respect to both data and information exchange and joint activities in areas of mutual interest.

space applications and so on). Achieving this in turn calls for changes in the *organisational structure* of space activities, including enhanced participation of private industries. And this is in turn related to the introduction of new *software* (i.e. a comprehensive space legislation and regulatory system for licensing, IPR, etc.). It should go without saying that such regulatory mechanisms can only be put in place in an optimal manner if there is an overarching policy vision spelled out at the highest political level. Hence, the bottom line of India's Space Stack clearly lies in the articulation of a sound national space policy that would allow the different layers of the space ecosystem to move forward.

Finally, and closely related to this aspect, the adoption of a space policy is coming to the fore because of the ever-growing need to provide a clear picture of the country's long-term space objectives that can guide in prioritising the development of capabilities and crafting international partnerships to reach the identified goals. The articulation of these interests, needs and priorities for the future is key to ensuring steady development of the space programme and of the country as a whole. While it is true that various policy documents such as Vision 2020 or the articulation of a space agenda by top ISRO leadership on various occasions already elucidate – as Ajey Lele argues – India's expectation for the future (Lele, India's policy for outer space, 2017b), it is for many also clear that India's "space policy must be driven and announced by the national political leadership rather than being left to the discretion of individual bureaucracies such as ISRO or Indian military services. Each organization will have its own interest, and these sometimes competing and even conflicting interests can only be shaped into a national policy by the political leadership" (Rajagopalan 2017a, b). The articulation of general principles and objectives in a national space policy could then be used as the foundation for lower-level government policies in specific sectors such as civil, military or commercial space by the responsible organisations. This broad articulation is still lacking. As also recognised by India's National Institute of Advanced Studies (NIAS):

> Today, Indian Space Programme is at a cusp with the need for a Long-term Strategy for Space – basically creating a roadmap that will look 30-50 years ahead and address several key questions in the public domain. Indian space needs to orient for a quantum jump in technological growth, adopt organisational models and collaborative strategies that will ensure economic efficiency and position a vibrant private sector. Important questions are being raised on the public and national consequences for Indian human space-flight and planetary exploration programme; how India must quest for a larger share and role in global space market; strategies to deal with changing political and economic environments and focused imperatives of international cooperation (Murthi and Rao 2015).

Before elaborating on what these goals could be, what directions the programme might take and what future changes might be introduced, a reflection on the external dimension of that Indian space programme – specifically India's relations with the major spacefaring nations – needs to be provided. This will be the objective of the following chapter.

References

Aliberti, M. (2015). *When China goes to the moon*. Vienna: Springer.

Annadurai, M. (2017). *South Asian satellite – A new approach to regional co-operation*. 60th Session of the United Nations Committee on the Peaceful uses of outer space. Vienna: UNOOSA.

Bagchi, I. (2011, March 6). India working on tech to defend satellites. *Times of India*.

Bajaj, V. (2010, October 15). In Mumbai, adviser to Obama extols India's economic model. Retrieved May 10, 2017, from *The New York Times*: http://www.nytimes.com/2010/10/16/business/global/16summers.html

Bommakanti, K. (2017). The significance of an Indian direct ascent kinetic capability. *ORF Space Alert, 2*(2).

Bonnefoy, A. (2014). *Humanitarian telemedicine*. Vienna: European space policy institute (ESPI).

Brown, P. J. (2010a, January 22). India's targets China's satellites. Retrieved November 18, 2016, from *Asia Times*. http://www.atimes.com/atimes/South_Asia/LA22Df01.html

Brown, P. J. (2010b, May 1). India's space program takes a hit. Retrieved November 10, 2016, from *Asia Times*. http://www.atimes.com/atimes/South_Asia/LE01Df01.html

Chandrashekar, S. (2016). Space, war, and deterrence: A strategy for India. *Astropolitics: The International Journal of Space Politics and Policy, 14*(2–3), 153–157.

Chatterjee Miller, M. (2013, May/June). India's feeble foreign policy. *Foreign Affairs, 92*(3), 14–19.

Cohen, S. (2001). *India: Emerging power*. Washington, DC: Brookings Institution Press.

Dutta, V. (2012, December 12). Apollo enters Myanmar with telemedicine service. *The Economic Times*.

Ganguly, S. (2016, September 19). India after nonalignment. *Foreign Affairs*.

Ghosh, A. (2009). *India's foreign policy*. New Delhi: Dorling Kindersley.

Global Times. (2014, February 18). India uncertain as Abe looks for anti-China alliance. *Global Times*.

Gopalaswamy, B. (2013). India's perspective. In A. M. Tellis (Ed.), *Crux of Asia. China, India and the emerging global order*. Washington, DC: Carnegie Endowment for International Peace.

Headquarters Integrated Defence Staff – Ministry of Defence. (2013). *The technology perspective and capability roadmap*. Retrieved September 10, 2016, from Ministry of Defence. http://mod.gov.in/writereaddata/TPCR13.pdf

IndoGenius. (2016a, August). *India's election calendar. The importance of India*. IndoGenius.

IndoGenius. (2016b, August). *India engages. The importance of India*. IndoGenius.

Jayaraj, C. (2004). *India's Space policy and institutions. United Nations/Republic of Korea workshop on Space Law. United Nations treaties on outer space: Actions at the national level*. New York: United Nations.

Kaplan, R. (2013). *The revenge of geography. What the map tell us about future conflicts and the battle against faith*. New York: Random House Trade Paperbacks.

Karan, P. P. (1953, December). India's role in geopolitics. *India Quarterly*, p 160.

Kissinger, H. (2014). *World order*. New York: Penguin Books.

Kumaraswamy, P. (2012). Looking west 2: Beyond the Gulf. In D. Scott (Ed.), *Handbook of India's International relations* (pp. 179–188). London: Routledge.

Lee, J. (2010, June 4). *India's rise helps complement U.S. interests*. Retrieved January 7, 2017, from Hudson Institute. https://hudson.org/research/7051-india-rsquo-s-rise-helps-complement-u-s-interests

Lele, A. (2013). *Asian space race: Rhetoric of reality?* New Delhi: Springer.

Lele, A. (2016). Power dynamics of India's space programme. *Astropolitics: The International Journal of Space Politics and Policy, 14*(2), 120–134.

Lele, A. (2017a). *Fifty years of the outer space treaty. Tracing the journey*. New Delhi: Pentagon Press.

Lele, A. (2017b). India's policy for outer space. *Space Policy, 39–40*, 26–32.

Lele, A. (2017c). India's strategic space Programme: From apprehensive beginner to ardent opera-tor. In R. P. Rajagopalan (Ed.), *Space India 2.0. Commerce, policy, security and governance perspectives* (pp. 179–192). New Delhi: Observer Research Foundation.

Listner, M. (2011, March 28). *India's ABM test: A validated ASAT capability or a paper tiger.* Retrieved March 30, 2017, from The Space Review. http://www.thespacereview.com/article/1807/1

Meredith, R. (2008). *The elephant and the dragon: The rise of India and China and what it means for all of us.* New York: Norton.

Mohan, C. R. (2013). Changing global order: India's perspective. In A. J. Tellis & S. Mirski (Eds.), *Crux of Asia: China, India and the emerging global order.* Washington, DC: Carnegie Endowment for International Peace.

Mohan, R. (2017, February 10). *Delhi, Tokyo, Canberra.* Retrieved February 11, 2017, from *The Indian Express.* http://indianexpress.com/article/opinion/columns/tpp-donald-trumpmalcolm-turnbullphone-call-leak-islamic-state-canberra-delhi-tokyo-4516471/

Moltz, J. C. (2012). *Asia's space race. National Motivations, regional rivalries, and international risks.* New York: Columbia University Press.

Murthi, S., & Rao, M. (2015). Future Indian (new) Space – Contours of a National Space Policy that positions a new public-private regime. In *3rd Manfred Lachs international conference on newspace commercialisation and the law.* Montreal: NIAS.

Naidu, S. (2009). India engagement in Africa: Self-interest or mutual partnership? In R. Southhall & H. Melber (Eds.), *The new scramble for Africa: Imperialism, development in Africa.* Scottsville: University of Kwua Zulu Natal Press.

Nye, J. (2004). *Soft power: The means to success in world politics.* New York: Public Affairs.

Ollapally, D. M., & Rajagopalan, R. (2013). India: Foreign policy perspectives of an ambiguous power. In H. Nau & D. Ollapally (Eds.), *Worldviews of aspiring powers: Domestic foreign policy debates in China, India, Iran, Japan and Russia.* New Delhi: Oxford University Press.

Othman, M. (2017). National space policy and administration. In C. Johnson (Ed.), *Handbook for new actors in space.* Denver: Secure World Foundation.

Pan-African E-Network. (2016). *About the project.* Retrieved January 17, 2017, from Pan-African E-network Project. http://www.panafricanenetwork.com

Pandit, R. (2008, June 17). *Space command must to check China.* Retrieved March 28, 2017, from *Times of India.* http://timesofindia.indiatimes.com/india/Space-command-must-to-check-China/articleshow/3135817.cms

Paracha, S. (2013). Military dimensions of the Indian space program. *Astropolitics: The International Journal of Space Politics and Policy, 11*(3), 156–186.

Pardesi, M. S. (2005). *Deducing India's grand strategy of regional hegemony from historical and conceptual perspectives.* Singapore: Institute of Defense and Strategic Studies. Retrieved from http://www.rsis.edu.sg/publications/WorkingPapers/WP76.pdf

Paul, T. (2014, April). India's soft power in a globalizing world. *Current History*, 157–162.

Pillai, C. M. (2016, April 25). *India as the pivotal power of the 21st century security order.* Retrieved January 12, 2017, from Center for International Maritime Security.

Prasad, M. (2015). ISRO and International Cooperation. In M. Rao (Ed.), *From fishing Hamlet to Red Planet: India's space journey.* New Delhi: Harper Collins.

Prasad, N. (2016). Diversification of the Indian space programme in the past decade: Perspectives on implications and challenges. *Space Policy, 36*, 38–45.

Prasad, N., & Rajagopalan, R. P. (2017). Creation of a defence space agency. A new chapter in exploring India's space Security. In S. Singh & P. Das (Eds.), *Defence Primer 2017. Today's capabilities, tomorrow's conflicts.* New Delhi: Observer Research Foundation.

Rajagopalan, R. P. (2011, October). Debate on space code of conduct: An Indian perspective. *ORF Occasional Paper, 26.*

Rajagopalan, R. P. (2013, October 13). Synergies in space: The case for an Indian Aerospace Command. *ORF Issue Brief, 59.*

Rajagopalan, R. P. (2017a). Need for an Indian military space policy. In R. P. Rajagopalan (Ed.), *Space India 2.0 commerce, policy, security and governance perspectives* (pp. 199–212). New Delhi: Observer Research Foundation.

Rajagopalan, R. P. (2017b, January). The need for India's space policy is real. *ORF Space Alert, 5*(1), 7–8.

Rajagopalan, R., & Sahni, V. (2008). India and the great powers: Strategic imperatives, normative necessities. *South Asian Survey, 15*(1), 5–32.

Reddy, V. S. (2017). Exploring space as an instrument in India's foreign policy and diplomacy. In R. P. Rajagopalan & P. Narayan (Eds.), *Space India 2.0 commerce, policy, security and governance perspectives* (pp. 165–176). New Delhi: Observer Research Foundation.

Reuters. (2016, January 25). *India to build satellite tracking station in Vietnam that offers eye on China*. Retrieved from *Reuters*. http://in.reuters.com/article/india-vietnam-satellite-china-idINKCN0V309W

Sachdeva, G. S. (2016). Space doctrine of India. *Astropolitics: The International Journal of Space Politics and Policy, 14*(2–3), 104–119.

Samson, V. (2010, May 10). India' missile defense/ASAT nexus. Retrieved from *The Space Review*. http://www.thespacereview.com/article/1621/1

Samson, V., & Bharath, G. (2011). Introduction. *India Review, 10*(4), 351–353.

Scott, D. (2012). *Handbook of India's international relations*. New York: Routledge.

Sheehan, M. (2007). *The international politics of space*. New York: Routledge.

Singh, H. (2009). *India's strategic culture: The impact of geography*. New Delhi: KW Publishers.

Space Daily. (2007, January 28). *India to set up Aerospace Defence Command*. Retrieved April 3, 2016, from *Space Daily*. http://www.spacedaily.com/reports/India_To_Set_Up_Aerospace_Defence_Command_999.html

Stepan, A., Linz, J., & Jadav, Y. (2010). The rise of the state-nation. *Journal of Democracy, 21*(3).

Tanham, G. (1992). *Indian strategic thought. An interpretative essay*. Santa Monica, CA: RAND Corporation.

Tellis, A. J. (2015). *Unity in difference. Overcoming the U.S.- India divide*. Washington, DC: Carnegie Endowment for International Peace.

Thomas, R. G. (1986). India's nuclear and space programs: Defense or development? *World Politics, 38*(2).

Unnithan, S. (2012, April 27). *India has all the building block for an anti-satellite capability*. Retrieved March 18, 2017, from *India Today*. http://indiatoday.intoday.in/story/agni-v-drdo-chief-dr-vijay-kumar-saraswat-interview/1/186248.html

Vasani, H. (2016, June 14). *India's anti-satellite weapons*. Retrieved June 15, 2016, from *The Diplomat*. http://thediplomat.com/2016/06/indias-anti-satellite-weapons/

Chapter 6
Cooperation and Competition in India's Space Relations

This chapter assesses the historical evolution and current status of India's relations with the major spacefaring nations in the context of the changing uses of space technology and foreign policy attitudes highlighted in the previous chapters. What clearly emerges is that these evolving attitudes are influencing the evolution of New Delhi's space relations, and vice versa. This is evident when considering the maturing relations with the United States or the emerging competition with China that moves in parallel to an emerging partnership with Japan, Israel and Australia in the context of Asian security dilemmas. While the Indian space programme entertains cooperative relations with over 30 countries, the focus of this chapter is on the country's space relations with the major spacefaring nations, namely, the United States, Russia, China, Japan and Europe, at both pan-European and at individual national level.

6.1 India and the Space Juggernauts

As discussed in Chap. 2, India's relations with the two "masters of all things space" date back to the beginning of the Space Age. Both Washington and Moscow have been key partners of India in its space endeavour, but New Delhi's cooperation with them proves to have been moving in diametrically opposed directions. Indeed, whereas ties with the United States progressively re-surged after two decades of geopolitical estrangement, those with Russia seem to have started witnessing a relative decline. As explained below, this is primarily the inevitable by-product of the evolved techno-political landscape of the twenty-first century and of India's changing needs and objectives with respect to its space cooperation.

© Springer International Publishing AG 2018
M. Aliberti, *India in Space: Between Utility and Geopolitics*, Studies in Space Policy 14, https://doi.org/10.1007/978-3-319-71652-7_6

6.1.1 India–US Maturing Space Relations

Early Cooperation

Indo-American cooperation in space dates back to the early 1960s. While the Soviet Union assisted India's very first satellite programme in the 1970s, in 1963, the American Nike Apache was the first ever rocket launched from TERLS. Indian scientists andengineers indeed greatly benefited from the assistance offered by NASA in regard to building up their own space programme to where it stands today. The first decades of the ISRO–NASA partnership furthermore notably included the joint SITE project using the US communication satellite ATS-6 to help develop nearly two and a half thousand villages across six states in rural areas of India through educational broadcasts.[1]

In the early 1970s, NASA's remote sensing satellite Landsat came to the benefit of India, which entered into an agreement with the United States to access and utilise data collected by the NASA satellite. A full ground facility near Hyderabad was provided by the United States. Such remote sensing technology was scarce at that time, and India's access to it is said to have laid the foundations for the IRS system, while the above-mentioned SITE is to be credited for opening doors to the realisation of the Indian communication satellite series INSAT. In fact, the entire INSAT-1 series was outsourced by ISRO and produced by Ford Aerospace in the United States. The INSAT series of satellites – the first two of which were launched by an American Delta Launch Vehicle and Challenger Shuttle – completely revolutionised Indian broadcasting, meteorology and telecommunications. The relationship even extended into the training of Indian scientists and astronauts in the mid-1980s, which ultimately was discontinued after the Challenger disaster of 1986. After roughly two decades of cooperation vital to the development of the Indian space programme, Indo-American relations stalled (Prasad 2015).

Geopolitical Estrangement

Several factors contributed to the cooling of relations between NASA and ISRO. As mentioned, the 1986 Challenger accident constituted a major blow to the US space programme and forced it to direct its focus inwards. ISRO turned to the French Ariane-3 to launch the third INSAT-1 series satellite in 1988 before returning to the American Delta 4925 rocket to deliver the fourth and final INSAT-1D into orbit in 1990. The INSAT-2 series would be, however, launched with European Ariane (Indian Space Research Organisation 2017a, b). Besides NASA's internal challenges, another factor for downward-spiralling cooperation relations was the huge gap in technological advancement between NASA and ISRO. The Americans were focussed on a fundamentally different path than India was, leading to the two nations' space partnership growing apart.

Geopolitical tension grew in a series of conflicts involving the Soviet Union, Afghanistan, the United States and Indian arch-nemesis Pakistan. When the Soviet

---

[1] It is interesting to note that India initially asked CNES to build a joint communication satellite. When the proposal did not pan out, Virkram turned to NASA, which provided access to its ATS-6, paving the way for the SITE project in 1975–1976 (Prasad 2015).

Union took military action in Afghanistan late in 1979, the United States stood with Pakistan to support Afghan rebel forces against Soviet insurgence. India's leader Indira Gandhi fostered a closer relationship with the Soviet Union during these days, in part resulting in the first Indian traveling to outer space as part of a Soviet-led mission in 1984 (Moltz 2012).

Additionally, the United States grew worried about their facilitating – or anybody else for that matter facilitating – the Indians' acquisition of technology that could be used for military and especially for nuclear weapon delivery purposes. A more powerful propulsion system certainly had vast civil and commercial applications as it could have enabled payload launches straight into GEO and thus eliminate dependence on foreign launch services. However, it could also have been used to develop ICBM's and thus would have presented a dangerously perceived shift from civil to military application of technology. In fact, India is, as said, one of the few nations that shifted technologies from civil applications to military applications rather than the other way around as most did and do. In light of concerns regarding the proliferation of ballistic missile technology, the Missile Technology Control Regime (MTCR) was established in 1987 by the United States and six other Western nations, binding the now 35 member countries to certain restrictions in regard to missiles and unmanned aerial vehicle development. India was the 35th member to join just last year, 30 years after the genesis of the MTCR (Missile Technology Control Regime, n.d.).

The soon-to-be Russian Federation was to provide India with cryogenic propulsion technology knowledge in the early 1990s, a transfer that did not in fact occur as the United States intervened. The United States and other Western nations refrained from cooperating with ISRO as they feared dire consequences resulting from an Indian acquisition of such technology. To this day, India is in pursuit of a more powerful launch vehicle system than the existing GSLV and PSLV models, a feat they could have achieved much sooner had there not been a blockade of technology transfer and discontinuance of cooperation led by the United States in the 1980s and 1990s. In the near future, GSLV Mk III is expected to be that vehicle for ISRO, enabling a substantial increase in India's single-launch capacity once the technical aspects are finalised (Prasad 2015).

India, in the face of persistent absence of cooperation with the United States, had been relying on the other global space superpower, the USSR, for space cooperation until this synergy was interrupted by the 1991 collapse of the Soviet Union. This occurred around the same time as the cryogenic engine technology exchange between the Soviet Union and India was foiled by US intervention and their sanctioning of Russian agency Glavkosmos, which was to transfer the propulsion technology. It is unclear exactly if or how much technology was in fact transferred to India before the deal between ISRO and Glavkosmos was revised to include the complete products without the debated cryogenic engine technology transfer amidst US pressure.

The Indo-American relationship did not improve much from the stagnant relationship of the 1980s, throughout the Carter, Reagan and eventually the Clinton administrations. President Clinton's hard-line non-proliferation agenda found

India – which to this day has not signed the non-proliferation treaty – in its cross-hairs. Commercial cooperation did however exist between ISRO and the US private company EOSAT, which agreed to market Indian remote sensing products world-wide as India grew increasingly keen to engage in international trade following the liberalisation of their economy in the 1990s and of course the development of indig-enous technology. It seems in analysing the long string of events throughout the 1980s and 1990s – as well as in several other situations that shaped the fundamental nature of India's space programme advancements or lack thereof – geopolitical fac-tors played a paramount role in determining the outcome for ISRO (Moltz 2012).

A New Beginning
After the turn of the millennium, geopolitics continued to strongly influence India–US space relations, but this time in a much more positive manner. Following the terrorists' attacks of September 2001, the administration of George W. Bush dra-matically changed US policy on India. After all, both nations shared foundational democratic values and faced the shared threat of terrorism in ways they had never witnessed before. Indian PM Vajpayee visited the United States the same year of the attacks and returned to India with brand new plans for Indo-American cooperation, including space cooperation. A clear primary concern of the United States for decades was the global threat of nuclear weapon proliferation, until the new century brought upon new challenges that, among a plethora of other changes, shifted America's approach in the global geopolitical arena. More specifically, the Indo-US High-Technology Group was founded in 2002, followed by the Next Steps in Strategic Partnership (NSSP) agreement 2 years later which included nuclear energy partnership and new, revived Indo-American cooperation in space (Correll 2006).

India and the United States blended well thereafter, seeing nearly an 80% increase in trade volume between the two nations in the first 8 years of the new mil-lennium and adding 1,6 million employees to the then $60 billion strong Indian IT sector which made up 6% of India's GDP. The two space programmes engaged in a partnership that would enable the United States to add two American payloads to the Chayandraan-1 lunar mission in 2008. Transitioning back to a fruitful space partnership after two decades of estrangement did not come without complications of course, as US export regulations limited possible exchanges before separate agreements cleared the barriers. Furthermore, the Indian space programme of the new millennium was quite different from the past in that it manifested interest in carrying out space science and exploration missions alongside the historically strong focus on civil space applications. The Chandrayaan-1 lunar probe was ini-tially successful in data collection before the spacecraft crashed onto the moon's surface in 2008, thus falling short of the half-way mark of the mission's intended duration and becoming a testament to ISRO's youthfulness in such kinds of endeav-ours (Verma 2016). Nonetheless, India for once had a taste of what such missions can do for the country and space programme in regard to acquiring clout in the scientific and geopolitical global community. In addition, Indo-US cooperation on Chandrayaan-1 triggered the signature of an enlarged cooperation framework agree-ment between NASA and ISRO in 2008. The concomitant placing of India among the group of closest US allies for export control opened a possibility to accelerate ITAR-related procedures.

It is also worth mentioning, however, that ISRO–NASA cooperation on *Chandrayaan* lunar mission raised some apprehensions in the United States about the possible transfer of implicit knowledge skills in the form of payload integration assistance that would benefit India's military in terms, for instance, of multiple warhead integration capacity (Bommakanti 2009).

Recent and Future Collaboration
Be this as it may, under President Barack Obama, export control regulations were further eased, resulting in some of the main space centres of ISRO no longer needing licenses to engage in exchanges with US entities. Removing such hurdles in a bilateral relationship is key for enabling exchange not only functionally but also diplomatically by instilling trust in the counterpart. In a 2010 India visit, President Obama and the Indian side also held talks regarding future cooperation – in particular in human spaceflight, space science and commercial launches of US payloads on Indian launchers (Moltz 2012). Commercial launches did indeed take place, most notably so in early 2017 when ISRO's PSLV launched 96 American satellites as part of a record-breaking single launch of 104 satellites (Indian Space Research Organisation 2017a, b). All of India's commercial launch services in the 3 years leading up to 2016 are claimed to have earned India a revenue total of $86 million.

As one of the results of the India–US Civil Space Joint Working Group – established in 2005 with the goal of creating a dedicated platform for discussions on future cooperation – ISRO and NASA joined forces in 2014 to begin preparing for the 2021-aimed NASA–ISRO Synthetic Aperture Radar (NISAR) mission, to be launched using India's GSLV. NISAR's remote sensing application is expected to yield better understanding of natural occurrences involving Earth such as natural disasters and environmental changes (National Aeronautics Space Administration 2016). Furthermore, the Jet Propulsion Laboratory at the California Institute of Technology – which will also cooperate with India in the NISAR mission – played a key role in assisting India's MOM, which almost parallel with the American MAVEN mission arrived in Martian orbit in 2014.

Other collaborative areas include the Laser Interferometer Gravitational-Wave Observatory (LIGO), where India and the United States collaborate on a multilateral platform for scientific advancement, astronautic partnership, plans to realise manned travel to Mars in the upcoming decades and traveling even deeper into space with the aim of exploring our solar system and other untapped territory. Cooperation on navigation satellite technology between India and the United States is comprised of exchange of satellite technology – which in part greatly benefits India's GAGAN – and closer collaboration to make IRNSS and GPS more compatible with each other by using the same signal for civil applications. In 2016 a memorandum of understanding concerning the sharing of EO data was signed by the United States and India, which will allow for an exchange of information gathered through the United States' LANDSAT-8 and India's RESOURCESAT-2. Over the last years several working groups were also formed to enhance bilateral communications. The United States furthermore has embraced India as a major defence ally in a joint statement released in June 2016. Such a close relationship on defence and security matters has not been attained between Europe and India, by contrast.

Besides the other manifold cooperative efforts between NASA and ISRO, one key area that is perhaps overlooked is the immense value the United States could bring India in its pursuit to involve private industry in space activity. The commercial space industry – as pioneered and exemplified by the United States – will inevitably grow more prevalent as time passes and is thus an unavoidable reality space programme ought to pay close attention to in the imminent future. Aligning themselves with the United States in this regard could be a solid strategic move for the future of Indian space endeavours (Moltz 2012).

6.1.2 A Slowly Declining Relationship with Russia?

To say that Russia – or the former Soviet Union – played an important role in the advancement of the Indian space programme would be an understatement. The launch of India's very first satellite – the Aryabhata – in the mid-1970s was carried out using a Soviet Kosmos launch vehicle, while almost a decade later a Soviet-led Intercosmos mission put the first, and to this day only, Indian citizen in space. The quasi-alliance between two of Asia's most influential powers started in 1962, with an agreement for launching Soviet rockets from Thumba, but took full shape in the 1970s.

Soviet Guidance in Space
In the early days of Indo-Russian space cooperation, the USSR's lightweight M-100 sounding rocket was utilised for meteorological observation missions and launched from India's Thumba Equatorial Rocket Launching Station. TERLS would wind up as the number one foreign launch pad for the Soviet Union. India greatly benefitted from the meteorological data gathered, which improved understanding of monsoons. Studying natural phenomena such as monsoons, which directly impact agriculture and thus the economic standing of many Indians, was and is key to the Indian space programme, well in line with ISRO's long-held focus on practical applications and improvements of life quality on Indian soil.

Placing launch facilities in foreign countries was more than a mere outsourcing of part of the Soviet space programme; in fact the two global superpowers and rivals at the time – the Soviet Union and the United States – undoubtedly did so at least in part to strengthen political alliances and spread their space dominance by taking allies under their wings. As was common procedure between the two cold war antagonists, the USSR attempted to one-up the US outsourcing of satellite launches by creating the Interkosmos and Intersputnik programmes in 1970, thus tying together nine countries east of the iron curtain in cooperation on the Soviet Union's own satellite launches, designs and operations.

With such close Soviet partnership and guidance, India put its first indigenous national satellite into orbit in 1975. The Indian part in this collaboration was the production of the satellite and its technology, while the Soviet Union made sure to consult their Indian partners and handle the launch as well as vital related systems India had not yet mastered. The partnership as well as the successful launch and placement of the satellite in orbit was clearly at the forefront of this Aryabhata satellite project, as opposed to the functional benefit of this satellite technology which was a rather simple collection of radiation data from the Earth's ionosphere. While signals from the Aryabhata were lost just days after its successful launch, it set the stage for the next Indian satellites Bhaskara I, Bhaskara II and Rohini satellites, some of which were placed in orbit by India's first indigenous Satellite Launch Vehicle SLV-3.

Following the same cooperation pattern used with satellites, the Soviet Union also made it possible for non-Soviets from allied nations to partake in manned spaceflight, thus expanding their brand beyond USSR borders and asserting their uncontested human spaceflight dominance. In this way the first and to this day only Indian cosmonaut Rakesh Sharma travelled to space aboard the Soviet Soyuz T-11 in 1984.

Many firsts for the Indian space programme were evidently facilitated by the Soviet Union. This prolific partnership however would not last forever. Indian advancement towards self-sufficiency in space as well as the dissolution of the economically plagued Soviet Union and resulting programme cuts led to weaker space cooperation between India and Russia in the 1990s and 2000s. Apart from some collaboration utilising Russian launch technology when Indian capacity had not yet reached the sufficient level to carry out particular missions, the partnership turned rather cold.

One geopolitical spat involving Russia and India in particular stands out in regard to space. During the last years of existence of the Soviet Union, India sought to clear a vital hurdle that would allow them to independently launch into GSO: a more powerful launch vehicle engine. As of today, India seems to be putting the finishing touches on its own cryogenic engine technology for the next breakthrough launch vehicle, the GSLV Mk. III. Initially, Russia was to provide India with the technological know-how it desired; however, the United States saw potential for misuse of such technology in connection with India's nuclear weapon arsenal and tensions with neighbouring Pakistan, which also had a nuclear programme. Upon American intervention, Russia only provided the finished product to India rather than technological insight that would have allowed India build its own cryogenic engines and heavy launch vehicle. In retrospect, this US blockade of the initial Indo-Russian technology transfer contract of 1992 effectively costed ISRO a quarter century of technological advancement (as ISRO introduced the GSLV Mk. III launcher equipped with an Indian-built cryogenic engine in only 2017) and also created serious burdens on the previously close ties between Roscosmos and ISRO.

Recent Cooperation

Relations between the two respective space programmes seemed to recover momentum in the course of the 2000s, greatly supported by the solidification of the broader political relations that followed the signature of a strategic partnership agreement by Russian President Putin and Indian Prime Minister Vajpayee in 2000. Groundbreaking cooperation plans were envisaged by the respective space agencies in three key domains, namely satellite navigation, space exploration, and human spaceflight. Because these areas were brand new domains for India, the envisaged cooperation even fed the impression that Roscosmos could become ISRO's most dependable partner in the twenty-first century. However, all these attempts to cooperate ended rather disappointingly. In the area of navigation, India and Russia signed several cooperation agreements on GLONASS between 2005 and 2008, envisaging Indian launches of GLONASS-M satellites onboard the GSLV, as well as joint development of the future GLONASS-K satellites and of users' equipment (Mathieu 2008). Through these agreements, India sought to gain valuable expertise in the area of satellite navigation as well as preferential access to data, while Russia intended to reduce the financial burners associated with the redeployment of GLONASS and to make India's augmentation system reliant on GLONASS rather than the American GPS. The agreements, however, failed to materialised, with Russia eventually deciding to launch GLONASS satellites on its own. The envisaged partnership on space science and exploration turned to be equally ill-fated. In addition to the utilisation of Russian ground stations to support India's first exploration mission to the Moon, cooperation in this field foresaw the launch of an Indian payload on-board the Russian Coronas-Photon mission in 2009 and a Russian participation in ISRO's second lunar mission Chandrayaan-2. In the first case, IRSO provided RT-2 gamma-telescope to investigate the processes of free energy accumulation in the Sun's atmosphere, but the mission failed less than one year after the launch. As for Chrandrayaan-2 lunar rover mission, the Russian participation consisted in providing the landing module building on the state-of-the art technologies developed for the Fobos-Grunt mission. However, as detailed by Vladimir Korovkin, "the Russian party did not provide the landing module in time, rescheduling the delivery first for 2013 and then for 2016. [Although some] cited the failure of Fobos-Grunt mission in 2011 to be part of the reason, though the mission failed due to an unsuccessful launch and never managed to test the landing device" (Korovkin 2017). Ultimately, in 2015 India decided to reschedule the mission in 2018 but to follow through without Russian participation, and so continued on its increasingly autonomous path with regards to its space programme.

A third area of cooperation that Russia and India envisaged during the 2000s was human spaceflight. Following a working group on space cooperation with Roscosmos, an agreement was reached between the two countries in December 2008 for the Indian manned spacecraft to be built following the trusted Russian Soyuz design, thus echoing the path of China. In that context, India also considered sending one of its citizens into space on board a Russian spacecraft to acquire the skills necessary for future manned space missions, and expressed interest in participating in the development of a new Russian manned spacecraft. Also in this case, however, the plans never materialised. In part because of this failed cooperation and in part because of the much-delayed development of the GSLV-Mk III (a delay that can be once again motivated by the missed transfer of cryogenic engine technology by Russia in the 1990s), the human spaceflight programme in India had to be postponed indefinitely.

From an overall perspective, it is safe to argue that all these disappointing cooperation experiences have certainly contributed to a diminished interest in closer cooperation, at least from the Indian perspective. Not surprisingly, little interaction has taken place over the past few years between Roscosmos and ISRO. Remarkably, not only have the two partners grown apart, but they also started to compete in the commercial arena, particularly in the launch service market. Today, India is considered one of the most potentially prolific commercial launch providers, as proven by its 2017 single launch of 104 mostly foreign satellites, a situation that now contributes to threaten Russia's established position in this market. While the defence and energy ties between Russia and India remain strong, especially so in light of Western sanctions against Russia, space cooperation is on a decline in comparison to the close partnership of the past.[2] India has become a space power of its own and as outlined has in many ways grown self-sufficient. This, however, cannot be taken to imply that there are no longer avenues for cooperation between Moscow and New Delhi. Indeed, some space-related cooperation remains, for instance viable in connection with defence industry, an area where Russia can still leverage its valuable expertise to engage in mutually beneficial undertakings with India.

In an effort to boost India's roughly $330 million defence exports, BrahMos Aerospace has been chosen to be India's commercial arm for Indian defence products sold abroad. BrahMos is partly owned by the Indian Defence Research Development Organization (DRDO) and partly by the Russian NPO Mashinostroyeniya, the former of which retains a majority share with 50.5% equity. The idea of having a separate entity – namely, BrahMos that handles international commercial activity for the DRDO – came as a result of increased global demand for Indian products after a series of expositions and aims to skyrocket Indian defence exports to US$2 billion in 2 years. Gaining significant market share over the medium to long-term however may prove difficult considering the dominance of well-established industry giants (Sputnik News 2017). Just days after news of the new-found partnership broke, India successfully test launched the BrahMos Extended Range supersonic missile off its eastern coast. Considered a milestone step for Indian defence, it now extends India's striking capability well beyond its current 180 miles (around 290 km). Complete with a guidance and stealth technology, the two-stage missile reached 2.8 Mach during the test flight. Touting BrahMos' as the global pinnacle of supersonic cruise missile system providers, great optimism and confidence seems to be spreading throughout the Indian defence industry in light of recent achievements (Maass 2017).

Just a month prior to the announcement of the BrahMos partnership, in February 2017 India and Russia agreed to fortify Indian defence capabilities through a GLONASS ground station in India. The increased navigation accuracy and coverage means better wartime command for Indian forces and secondarily brings with it civil applications. The vast Russian defence technology used by Indian armed forces

---

[2] In 2016, several energy and defence deals worth billions of dollars were closed at the 2016 annual Indo-Russian summit (Busvine and Pinchuk 2016). Closer alliance with influential Asian partners such as China, Turkey and India were vital to the health of the Russian economy at the time and therefore somewhat inevitable given the sanctions of Western economic powers on Russia in connection to the annexation of Crimea in 2014.

is compatible with GLONASS, thus making this a sensible move from a technological perspective. India has its own navigation satellite system IRNSS; however, this is a regional system as opposed to the global systems GLONASS and GPS. Perhaps it is not too far-fetched to consider that this improved navigation capability by the addition of GLONASS signal to the already operational IRNSS was a key move to bolster technological capabilities in preparation for the partnership with supersonic cruise missile company BrahMos.

6.2 India and the Two Asian Space Giants

This section provides a reflection on India's past and current space relations with the other two major spacefaring nations of Asia, namely, Japan and China.[3] The analysis is embedded in the broader context of the evolving regional dynamics and diversification of the Indian space programme discussed throughout Chap. 5.

6.2.1 Lack of Cooperation or Confrontation?

One of the most striking features of Indian space relations is the few cooperative efforts that have been taking place within the Asian context during the past 50 years. While this can be in part motivated with the limited space capacities of the large majority of Asian states, the limited intra-Asian space cooperation was more importantly connected with the wider political dynamics in the region, which can be described as having been problematic at best. The region is still haunted by the remnants of post-World War II early dynamics: the Korean and Taiwan issues, the separation between India and Pakistan, the numerous territorial disputes and the persistence of the US "hub and spokes" system, which for a long time has inhibited the development of multilateral relations. These enduring historical divisions and geopolitical rivalries still represent a potential threat and have generally prevented the unfolding of cooperative undertakings in space (Aliberti 2013).

During India's early efforts in space, an important exception was its cooperation with Japan, the most advanced Asian nation at the time. Hideo Hitokawa, the father of the Japanese space programme, was in fact employed as an adviser to the Indian space programme, on the invitation of Prime Minister Indira Gandhi (Prasad 2015). He also hosted Indian scientists at the Institute of Space and Aeronautical Science (ISAS), University of Tokyo, to gain first-hand knowledge on sounding rockets and solid propulsion technology. And it was Itokawa who proposed to Sarabhai the name of Rohini for India's first satellites (Harvey et al. 2010).[4] Yet, despite these

---

[3] This section draws on the research conducted for the book "When China Goes to the Moon…" (Aliberti 2015).

[4] Interestingly, Hideo Itokawa wrote of this period in his book "Third Road – India, Japan and Entropy", which was published in 1994.

early connections between the two countries, in subsequent years India and Japan did not cross paths for cooperation in space. As explained by Kazuto Suzuki, this was mainly because of the huge gap between the philosophy of space development of India and Japan. While the former was conducting activities under the tenet of societal development, the latter was pursuing a strategy of catching-up to achieve first-class status with the United States.

What is perhaps more remarkable than this lack of cooperation, however, is that according to several analysts Tokyo and New Delhi have in recent years started to compete against each other in the space arena. Admittedly, this competition is not seen as a direct confrontation, but the indirect – yet inevitable – by-product of their respective reactions to China's rise in space, which would have eventually ignited an "intra-Asian space race" (Moltz 2012; Lele 2013).

The idea of an Asian space race has gained currency in the past decade, prompted by the reorientations that China's space achievements have caused in the space policy posture of several space powers, including India and Japan.

In Japan, just 1 week after the Shenzhou-5 mission, on 22 October 2003, the Ministry of Education, Science and Technology set up a commission to review Japanese space aims for the following 20 years, including long-term participation in the ISS and manned spaceflight (Sheehan 2007). The conclusions of the commission's review were issued in a policy proposal, released by Japan's Aerospace Exploration Agency (JAXA) in April 2005. Entitled *JAXA Vision 2025*, the document set forth ambitious plans for an autonomous manned spaceflight programme. Notably, it also included a long-term plan for a human landing on the Moon by 2025, a task to be achieved in close collaboration with NASA (Japan Aerospace Exploration Agency 2005).

The following year (in November 2006), ISRO, which in 2003 had already set up several study groups on space exploration, also produced a formal proposal for the development of a human spaceflight programme. The proposal was presented to Indian Prime Minister Mamohan Singh on 17 October 2006, and, on the latter's advice, it was submitted by ISRO Chairman Gopalan Madhavan Nair to a cross section of the scientific community that met in a brainstorming session in Bangalore on 7 November 2006 (Peter 2008). The conference agreed to immediately initiate a human spaceflight programme and to autonomously launch its first manned flight by 2014 and land an Indian astronaut on the Moon by 2020 (Jayaraman 2006). The decision represented a major change in Indian space policy which, since its inception, had declined any potential involvement in human spaceflight. As many analysts have argued, although India would insist that its "manned programme was a logical next step for the Indian space programme, and not a reaction to the emergence of the Chinese manned programme, the shift in policy was so dramatic in comparison to the almost ideological opposition to manned flights, that the timing seems hardly coincidental" (Sheehan 2007).

Between 2007 and 2008, with the almost simultaneous launch of the Kaguya-1, Chang'e-1 and Chandrayaan-1 lunar orbiters by Japan, China and India, it appeared that the Asian space race had already started. The proximity of the launching dates and the nature of the missions fed the impression that the three countries were competing against each other, a competition that would later extend to manned space capabilities and, quite likely, end up in a space arms race. It should, however, be acknowledged that the evolution of the policy postures of Japan and India demonstrate that the recent and current dynamics cannot be truly understood as a space race (Aliberti 2015).

6.2.2   Human Spaceflight and Exploration Dynamics

In the field of space exploration and human spaceflight, the dynamics of the past 10 years have drastically downsized the prospect of a one-upmanship contest in space. If it is true that Japan, China and India each launched almost concurrent lunar probes between 2007 and 2008 and that both JAXA and ISRO announced ambitious plans for robotic and human space exploration, more or less openly, intended to counterbalance China's rise, very little has, in fact, followed.

In Japan, the lunar exploration programme has encountered political setbacks that have compromised its steady implementation.[5] Even more ill-fated was the proposal for the implementation of an autonomous human spaceflight programme contained in JAXA's *Vision 2025*. In spite of the mounting excitement among the Japanese public generated by the release of the document, as early as September 2008 (i.e. a few days after China's first spacewalk) Kawamura Takeo, then chief secretary of the Japanese Cabinet and a leading figure within Japanese space policy planning, declared that no Japanese autonomous manned space programme was in sight, beyond the long-term involvement in the ISS programme (Kyodo News Service 2008). The decision to refrain from manned spaceflight was clearly carried out in the space budget allocations of the following years. Since 2008, the Space Policy Commission has given the lowest priority to manned spaceflight in the budget plan, limiting funds to the completion flight of the H-II Transfer Vehicle (HTV) and to the full utilisation of the Japanese experimental module KIBO onboard the ISS (Suzuki 2013).

As for India, although the manned spaceflight programme initially received stronger political backing, and initial funding began in April 2007 (Peter 2009), also in this case early expectations had to be reconciled with reality, as the financial commitment to human spaceflight has remained rather limited. In addition, cooperation with Russia in the development of a Soyuz-based manned capsule has not gone as planned (see Sect. 6.1.2), while the development of the enhanced GSLV Mark III launch vehicle encountered considerable delays. Probably as a result of these adverse developments, in August 2013 ISRO Chairman Koppilil Radhakrishnan

---

[5] In spite of the resounding success of the Kaguya-1 mission, policy-makers have in fact provided little political backing and financial commitment to the programme. Because of the climate of severe budgetary pressures affecting Japan's economy since 2007 on the one side, and the termination of U.S. plans on the other, the follow-on Kaguya-2 moon lander scheduled for launch in 2010 and tactically intended to anticipate the 2013 Chinese Chang'e-3 mission, was suspended and has not yet received the final "green light" from the government. The mission is now scheduled for 2018, i.e. even after China's Chang'e-5 sample return mission.

announced that a human spaceflight mission was not an ISRO priority (NDTV 2013),[6] a position that had already been clear since the release of the 12th Five-Year Plan (2012–2017), in which human spaceflight did not figure among the list of projects to be implemented over the period.

Although in his announcement Mr. Radhakrishnan was keen to emphasise that the country would not be starting from scratch should a decision for human spaceflight eventually be taken, he avoided stating a definite time frame for the potential launch, limiting himself to specifying: "We are not going to see the human spaceflight as a programme in the 12th Five-Year Plan. *We will see maybe later*" (NDTV 2013). As a result, it is now very difficult to make the case for one-upmanship with China. Even in the event of an abrupt volte-face in the current Indian space agenda, it can be projected that the country could make its first steps only by the end of the current decade or early in the next. By that time China will already have started to assemble the core module of its space station in orbit and will likely have mastered the crucial technologies to reach the ultimate target of a moon landing (Aliberti 2015).

Fully aware of its short-term inability to compete directly with China in this arena, India appeared to have been looking for alternative paths to gain international prestige, including a much-heralded reorientation of its space programme towards Mars exploration, which gave it the opportunity of achieving a "space first", at least in comparison to China. The launch of a Mars Orbiter Mission in November 2013 could be seen in this light (Lele 2014).

All in all, although the inclusion of exploration in India's space portfolio suggests that prestige considerations have become an important factor, they have not yet come to dominate the development rationale of the space programme to ignite a possible competition with China in the field of space exploration. An examination of the priorities and goals listed in the 12th Five-Year Plan shows a sharp contrast between the great attention devoted to the "societal use of space" and the scant attention paid to *grandeur* projects.

6.2.3   Military Space Competition?

Together with space exploration, a second field of potential competition between India, Japan and China is that of military space. Over the past 10 years, China's fast and substantial build-up in military space capabilities has certainly changed the strategic balance in the Asian chessboard, urging both Tokyo and New Delhi to respond, but also in this arena, it remains difficult to make the case for a space arms race between the three Asian giants.

As for Japan, although its 2008 Basic Space Law made remarkable openings to the military uses of space, Japan's Ministry of Defence has made it clear that it is not

---

[6]The position was also clarified in a personal interview with Ajej Lele, where he stressed, "As of today, a human mission is not in our space agenda. We are in a very early phase of developing a few critical technologies required for realising a human mission". Quoted from: (Lele 2014).

willing to invest much in military space capabilities – let alone compete with China in a space arms race (Suzuki 2013).[7] Manifestly unwilling – and unable – to keep pace in an arms race with China, given the resource difficulty in competing with a giant and the inherent anti-militarism of the Japanese population, the country is trying to gain strategic advantage by forging more or less formal mechanisms to contain the expansion of China's power. These manoeuvres have included an upgrading of the US–Japan security alliance and the proposal for the so-called Asia Democratic Security Diamond, an informal alliance of Asia-Pacific democracies (Japan, India, Australia and the United States) intended to hedge against China's power projection.[8] Regardless of the result of these "earthly manoeuvres", it appears highly unlikely that space will become an arena of military confrontation between Japan and China.

As for India, it is undeniable that since the ill-famed Chinese ASAT test of 2007, the country has expanded its interest in the development of both defensive and offensive space capacities. India's armed force clearly maintain that China's military space capabilities pose a main threat to the country and have been urging the swift development of robust retaliatory capabilities to protect its space assets through deterrence.

Although the open recognition of Indian efforts to develop ASAT capabilities made by DRDO Director General V.K. Saraswat in January 2010 (Brown 2010) fed the general impression that India did not want to lose strategic ground vis-à-vis China's defence build-up and that the prospect of a military space race was hence just a matter of time, the reality is that the armed forces have continued to suffer from a lack of strong governmental support for their efforts (see Sect. 5.3). This position can be explained with the fact that Indian policy-makers are extremely wary of deterring China's military space rise in ways that they would ultimately be difficult to support. Concretely, the disquiet of Indian policy-makers is that a possible green light to the development of offensive space capabilities may inadvertently create the exact thing that they want to avoid: a military escalation and possible conflict in or about space with China.

---

[7] There are multiple reasons for this reluctance. First, anti-militarism remains a deep-rooted attitude within a number of Japan's parliamentary factions and within the nation as a whole. Second, considering that the government is confronted with important budgetary constraints that affects the defence budget, "there is no luxury to increase spending for the unfamiliar domain of space" (Suzuki 2013). Furthermore, the MoD has little experience and almost no staff or technical expertise in space technology. This makes it dependent on JAXA for expertise. As argued by Kazuto Suzuki, "Given the secretive nature of the MoD, it would not be acceptable to depend on a civilian agency to develop military sensitive technology. So instead of relying on JAXA the MoD has to date chosen not to invest so much in space" (Ibid). Finally, many analysts have noted the importance of Japan's alliance with the United States and argued that these ties mean that Japan does not have to become a military space power (and compete in an arms race in space) in order to maintain its security. See Suzuki (2013) and Aliberti (2015).

[8] This strategic initiative was launched already during Mr. Abe's first tenure as prime minister (September 2006–September 2007) and named the Quadrilateral Security Dialogue (Japan, the United States, India and Australia) as a solution to the maritime disputes involving China. The new strategy is based on "three pillars: (1) reinvigorating the US–Japan alliance, (2) a reintroduction of the United Kingdom and France to Asia's international security realm, and (3) bolstering international cooperation between key democracies in the Indo-Pacific, such as India and Australia" (Miller 2013).

There are, first of all, structural impediments that work against the case for a military space race. For New Delhi's politicians, it is evident that China's "comprehensive national power" exceeds India's by a very large margin. With a GDP that is one-fifth that of China's, as well as a military and space budgets that are a tiny portion of the Chinese ones, fomenting a space arms race is the least of the strategically desirable options for New Delhi. Indian leaders seem to be fully aware that for the immediate future China will remain a distant peer, and this gap compels India to keep a prudent approach vis-à-vis China, gradually and methodically building up its military capabilities but in the meantime carefully postponing the strategic contest for another day (Aliberti 2015).

In addition to these strategic considerations, it should be noted that current Sino-Indian relations give rise not just to pessimism for the future. While their interplay remains enigmatic at best, both New Delhi and Beijing have in recent years adopted a more practical and positive stance towards each other. After all, as the leaders of the two countries have repeatedly affirmed, Asia is vast enough to allow the peaceful rise of both China and India, and it is in their interests to maintain a stable environment and a cooperative relationship. Today China and India are more "politically and economically engaged than at any time in recent history" (Smith 2014).

Beside the ever-increasing economic exchanges and a maturing political dialogue, the Chinese and Indian armies have also started periodically to hold joint military exercises (they held their first joint exercises in 2007; these were followed by two more in 2008 and 2013), thus significantly expanding the level of cooperation (Smith 2014). It is also significant that India has joined the Shanghai Cooperation Organisation (SCO) as full member in June 2017 and has continued to gently decline Tokyo's diplomatic offer of the formation of a Democratic Security Diamond against China (Global Times 2014). Indeed, although willing to increase its cooperative relations with Tokyo as well as to gain strategic advantage vis-à-vis China, New Delhi has been visibly reluctant to take part in initiatives that would limit the freedom of manoeuvre of its diplomacy and generate undesirable rifts in its relationship with Beijing.

6.2.4   Diplomatic and Commercial Rivalry

A third issue area of regional competition between the three Asian giants has been seen in the provision of international space services or public goods, with the aim of extending their political influence over the region. In this respect, whereas Japan and China are indeed competing through regional cooperative initiatives such as the Japanese-led Asia-Pacific Regional Space Agency Forum (APRSAF) and the Chinese-led Asia-Pacific Space Cooperation Organisation (APSCO) or the provision of launch package agreements to developing countries, India has been apparently standing aside because of its weaker position. There is no Indian-led regional organisation that can be compared to APRSAF or APSCO. Admittedly, ISRO hosts the headquarters of the Centre for Space Science and Technology Education in Asia and the Pacific (CSSTEAP), which was established in 1995 in response to a resolution of the UN General Assembly recommending the creation of centres for space education in developing countries. The CSSTEAP has limited scope, however, and less ambitious targets than APSCO and APRSAF. Hence, it is unlikely to play a major role in giving political shape to cooperative undertakings in the region. CSSTEAP's core objectives are in fact limited to educational activities in space science and technology, a task that figures only among the marginal activities of APSCO and APRSAF.[9] It is also worth mentioning that India has been cooperating with Japan in the Sentinel Asia project for disaster prevention within the framework of APRAF (Horikawa 2017).

Although India's position remains still weak in comparison to Japan and China, some recent initiatives, such as the construction and launch of a communication satellite for South Asian nations discussed in Sect. 5.2.2, suggest that India's space diplomacy is progressively acquiring a more ambitious profile and that it will seek to reach a dominant position in the South Asian context, where it already exercises more influence as compared to China and Japan.

What is also interesting to note is that India is moving ahead in the commercial space market, particularly the launch sector, an area where it has some competitive advantages vis-à-vis Japan, whose services are expensive, and China, still strongly affected by the ITAR regime.[10] Remarkably, in February 2017, India's PSLV orbited an impressive 104 satellites in a single launch – a world record that further enhanced India's image as the leading Asian power in the commercial space markets. Competition in this arena, however, is subject to global, rather than regional, dynamics and cannot therefore be labelled a space race. If we think of it as a race, then, in the words of Ayej Lele, "it is a form of global space race, where every state is looking for a share".[11]

All in all, whereas regional space dynamics does not seem particularly promising from a cooperation perspective, the narrative of an intra-Asian space race does not prove particularly enlightening. Chinese efforts (be them in the area of civil or military space activities) clearly have global, rather than regional, targets (Aliberti 2015). It appears also evident that India does not aim to directly challenge China, let alone Japan, with head-to-head endeavours, but rather achieve an increased influence in the regional context through cooperative space undertakings and the provision of services. In addition, the current consolidation of India–Japan political ties, the expansion of India's space capabilities and the broadening of Japan's space policy are seen to be opening up important cooperation opportunities between IRSO and JAXA in such fields as Earth observation, satellite navigation, space science

---

[9] The main task of the Centre is to develop the skills and knowledge of university educators, environmental research scientists and project personnel in the design, development and application of space science and technology for subsequent application in national and regional development and environment management.

[10] While also India has in the past been subject to ITAR restrictions, since the early 2000s these measures have been progressively removed. Further in 2016, India and the United States reached an understanding under which India would receive license-free access to a wide range of dual-use technologies (Samson 2017).

[11] Quoted from Moltz (2012)

and planetary exploration (Horikawa 2017; Suzuki 2017). It remains to be seen whether such closer cooperation prospects will also extend in the security field and transform into a fully-fledge partnership to balance the power of China.

6.3 Relations with Europe: A Tale of Missed Opportunities?

As in the case of the United States and Russia, there is a long-standing record of cooperation between India and a range of European stakeholders, be it at national or pan-European level. While such cooperation has in many cases been key for the subsequent emergence of India in the space arena, it appears to have been insufficiently leveraged in a coordinated fashion on the European side, primarily due to the fragmented nature of European governance and the feeble connections established between its space policy and foreign policy. With the sole exception of France, and more recently Germany, cooperation efforts have been driven by programmatic needs rather than general foreign policy objectives.

6.3.1 Europe–India Space Cooperation at Pan-European Level

All the three major pan-European institutions involved in space matters, namely, the European Space Agency (ESA), the European Organisation for the Exploitation of Meteorological Satellites (EUMETSAT) and the EU, have entertained cooperative relations with India. While in many respects such cooperation has proved highly beneficial and reliable, as explained below, it is undeniable that it has not been ground-breaking especially if compared to other cooperative ventures both sides have developed with third countries.

ESA–ISRO Erratic Cooperation
The European Space Agency, as the primary European *mechanism* and *actor* of international space cooperation, has a 40-year-long relation with its Indian counterpart. The first cooperation agreement with ISRO was signed in 1978, following the opening consultation meetings in 1977 (Bigot 2017).

The first major milestone in the development of cooperative relations between the two organisations was achieved when, among 72 competing proposals, ESA selected a project from ISRO to develop and provide an experimental telecommunication satellite for the third qualification flight of Ariane-1 in July 1981 (Vasagam 2015). Developed in just 2 years also thanks to ESA assistance as a sandwich passenger-carrying Meteosat-2 on top and the Capsule Ariane Technologique (CAT) module below, the Ariane Passenger Payload Experiment (APPLE) was India's first telecommunication satellite. Besides giving ISRO valuable hands-on experience in "the design, fabrication, launch and post launch mission manoeuvres for placing and maintaining such a satellite in GSO", it enabled experimentation of the advanced communications technology for a number of purposes, including tracking Indian railways wagons and telemedicine (Vasagam 2015; Blamont 2017). Even more importantly, the launch of APPLE paved the way for greater Indo-European cooperation in the field of orbital launch services. Indeed, since 1981, India has turned more than 20 times to Arianespace for the launch of its heavy GEO satellites (including for its first dedicated military satellite GSAT-7 on 30 August 2013),[12] making ISRO one of Arianespace's most important clients. On its side, ISRO also launched ESA's PROBA satellite onboard its PSLV in October 2001, in addition to a number of European small satellites (see Sect 3 of Annex for the list of foreign satellites launched by India).

Since the early 1990s, ESA and ISRO have also cooperated in the field of Earth observation. More specifically, in 1991 ESA and the National Remote Sensing Agency (an autonomous body under DOS that has now integrated within ISRO and been renamed NRSC) signed a memorandum of understanding "for the direct reception, processing archiving and distribution of ERS-1 SAR data at the Shadnagar ground station 50 km away from Hyderabad" (Bigot 2017). Such cooperation was subsequently formalised in the 1993 cooperation agreement and extended in 1995 to cover also the data from the ERS-2 satellite. To strengthen and complement the provision of these data, many ISRO scientists also received training in ESA's laboratories.

In the following years, cooperation in this field was further enhanced with the exchange of data between Indian EO satellites such as Ressourcesat-1 and RISAT-1 and ESA's Envisat and Soil Moisture and Ocean Salinity mission (SMOS)'s satellites (European Space Agency 2007). Furthermore, when renewing its cooperation agreement with ESA on 9 January 2002,[13] ISRO also signed the International Charter on "Space and Major Disasters" that was initiated by ESA and the French Space Agency 2 years before, thereby enabling the provision of Indian remote sensing data for the purposes of disaster relief (Indian Space Research Organisation 2002; Defence Aerospace 2002).

While the primary objective of the 2002 cooperation agreement was to renew and expand ongoing cooperative activities in the field of satellite applications, particularly Earth observation, it was also intended as an enabler for both ESA and ISRO to carry out new programmes of common interest in areas such as space, life and material sciences. Indeed, 3 years later (in June 2005), a new agreement was reached between the two agencies regarding cooperation on India's first lunar mission, Chandrayaan-1. Under the agreement, ESA would provide technical expertise in areas such as flight dynamics and mission analysis, tools inherited from its

---

[12] As noted by Lele, this specific arrangement "should not be viewed only as a commercial activity but also as one that demonstrates India's faith in the French administration, where they are depending on a foreign agency for the launch of a strategic system into space" (Lele, Space Collaboration Beteen India and France – Towards a New Era, 2015).

[13] This umbrella agreement enables both ISRO and ESA to carry out programmes of common interest in space science and applications including communication, remote sensing for monitoring the environment and corresponding data processing, meteorology and navigation and life and material sciences under microgravity conditions.

SMART-1 mission of 2003 (European Space Agency 2006) as well as three scientific instruments to be integrated onboard the Chandrayaan-1 spacecraft (European Space Agency 2005). These ESA-furnished instruments, a low X-ray spectrometer, sub-keV atom reflecting analyser, and a near infrared spectrometer, would be built by research institutes in ESA member states, respectively, in the United Kingdom, Sweden and Germany, and integrated within the spacecraft by ISRO.

The mission, successfully launched in October 2008, represented "an important step forward in the cooperation between the two agencies and fostered increased collaboration between European and scientific communities with excellent scientific results from the analysis of the data collected so far" (Bigot 2017). Yet, in spite of this important milestone, in the following years ESA–ISRO relations have curiously stagnated.

Even though in January 2007 the two agencies renewed the 2002 cooperation agreement for an additional 5 years and at the 58th International Astronautical Congress (IAC) held the same year in Hyderabad both ESA Director General Jean-Jacques Dordain and ISRO Chairman G. Madhavan Nair expressed a keen interest in tightening relations outside the field of space and Earth sciences (European Space Agency 2007), little did in fact follow in practical terms. Throughout the tenure of Jean-Jacques Dordain, relations with ISRO no longer figured among ESA's cooperation priorities, and on ISRO's side too, marginal efforts were put in place to cultivate its relations with the Paris-based organisation.

A change in posture has only recently materialised with the new directorship of Joan Dietrich Wörner, who has since the beginning expressed a strong interest in revitalising ESA–ISRO space relations. Already before his appointment as ESA director general in an interview with the Speigel, Wörner emphasised the importance of having India onboard major international space projects (Der Spiegel 2015). While in the interview he proposed lifting the access restriction to conduct experiments onboard the ISS, the real underlying target for this enhanced cooperation with India (as well as China) was the well-known Moon Village's proposal he was about to launch (Reuters 2015).

After a series of bilateral discussions and meetings, the ESA–ISRO cooperation agreement was renewed in January 2017 for five additional years, making it valid until January 2022. In addition to space science and applications in the field of telecommunications, remote sensing and corresponding data processing, meteorology, navigation as well as microgravity experiments, Article 2 of this new umbrella agreement identifies "cooperation on satellite development" as a possible field of collaboration to be formalised through mutual agreements. In its Article 5, the agreement also mentions the possibility of setting up "joint working groups" to examine and define new cooperative programmes.

Consistent with this provision, following their meeting in Paris in October 2016, the ESA director general and ISRO chairman agreed to constitute a dedicated joint working group to study possible new cooperation opportunities between the two agencies. The joint working group is now mandated to report back to the two agency heads with specific recommendations regarding specific joint projects of interest (by mid-2017), and the status of progress made by the joint working group will be reviewed by the agency heads and a plan of action for future cooperation will be proposed.

EUMETSAT–ISRO Emerging Partnership

Apart from ESA, another pan-European institution boasting relevant experience in cooperating with India is the European Organisation for the Exploitation of Meteorological Satellites (EUMETSAT). While its cooperation record is more recent as compared to that of ESA, it has, in fact, followed a more structured approach that is now evolving into a very fruitful partnership.

Following initial contacts with the Indian Meteorological Department (IMD), the first cooperation agreement was signed with ISRO in November 2000.[14] The agreement was subsequently renewed in December 2003, December 2008 and more recently in January 2014, progressively expanding the scope and the quality of cooperation.

The primary focus of this cooperation has been on satellite data and product exchange of the parties' respective meteorological satellites in support of weather analysis and forecasting. This includes or has included meteorological data from ISRO's INSAT 3A, Kalpana and INSAT 3D satellite series and EUMETSAT's Metop series of polar orbiting satellites, Meteosat geostationary satellites and the Jason Altimetry Mission. Data processing and dissemination chains have been established at both Shadnagar ground station, near Hyderabad, and at EUMESAT's headquarters in Darmstadt, and data products have been made available via EUMETCast, a multiservice dissemination system capable of delivering products within 5 min of processing. This successful cooperation in the field of meteorology has been subsequently extended to also cover calibration/validation, data processing and reprocessing methodology and satellite data applications, including numerical weather prediction, as well as training, thanks to the creation of a visiting scientists scheme (EUMETSAT 2017a, b, c).

A more recent area of cooperation between EUMETSAT and ISRO is oceanography. To be sure, this cooperation already finds it roots in the late 1990s with the Indian Ocean Experiment (INDOEX), an international field experiment (with the participation of France, Germany, Sweden, the United Kingdom, the Netherlands, India and the United States) dedicated to study the transport of aerosols and trace constituents issued from the Indian subcontinent, and their interaction with clouds and radiations (University Corporation for Atmospheric Research 2017).

To support INDOEX, EUMESAT established a dedicated service, the Indian Ocean Data Coverage (IODC), by placing its Meteosat-5 satellite at 63°E, thereby enabling the provision valuable imagery of the Indian Ocean area during the duration of the experiment. Meteosat-5 started operational service from this position on 1 July 1998, but even after the successful conclusion of INDOEX, it continued to offer the IODC service on a best effort basis until its date of re-orbiting on 16 April 2007 (EUMETSAT 2017a, b, c).

In light of the valuable information provided through the IODC service, particularly in terms of tropical cyclone monitoring, EUMETSAT decided to continue the

---

[14] It is interesting to note that in cooperating with India, EUMETSAT has entertained relations with a space agency, rather than with a meteorological administration, as in the case of other cooperative ventures with third countries.

service through its Meteosat-7 at 57°E,[15] while in furtherance of the agreement renewed in December 2008, ISRO eventually decided to enable EUMETSAT's access and redistribution rights for its Oceansat-2 mission launched on 23 September 2009 (Indian Space Research Organisation 2012).

As a result, the ocean wind data from the scatterometer onboard ISRO's Oceansat-2 satellite were made available via EUMETCast. Coupled with the ocean wind data from the Metop satellites, this exchange of data not only "greatly increased the global coverage and accuracy of data products and services made available for users"; it also significantly contributed to generating trust and a real partnership between the two organisations (EUMETSAT 2014).

Building on this fruitful collaboration, in their 2014 agreement ISRO and EUMETSAT consistently envisaged enhancing the mechanisms for reciprocal bilateral data and products exchange from the parties' respective meteorological and ocean satellites in support of weather analysis and forecasting as well as climate research[16] and agreed to hold annual bilateral meeting to review the status of cooperation as well as regular technical workshops followed by a videoconference at director level every 6 months. According to both EUMETSAT and ISRO, these meetings have proved particularly valuable for discussing and identifying long-term cooperation plans between the two organisations.

More specifically, at the bilateral meeting of 5 April 2016, the two parties discussed a number of important matters of common interest, including long-term cooperation plans for the IODC service, the possibility of ISRO accessing data from the EU Copernicus Sentinels operated by EUMETSAT as well as the possibility of EUMETSAT obtaining access and data redistribution rights from ISRO's Scatsat-1 mission, so as to complement its oceanography satellite programmes.[17]

While a possible cooperation with respect to the EU Copernicus Sentinels remains subject to an agreement between ISRO and the European Commission, access to India's new scatterometer mission was agreed upon and subsequently formalised through an exchange of letters between EUMETSAT Director General Alain Ratier and ISRO Chairman Kiran Kumar in the summer of 2016,[18] hence continuing the very successful service started with Oceansat-2 (EUMETSAT 2016).

---

[15] Meteosat-5 provided the service at 63°E from 1 July 1998 to 16 April 2007, and Meteosat-7 provided the service at 57°E from 5 December 2006 to 31 March 2017.

[16] Article 1 of the EUMETSAT–ISRO Agreement on "Cooperation in Exchange, Redistribution and Utilisation of Data and Products from Meteorological and Ocean Satellites in Support of Weather Analysis and Forecasting and Other Related Areas"

[17] ISRO's Scatsat-1 mission was launched on 26 September 2016 by a PSLV-C35 rocket. The planned mission duration is of 5 years.

[18] Pursuant this exchange of letters, Article 3.3. of the Agreement on Cooperation in Exchange, Redistribution and Utilisation of Data and Products from Meteorological and Ocean Satellites in Support of Weather Analysis and Forecasting and Other Related Areas signed in 2014 was amended to include Scatsat-1. The article now reads as follows: "'ISRO Data' means all data generated by the meteorological payloads of ISRO's Geostationary satellites (INSAT 3A, Kalpana and INSAT 3D series) and by the meteorological and ocean payloads of ISRO's Low Earth Orbiting satellites (Oceansat series, SCATSat −1, Megha-Tropiques and SARAL)".

In addition to their bilateral cooperation, EUMETSAT and ISRO have also put in place a tripartite agreement with CNES to enable data access to the Indo-French Megha-Tropiques SARAL satellites by the European user community (see also Sect. 7.2).

Within this framework, EUMETSAT supported the SARAL (Satellite with Argos and Altika) ocean altimetry mission by hosting and operating the European processing centre. As reported by the Darmstadt-based organisation,

> this trilateral cooperation delivered its first benefits to users in 2013 with the start of real time dissemination of altimeter products from the SARAL satellite, after full validation of the EUMETSAT ground segment elements. Ocean altimetry data from SARAL complements Jason-2 data as SARAL follows a different orbit and the combination of their data provides better coverage and sampling of the global ocean circulation (EUMETSAT 2014).

As for Megha-Tropiques (a mission to map atmospheric distribution of water vapour, clouds, rainfall and water evaporation in the tropical belt), EUMETSAT and CNES "implemented the network connectivity and software needed to acquire and redistribute the data from the SAPHIR humidity sounder on-board the ISRO-CNES Megha-Tropiques satellite", which are now disseminated to meteorological agencies in near real time through EUMETCast.

Apart from this trilateral cooperation, EUMETSAT and IRSO continue to cooperate within a number of multilateral frameworks such as the World Meteorological Organisation (WMO), the Committee on Earth Observation Satellites (CEOS), the Intergovernmental Group on Earth Observation (GEO) as well as the Coordination Group for Meteorological Satellites (CGMS). Here, the two parties have pursued the implementation of an international strategy agreed among the CGMS members and involving satellites from ISRO, the China Meteorological Administration and Russia's ROSHYDROMET to secure the most appropriate coverage of the Indian Ocean region and the successful continuation of the IODC service (EUMETSAT 2016). An overview of the current IODC data dissemination put in place by the CGMS members is given in Table 6.1.

As part of this international strategy, and in view of the planned deorbiting of Meteosat-7 in March 2017, EUMETSAT envisaged the possibility of moving its Meteosat-8 satellite to 41.5°E, and following the eventual authorisation by EUMETSAT Council in June 2016, Meteosat-8 took over from Meteosat-7 as the prime satellite providing the IODC service on 1 February 2017 (EUMETSAT 2017a, b, c).

All in all, relations between EUMETSAT and ISRO can be assessed as being reliable, mutually beneficial and time-tested. While their cooperation certainly remains far from being ground-breaking, particularly as compared to the deeper relations EUMETSAT has with the National Oceanic and Atmospheric Administration (NOAA), it appears nonetheless to be progressively evolving in a similar direction, with more ambitious forms of collaboration, including the planning of joint missions, possibly becoming a reality in the near term.

Table 6.1 Indian Ocean data coverage

		Data reception (users)			
		EUMETSAT	India	China	Russia
Data provider	EUMETSAT	EUMETCast-Europe	EUMETCast-Europe	CMACast	EUMETCast-Europe
	Meteosat-7	EUMETCast-Africa	CMACast	GTS (products)	CMACast
	Images	GTS (products)	GTS (products)		GTS (products)
	Products				
	INDIA	EUMETCast-Europe	Internet (MOSDAC)	Internet (MOSDAC)	Internet (MOSDAC)
	INSAT 3D	GTS (winds)	GTS (winds)	GTS (winds)	GTS (winds)
	Images				
	Winds				
	CHINA	EUMETCast-Europe	CMACast	CMACast	CMACast
	FY2-E		Beijing GISC	Beijing GISC	Beijing GISC
	Images				
	RUSSIA	EUMETCast-Europe	EUMETCast-Europe	Unknown	Password protected ftp server
	ELektro-L-N1/2				
	Images				

Source: Coordination Group for Meteorological Satellites

EU–India (Deficient) Cooperation

Similar to their broader political dynamics, EU–India space relations can be characterised as having been "long on general agreements, but short on actual implementation". Cooperation in the field of space technology was indeed mentioned as early as 1994 in the EU–India cooperation agreement. Even more importantly, it was specifically identified as a fully fledged component of the EU–India strategic partnership that was launched in November 2004[19] and included within the EU–India Joint Action Plan (JAP) endorsed at the Sixth EU–India Summit of 7 September 2005 (Council of the European Union 2005), In its section "Economic Sectoral Dialogues and Cooperation", the document more specifically stated:

---

[19] In the joint declaration that was released in conjunction with the launch of their strategic partnership on 8 November 2004, the leaders of the two unions emphasised their space cooperation on Galileo: "we welcome the progress in the on-going discussion on the EU-India Draft Cooperation Agreement on the Galileo satellite navigation project. It will ensure India's equitable participation in Galileo space, ground and user segments and will guarantee the availability of highest quality signals over the Indian territory. Considering that India has well proven capabilities in space, satellite and navigation related activities, the agreement will provide an important positive impulse for Indian and European industrial co-operation in many high-tech areas. [...] The EU and India both have very mature space programmes and a long history of working together through their respective space agencies ESA and ISRO in the peaceful exploration and use of outer space. The EU expresses its interest in the Indian unmanned lunar exploration mission Chandrayaan-1. We support and encourage the cooperation between ESA and ISRO".

Both India and Europe are at the cutting-edge of research in the field of Space Technology, and there is a wide scope for cooperation. With a view to promote collaboration and provide an appropriate environment for fruitful cooperation in the space sector, both parties will:

Support further collaboration and dialogue between Indian Space Research Organisation (ISRO), Department of Space (DOS) and the European Space Agency (ESA) and the European Commission, in areas such as earth observation and remote sensing for monitoring of natural resources and environment, communications, meteorology, navigation, life and material sciences under micro gravity conditions, space exploration, space sciences and any other area relevant to our respective Space programmes;

Jointly identify specific new areas/projects of cooperation between the respective space agencies for further discussion/implementation through the existing mechanism for technical cooperation (Council of the European Union 2005).

While the JAP suggested several cooperative ventures in a variety of fields, none of these eventually moved beyond the announcement phase. The failed cooperation on the Galileo programme offers the most eloquent case in point and substantiates the idea of a strategic partnership leading mainly to dialogue and commitment to further dialogue rather than to significant policy measures or breakthroughs (Malone 2011).

New Delhi had announced its intention to join this EU's satellite navigation flagship programme in the wake of the 2003 Iraq War in which the GPS was extensively used by the US military. Acquiring indigenous hands-on experience in the field of satellite navigation through equitable participation in the development of Galileo was for New Delhi a way to pave the way for future strategic autonomy. In addition, China had also announced the intention to join the Galileo programme, and India did not visibly want to lag behind its alleged space rival. Apart from these strategic objectives, India saw in the Galileo programme an opportunity to demonstrate its technological prowess in the space domain, to stimulate the growth of Indian private companies involved in the country's space programme and facilitate further Indo-European cooperation in high-tech areas. In fact, India's Department of Space had already envisaged linking Galileo to its own GAGAN augmentation programme in order to boost signals from Europe's new GNSS. As for the EU, cooperation with India was imagined as a means to share the burdens of this costly programme, reinforce the strategic dimension of its partnership with New Delhi and tap into India's potentially huge market for satellite navigation applications.

The extensive negotiations held since January 2004 eventually reached approval at the EU–India Summit of 7 September 2005, when the Draft Cooperation Agreement on the Galileo satellite navigation project was sealed in the presence of Indian Prime Minister Manmohan Singh, UK Prime Minister Tony Blair as EU President and the EC President José Manuel Barroso (European Commission 2005). Hailed as a major milestone for both the advancement of the Galileo project and the broader EU–India strategic partnership, the agreement made India the fourth country to join the Galileo programme, after China, Israel and Ukraine (European Commission 2005).[20] The total investment offered by India was around €300 million, an amount much larger than that of China (around €200 million) and roughly one-tenth of the initially estimated costs of the Galileo programme.

---

[20]At that time discussions were also underway with Argentina, Brazil, Morocco, Mexico, Norway, Chile, South Korea, Malaysia, Canada and Australia.

The details of India's participation in Galileo were to be finalised at the 7th EU–India Summit of 2006. At that meeting, however, cooperation already started to creak under its own weight. For one thing, India did not seem to appreciate the policy shift made by the Commission vis-à-vis the possible use of Galileo for military purposes. In a statement released just before the Helsinki Summit, EC Vice President Jacques Barrot had suggested that Galileo should be opened for military uses despite remaining a civilian project. Although this possibility did not come as a surprise, the commission's ambiguity and the still pending decision on Galileo's applications did not facilitate the negotiations at Helsinki.

In addition to that, Indian officials expressed a number of security concerns over the protective measures put in place for Galileo. More specifically, they feared that Galileo's Public Regulated Service (PRS) did not have sufficient firewalls to prevent it from being used by unauthorised users or other nations (read in particular China) participating in the project (Deshpande 2006). And this was a particularly delicate issue for New Delhi. As also commented by Wülbers,

> for India participating in Galileo meant sharing of strategic and military information that India has safeguarded with care. […] Sharing security information means for India a loss of its autonomy. For India, its autonomy has been very important since it is flanked by nuclear powers on both sides and prefers to be able to act according to its security interests rathen than be bound into arrangements that may involve information sharing with its perceived rivals (Wülbers 2011).

The core issue, however, remained the very definition of India's role within the programme, which eventually became subject to a "fundamental incompatibility". Since the onset of the negotiations, India had made it clear that it would join the programme as an equal partner – that is as a partner actively involved in its development and management – and not as a mere customer or end user. And the Joint Communication released at the end of Prime Minister Manmohan Singh's Summit meeting with EU leaders in November 2004 seemed to confirm this understanding. The communication affirmed that the cooperation agreement would "ensure India's *equitable* participation in Galileo space, ground and user segments, the availability of highest quality signals over the Indian territory as well as cooperation to establish regional augmentation systems based on EGNOS and GALILEO". But while the EC was particularly "keen on India's financial participation in the project, it had reservations about New Delhi's involvement in the operational aspects of the project" (Times of India 2004). An Indian senior official was then reported to comment: *"If we are putting in 300 million euros we must have a say in the control of the satellite, […] If we don't have access to their codes we can be denied access to Galileo's signals in times of war."* And while the EU had assured that such denial would occur only in case of a global war, the official continued: *"How do we know what is their definition of a global war. They can term an India-Pakistan war as global, saying it has nuclear connotations"* (Times of India 2004). For India, failing to access Galileo's encrypted signals and being rejected as a full partner simply called into question the strategic driver that had pushed it to join the project in the first place.

At the New Delhi Summit of 2007, none of these outstanding issues were eventually addressed. And while the new JAP endorsed in 2008 at the Marseille Summit acknowledged the difficulties faced in the finalisation of the agreement and stressed the need to further the dialogue between the two sides in this respect (European Union External Action Service 2008), after that India's participation in the Galileo programme was suddenly removed from EU–India public debates.

At the time, New Delhi's decision to withdraw from the project was primarily imputed to its intention to acquire an indigenous RNSS. However, there also appears to have been other and perhaps equally important motivations. These certainly include India's concerns about being caught between the crossfires of EU–US discussions on Galileo–GPS interoperability and hence the risk of compromising its blossoming cooperation with Washington, which had always expressed an evident opposition to the Galileo project, as well as New Delhi's frustration in dealing with the complex instructional structures set-up for the project. However, more importantly, there appears to have been India's disappointment in being rejected as a full partner and "co-decision-maker in the Galileo project, especially after it had been asked to invest 300 million euros in the project (100 million more than China)" (Vicziany 2015).

Although the EC did not appear much impressed by India's offer to invest a mere 300 million euros, particularly when the Galileo costs began to move upwards, that amount corresponded to more than half of the total annual budget invested by ISRO during the mid-2000s (see Sect. 3.1.2). Hence, when it became clear that India would not come onboard the Galileo project as a full partner, it decided to step back and to focus on its national system. In retrospect, India should have felt pleased to see that the total cost of its IRNSS programme (as of 2016) did not exceed €300 million and that also Chinese participation in Galileo eventually failed to take off.

As pungently commented by Marika Vicziany, "on a diplomatic level, the Galileo project has been a dramatic disappointment for the normative approach to international relations that the EU has been famous for. The project not only failed to keep the two emerging Asian powers on side, but it appears to have also insulted them […] The EU authorities appeared not to be too concerned about why India and China might be sensitive about the EU's unpredictable behaviour in disengaging with them, preferring to see the problems of the Galileo project as financial and managerial ones. Perhaps this indifference was driven by the EU's desperate search for ways of keeping this EU flagship alive? For India, being rejected as a full partner in the Galileo project simply underscored the widespread impression that only the United States was willing to deal with India as a serious political and security partner".

An attempt to circumvent this failed cooperation experiment and tap into India's GNSS market potential was put forwards in 2012 with the launch of the *GNSS Asia* project. First financed under the EU 7th Framework Programme (FP7) and then under Horizon 2020 and managed by the GSA, the GNSS Asia project focuses on strengthening downstream GNSS industrial cooperation activities between Europe and five Asian nations, namely, India, China, Taiwan, Korea and Japan – referred to as Asia-Pacific (European Global Navigation Satellite Systems Agency 2015).

GNSS Asia provides cost-free support services (including market analyses, identification of key opportunities, networking, etc.) with the overarching objectives of

(a) supporting industrial cooperation between European and Asian GNSS companies, with a special focus on applications and receivers/chipsets, (b) promoting European GNSS Galileo solutions, and (c) raising awareness of Galileo's applications and benefits (European Global Navigation Satellite Systems Agency 2015). In meeting those objectives, special emphasis has been placed on workshops, thematic events, round tables and seminars, particularly in view of the lack of awareness of GNSS technology among Indian researchers, solution providers and potential users (GNSS Asia 2016).

In India, the key objectives of GNSS Asia include the promotion of GNSS applications for infrastructure modernisation, most specifically road and rail improvements throughout the country, the compatibility of GAGAN with Europe's EGNOS system and the opportunity to leverage European expertise in satellite-based augmentation systems (SBAS) to support the expansion of GAGAN to sectors beyond its application in aviation (European Global Navigation Satellite Systems Agency 2014; GNSS Asia 2017).

As compared to the ground-breaking cooperation initially envisaged for the development of Galileo, the activities undertaken within the GNSS Asia framework are certainly less indicative of a strategic partnership. Their importance, however, should not be dismissed either. Besides providing a valuable thrust for possible GNSS industrial cooperation among European and Indian companies, these activities have also contributed to highlighting the need to put space back on the agenda of EU–India relations.

Consistent with the broader effort to revitalise the EU–India strategic partnership announced at the 13th EU–India Summit of 2016, the EU–India Agenda for Action 2020 indeed identified space as a sector of policy dialogue and cooperation in its own right. The Agenda in particular recommended enhancing "space cooperation including earth observation and satellite navigation for the strengthening of interaction between the Indian Regional Navigation Satellite System and EU's Galileo as well as joint scientific payloads" (Council of the European Union 2016).

Although the adoption of this vision document as a common roadmap to jointly guide relations in the period 2016–2020 can certainly provide new impetus to enhance collaboration in the space sector, it appears already quite evident that much cooperation potential has largely remained unexplored. By simply considering space as one of the sectors of policy dialogue and cooperation, the Agenda fails to acknowledge the wide range of opportunities that space could provide for achieving a multitude of shared policy objectives. Indeed, for each of the sectors where the EU and India have committed to working together, namely, climate change, urban development, research and innovation, information and communications technology, energy, environment, transport, the 2030 sustainable development coals and even security, there is an evident role that space assets and related applications could play in achieving those objectives. But none of these contributions is in fact acknowledged in Agenda 2020. As also lamented by Isabelle Sourbès-Verger in an article for the EU–India Think Tank Twinning Initiative, in the EU–India policy documents:

There is a wide range of topics to be observed, including trade and investment, science and technology, economic and development cooperation and security. Space capabilities can be put to use for most of these areas, and yet these are not mentioned. This might stem from the belief that space programmes' technicalities ought to be directly handled by national agencies or ESA and not to be taken into consideration at the EU level, possibly for lack of suitable community powers. In fact, the capabilities of space systems could be better integrated in EU-India relations with an approach that includes the main challenges identified in the EU-India Dialogue. (Sourbès-Verger 2016)

Thanks to its unique transversal qualities, space could obviously provide a wide range of opportunities for closer EU–India cooperation. Thus far, the fact that none of these opportunities is mentioned within Agenda 2020 seems to suggest that the EU and India are once again missing the opportunity to move beyond rhetorical commitments to dialogue and cooperation and to transform general political objectives into actionable policy measures.

6.3.2   National and Industry-Level Cooperation

As compared to pan-European institutions, national space agencies in Europe have had some more substantial cooperative ventures with Indian stakeholders. As of 2017, bilateral agreements had been signed by France, Germany, Italy, the Netherlands, the United Kingdom, Sweden, Bulgaria, Hungary and Ukraine and have covered a broad scope of activities, spanning from joint missions and payload accommodation to commercial satellite launches and scientific experiments. An overview of these cooperative undertakings is offered in Table 6.2.

In terms of bilateral cooperation, particularly meaningful has proven the experience of the French space agency (CNES) – to which the next section is devoted, being one of ISRO's most important partners – and, to a much less extent, those of the German aerospace centre (DLR) and of the Italian space agency (ASI).

ASI has put in place a framework agreement for cooperating with ISRO since 2000 and has been one of the few agencies worldwide to have a payload hosted onboard an Indian national mission. ASI's Radio Occultation Sounder for the Atmosphere (ROSA) was indeed integrated within the scientific payload of ISRO's Oceansat-2 mission following an MoU signed by the two agencies in 2005.[21] Moreover, the launch of ASI's AGILE X-ray and gamma-ray satellite on 23 April 2007 was the first launch dedicated entirely to a foreign customer on a commercial basis (Harvey et al. 2010). This launch hence paved the way for the entry of the PSLV rocket in the international launch market as well as for the subsequent launch of a number of European small satellites (including NLS-1 and NLS-2 which remarkably were Austria's first satellites in orbit).

Whereas additional areas of potential ASI-IRSO cooperation (including networking of the respective ground stations for data reception and joint research pro-

---

[21] ROSA is a GPS occultation receiver intended to study the main physical parameter of the Earth's atmosphere. For more information on the ROSA-Oceansat-2 cooperation, see De Cosmo (2006).

Table 6.2 Main cooperation activities at the level of national space agencies

Category	Country/mission
Joint missions	France (Megha-Tropiques, SARAL)
Payload accommodation	Italy (ROSA)
	Germany (MEOSS, MOS, SIR-2)
	UK (CIXS)
	Sweden (SARA)
Ground station support	Norway (Svalbard, Tromso)
	Germany (Neustrelitz)
EO experiments and data sharing	Germany (airborne campaign, IRS data reception)
	France (Ka-Band propagation, SARAL, Megha-Tropiques)
Capacity building/tech transfer	France (viking engine
	Germany (scientists exchanges)
Satellite launch services	Italy (AGILE)
	France (SPOT-6, SPOT-7, Hylas-1)
	UK (STRAND-1, DMC-3, CBNT-1)
	Austria (NLS-1,2)
	Germany (BIRD, AISat, BIROS)
Bilateral meetings/workshops	Italy (ASI-ISRO Workshop on Space Research 2012)
	Germany (ISRO-DLR Cooperation Workshop 2015)
	France (Quality Workshop for ISRO at CNES 2016)

grammes combining X band SAR data of COSMO-SkyMed and C band SAR data of RISAT) had been identified by the two agencies as early as 2007 (Agenzia Spaziale Italiana 2014; Somasekhar 2007), the 2012 *Enrica Lexie* case exercised a negative impact on the finalisation of these cooperative efforts, which as of 2017 remain still on hold.[22]

As for Indo-German space relations, the first inter-agency agreement was signed by ISRO and the DFVLR (the aerospace centre of West Germany) as early as 1971 and more recently renewed in 1995. Early cooperative ventures during the 1970s primarily focused on the training of ISRO staff members at DFVLR for high-altitude tests of rocket motors as well as satellite systems and operations and on the supply of electronics and equipment from German companies (Prasad 2015).[23]

---

[22] The *Enrica Lexie* case refers to a still ongoing international dispute about two Indian fishermen killed off the coast of Kerala by two Italian marines onboard the Italian tanker Enrica Lexie. The case has caused serious diplomatic tensions between the two countries and has sparked a controversy over the legal jurisdiction and functional immunity between the two governments.

[23] The training activity was quite useful to ISRO and an average of eighty man-months of training was provided for a period of roughly 8 years from 1973 to 1981. The subjects of training included materials and structures, specifically composites of carbon fibre reinforced plastics. The training on the design of rocket propulsion test facilities resulted in commissioning of the High-Altitude Test Facility at SHAR in 1980. The training in the areas of mission planning and mission operations resulted in ISRO experts gaining additional knowledge in the area of orbit determination and satellite operations (Prasad 2015).

Increasingly supported by a sound political commitment on both sides, recent cooperation efforts between the two agencies have been mainly undertaken with respect to acquisition and commercial distribution in Europe of IRS EO data through the DLR ground station Neustrelitz, provision of components and equipment for Indian EO satellites, capacity building through the exchange of scientists and the organisation of thematic workshops and launch of German satellites by India's PSLV launcher, including the Automatic Identification System Nanosatellite (AISat) in June 2014, the Bispectral InfraRed Optical System (BIROS) in June 2016 and the Environmental Mapping and Analysis Programme (EnMAP) that is now scheduled for launch in 2018/2019.[24]

It is also important to mention that as a federal agency carrying out research and development activities in the field of aeronautics, energy, transport and security, DLR's cooperation with India has not been limited to the space sector but has more broadly encompassed joint efforts also in other domains such as renewable energy and urbanisation (DLR 2016). In addition, cooperative undertakings in the broader STI sector have been supported by an ad hoc Indo-German Science and Technology Centre (IGSTC) and by a High Technology Partnership Group (HTPG). Whereas the former was set up "to catalyse innovation centric R&D projects by synergising the strength of research/academic institutions and public/private industries from India and Germany" (Indo-German Science and Technology Centre 2017), the latter seeks to identify specific opportunities for high-technology trade and cooperation – also in the areas of manufacturing under the Make in India initiative including space – through regular meetings.[25] Interestingly, at its meeting in November 2016, the HTPG held a thematic workshop on space in which specific areas of future cooperation between DLR and ISRO were identified (Embassy of India 2016).

European Space Industry in India
In the framework of Europe–India space cooperation, attention should be also devoted to the European space industry involvement in the country's space programme. In this respect, it is worth noting that India has often partnered with foreign industries. Its major partners in the aerospace sector are European countries like the United Kingdom, and France, as illustrated in Fig. 6.1.

Consistently, the two major aerospace groups in Europe, Airbus and Thales, have a long-standing and strong presence in India (see Table 6.3).

Despite focusing on sectors such as public mobility and smart cities where the contributions of space assets prove particularly relevant, as well as on government initiatives such as *Make in India* that have opened up important avenues for space manufacturing in India, the Thales Group has not been particularly active in the Indian space programme.

---

[24] For a more detailed overview of Biros and EnMAP missions, see DLR 2016.

[25] The creation of this group goes back to the decision made by Federal Chancellor Angela Merkel and Indian Prime Minister Manmohan Singh during the Second Indo-German governmental consultations in April 2013 in Berlin, where both governments agreed to expand trade and cooperation in high technology (Ministry of External Affairs 2013).

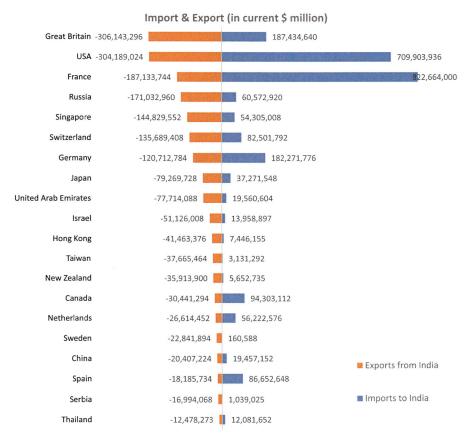

Fig. 6.1   India's main aerospace trade partners in 2014 (*Source*: OECD 2014)

Table 6.3   Airbus and Thales Group in India

	Thales	Airbus
Presence in India	Since 1953	Since 1960
Major sites	Delhi, Bangalore Hyderabad, Chennai, Mumbai	Delhi, Bangalore Hyderabad, Chennai, Mumbai
Direct employees	300+	500+
Core business	Defence, Mobility, Smart Cities	Aircraft, Helicopters,
Major partners/ suppliers	HAL, BEL, Samtel, LT&T Technology Services	HAL, ISRO, BEL, BDL
Academic partnership	Institute of Technology in Delhi (IIT Delhi); Indian Institute of Science (Bangalore), Indian Institute of Technology (Bombay)	Indian Institute of Management Bengaluru (IIM-B), Tata Institute of Fundamental Research Mumbai (TIFR)

By contrast, Airbus Defence and Space (ADS) has established itself as the prime foreign industrial partner of ISRO and the Indian space industry, also thanks to a decade-long alliance that EADS Astrium and Antrix/ISRO formalised through a cooperation agreement in February 2006.

The primary aim of this alliance has been "to jointly offer communications satellites in the market segment around 4kW of payload power and with a launch mass of 2–3 tons" (Indian Space Research Organisation 2006). Within this cooperation scheme, Airbus is specifically responsible for building the payload according to individual customer specifications, while Antrix builds the satellite's platform, so as to offer cost-effective solutions to telecommunications operators.

As noted in a previous ESPI study (Al-Ekabi 2016),

> the industrial partnership has proved advantageous for Airbus because it increases its level of competition in the global commercial market by pooling skilled low-cost labour from India, resulting in lower-cost satellites. This is a competitive advantage that Airbus has over its American counterparts who, at the moment, do not have comparable cooperation agreements with India.[26]

At the same time, the effective weight of this partnership should not be overestimated either. Indeed, while the two organisations have jointly designed and built two communication satellites for European customers, namely, W2M in 2008 for Eutelsat and HYLAS-1 in 2010 for Avanti Communications of London (Antrix Corporation 2015), following these two orders, the Airbus–Antrix venture has been absent from the commercial market, primarily because of technology transfer and product-liability issues as well as Antrix's difficulties in keeping up with India's domestic demand for telecommunications satellites. However, with Modi's efforts to modernise India's industrial base and ease doing business, a number of opportunities may open up in the future.[27]

Commercial satellite manufacturing is not the only area where ADS and Antrix/IRSO have cooperated. An additional key element of their decade-long partnership has in fact pertained to the launch Airbus-built satellites onboard ISRO's PSLV launcher (Diplomacy & Foreign Affairs 2013; Indian Space Research Organisation 2008). The agreement, signed on 30 September 2008 by ISRO/Antrix Chairman

---

[26] However, as further detailed by Al-Ekabi (2016), "the competitive edge that European satellite integrators have over their U.S. counterparts might be challenged due to the gradual increase in cooperation between India and the United States as a result of the loosening of U.S. government regulations on collaboration with that country. In 2008, the U.S. government eased sanctions on industrial cooperation by U.S. companies with India in the field of space and missile technology, sanctions that were originally put into force after Indian underground nuclear tests.240 ITAR regulations formed a second barrier that made it impossible for U.S. manufacturers to export U.S. satellite components and fully manufactured satellites to India, which impeded them from entering into industrial cooperation with ISRO/Antrix similar to the Airbus cooperation. Because of the 2014 changes in the ITAR regulations this barrier has been lifted. The access advantage European satellite integrators had is therefore reduced but no U.S. satellite integrator has so far established any form of cooperation agreement with ISRO or Antrix".

[27] Remarkably, following Modi's visit to France in April 2015, ADS announced that "links with Indian industry in this space sector will grow in the coming years through the cooperation on design and manufacturing of larger telecommunications satellites in India" (de Selding 2015).

Madhavan Nair, on the occasion of Manmohan Singh's state visit to France, has thus far led to the successful launch of the SPOT-6 and SPOT-7 Earth observation satellites, respectively, in 2012 and 2014, and of the three DMC-3 satellites built by Surrey Satellite Technology Ltd. (SSTL) – a UK-based satellite manufacturer and subsidiary of Airbus Defence and Space – in 2015. In addition, one should not overlook the fact that the large majority of Indian INSAT and GSAT communication satellites have been put into orbit by Arianespace, which is largely owned by Airbus Defence and Space through its joint venture with Safran (ASL). Moreover, over the years, many ADS products and components have been provided to ISRO for many undertakings, including its Mars Orbiter Mission and its IRNSS programme (Airbus 2017). The increasing importance that Airbus attaches to the Indian space market was recently evidenced by its presence, along with a number of smaller European space companies, at the 2016 Bengaluru Space Expo (Indian Space Research Organisation 2016).

European companies' presence in India is not only limited to the upstream sector, but also increasingly evident downstream, and specifically in the market for satellite telecommunications. Despite the restrictive regulations that have made it difficult for non-Indian satellite operators to do business in India, the two leading European satellite operators, EUTELSAT and SES, have achieved a valuable share of India's satellite telecommunications market through their broadband and broadcast services.

With seven satellites providing full coverage of the Asia-Pacific region, SES offers the whole set of telecom services in India, and it is particularly strong in its DTH market, where it provides transponder capacity to India's premier service providers such as Dish TV (14.9 million subscribers as of June 2016) and Airtel Digital TV (12.1 million subscribers as of June 2016) through its NSS-6, SES-8 and SES-7 satellites. NSS-6 alone, with its 42 transponders, provides the largest DTH platform in India (Television Post 2013; SES 2017). Similarly, EUTELSAT has been delivering comprehensive coverage and high-bandwidth capability for Indian broadcasters, video and Internet service providers and telecom operators, in particular thanks to its EUTELSAT 70B satellite, the transponders of which have been delivering more than 35 Indian TV channels (EUTELSAT 2017; Track Dish 2015).

6.3.3 India–France Partnership in Space

In the relatively modest list of Indo-European cooperative undertakings in space, France accounts for the vast majority of projects and, in fact, figures not only as India's oldest and closest European partner but also as one of India's most dependable space partners internationally.

The importance of Indo-French space cooperation most sharply emerges when looking at the peculiar role played by the Centre National D'Etudes Spatiales (CNES) in the very conceptualisation and development of the Indian space programme during its formative years, as well as at the sheer volume of agreements and activities put forwards in a wide range of areas, which more broadly reflect an unparalleled amount of economic and political commitment at the highest level. As aptly underlined by the CNES representative in India, Mathieu Weiss, on the occasion of the 50th anniversary of Indo-French space cooperation

> this partnership for and through space has often served as a bridgehead, feeding into each bilateral encounter, each summit and each meeting of the French-Indian strategic partnership. Even when tensions were at their highest during the U.S. embargo, France never renounced its attachment to the Indian subcontinent. These episodes of its history are deeply embedded in India's memory and it is this trust built over the years that underpins all of our bilateral exchanges, notably in space (Weiss 2014).

The unwavering alliance between ISRO and CNES started taking shape in 1962 when the newborn CNES helped the Indian National Committee for Space Research to set up the Thumba Equatorial Launching Station (TERLS), near Trivandrum. Just a year later, a sodium ejector payload fabricated by CNES laboratories was successfully launched aboard a US-provided Nike Apache sounding rocket, marking the beginning of India's space programme.

These initial seeds of cooperation were subsequently formalised through an MoU signed in May 1964 between CNES and DAE. Under the agreement, France provided valuable equipment to TERLS (including a COTAL radar) and more importantly licenced the manufacturing of its Bélier and Centaure sounding rockets to India, with the accompanying transfer of solid propulsion technologies. Overseen by a CNES team led by Jacques Blamont, then scientific and technical director of CNES, more than 50 rockets were built locally and launched from 1965 onwards under the patronage of the United Nations (see Fig. 6.2).

The indigenisation of the Centaure technology was a major boost for the development of the first all-Indian rocket, the Rohini, which in fact was largely derived from this French solid propulsion technology.

Fig. 6.2 France's Centaure rocket in India (Credit: CNES)

The early 1970s also saw the transfer of technologies for liquid propulsion systems. In 1972, under an agreement with the French company Société Européenne de Propulsion (SEP), ISRO acquired the technology of the Viking liquid engine that was being developed for the Ariane launch vehicle programme. In return, ISRO provided SEP with the services of 100 man-years of ISRO scientists and engineers for its work on the Ariane launch vehicle development (Prasad 2015).[28] The Viking engine, subsequently indigenised under the name "Vikas" and successfully tested in SEP facilities in 1985, had a great impact on the very configuration of the PSLV and GSLV launchers and hence remains a crucial staple of India's launch technology to this day (Blamont 2017; Bouhey-Klapisz 2017).

France's contributions to the development of the Indian space programme in its formative years were not limited to technology transfer in the field of propulsion. During the same years, assistance was also provided for the setting-up of India's space-based educational programme pursued by the father of the Indian space programme, Vikram Sarabhai. In 1970, he obtained a letter from India's Prime Minister Indira Gandhi, proposing to the French government a joint mission for a geostationary relay satellite for cultural and educational TV (Blamont 2017).

While this proposal was not accepted, following its launch in 1974, Symphonie was made available to Indian scientists for the conduct of important experiments by ISRO and the Posts and Telegraph Department of the Ministry of Telecommunications. These experiments were also accompanied by training and exchanges of scientists, procurement of components and equipment from French companies through the good offices of CNES and utilisation of French space facilities for conducting special tests.[29]

As remarked by Prasad, "the telecommunication application project with SYMPHONIE satellite laid the foundations for using satellite-based communications in India. This along with educational TV experiments using Application Technology Satellite (ATS) of the USA was the foundation on which the applications of INSAT series of satellites had grown in India" (Prasad 2015).

On the institutional front, in April 1972, CNES and ISRO convened to set up a joint commission to identify and put forwards cooperative projects. Since 1973, the ISRO–CNES Joint Commission has met regularly every year, acting on the recom-

---

[28] As detailed by Blamont (2017), "to acquire the Viking engine technology, ISRO engineers worked in all areas of development activities of the Ariane programme. They participated in design reviews, progress reviews and even had interaction with European industries. They received all detailed design drawings and documents, and participated in inspection and quality assurance of systems, subsystems and components. They were also part of assembly and integration, checkout and testing operations in SEP facilities. They had discussions with SEP specialists and received clarifications to understand the technology fully. Some 40 engineers, working under a five-year contract, participated in the technology acquisition program at Vernon and Brétigny in France". The ISRO Chairman created in 1980 a Liquid Propulsion Project (LPP) which organised three teams under the leadership of three SEP trained experts—the first team developed the system, the second was tasked to realise all hardware in India in association with Indian industries, and the third team was to establish all development facilities at Mahendragiri. Indian industries and academic institutions were associated in the development effort".

[29] Companies like Thomson, SAFT, Matra, Deutsch and SAGEM, among others, would later sign important commercial contracts with ISRO.

mendations of the joint working groups that were set up in various specialised areas such as launchers, telecommunications, remote sensing, etc. (Prasad 2015). Building on this fruitful work, in May 1975 CNES proposed replacing the existing inter-agency agreement with an intergovernmental agreement in order to widen the scope of Indo-French space cooperation and receive stronger commitment at the political level. The agreement, signed in 1977, laid the groundwork for subsequent assistance and training in the area of telemetry and satellite development (most notably for APPLE, for which CNES provided the thermal testing facilities at its Space Centre in Toulouse) as well as cooperative undertakings in the field of atmospheric and space research and application space technology.

While relations fell into a period of inactivity from the mid-1980s to the early 1990s, also owing to ISRO's choice of Ford Aerospace for the development of the INSAT satellite and the utilisation of US rockets for their launch, with the progressive emergence of India as a mature spacefaring nation, the very nature of the partnership evolved accordingly, moving from mere assistance and technology transfer to sound cooperative undertakings and commercial contracts. In 1993 the cooperation agreement was renewed with an enhanced scope (in areas such as remote sensing, meteorology and communications) and eventually upgraded into a fully fledged strategic partnership that marked a new level of engagement in 1998. Indeed, following the inclusion of space among the priority areas of the Indo-French strategic partnership, CNES and ISRO started envisaging more ambitious ventures, including projects for joint satellite development.

Just one year later, the two countries signed a statement of intent covering the joint development of an Earth observation satellite for studying tropical climatology: Megha-Tropiques (Lele 2015). Considering the new series of sanctions that had just been put in place following India's 1998 nuclear tests, the initiative proved particularly audacious. Understandably, it also underwent a lengthy and complex negotiation process that lasted for more than 10 years.

While the design had already been completed by the turn of the new century, development was suspended in 2002 due a financial shortfall at CNES (de Selding 2011). A new settlement between the two agencies was reached in November 2004, with the ISRO chairman agreeing to reorganise the mission with greater responsibility from ISRO, including the provision of an IRS-derived satellite platform in addition to the launch. But even then, discussions rambled for a few more years, also due to concerns expressed by the United States regarding the possible infringement of the International Traffic in Arms Regulations (ITAR).

The final green light was given only on 30 September 2008, within the framework of a new intergovernmental agreement on bilateral space cooperation signed by India's Prime Minister Manmohan Singh on occasion of his state visit to France.[30]

---

[30] Together with the framework agreement on space cooperation, India also signed an MoU between the Indian Institute of Space Science and Technology (IIST) of ISRO and Ecole Polytechnique (EP) at Paris for furthering collaboration in academic and research activities involving students and professionals and an agreement between ISRO/Antrix and Astrium enabling Astrium to offer attractive solutions in international markets for in-orbit delivery of its Earth observation satellites using the PSLV launch services from Antrix/ISRO (Indian Space Research Organisation 2008).

The agreement detailed a number of issue areas for future mutual cooperation, including:

- The study of climate change with EO satellites
- The development of microsatellites for scientific purposes
- Joint research and development activities
- Development of ground infrastructure for joint space missions
- Organisation of combined training programmes
- Exchange of technical and scientific staff

The new agreement more specifically announced the development of the Megha-Tropique flagship programme, as well as the development of a second ISRO–CNES joint mission, the Satellite for Argos and Altika (SARAL), dedicated to operational oceanography.

The operational details of these two missions were finalised during the 2010 visit of President Nicolas Sarkozy to India, during which the two countries also committed to furthering their efforts in the field. Eventually, Megha-Tropiques was successfully launched on 12 October 2011 while SARAL following on 25 February 2013. Box 6.1 offers a more detailed overview of these two satellites jointly developed and operated by CNES and ISRO.

The successful deployment and operations of Megha-Tropiques and SARAL provided an important basis for the celebration of 50 years of India–France cooperation in space. To mark this anniversary, which unsurprisingly was one of the major highlights of Indian Prime Minister Modi's first visit to France in April 2015, the

---

Box 6.1: CNES–ISRO Joint Satellite Missions

Megha-Tropiques

Megha-Tropiques (Megha meaning clouds in Sanskrit and Tropiques meaning tropics in French) is a joint Indo-French satellite dedicated to study the water cycles in the tropical belt. It is more specifically designed to deliver simultaneous measurements of the different states of water in the ocean-atmospheric systems of the tropics by three-dimensionally mapping distribution of water vapour, clouds, rainfall and water evaporation over this region.

This 1000-kg class satellite comprises a platform derived from ISRO's IRS and a set of three payloads: MADRAS, a microwave imaging radiometer providing data on the rain, clouds and water vapour properties; SAPHIR, a microwave sounding radiometer for the study of atmospheric water vapour; and SCARAB, a wideband radiometer for measuring radiation fluxes at the top of the atmosphere. ISRO provided the platform and the major part of the

(continued)

Box 6.1 (continued)

MADRAS instrument, in addition to the PSLV launch to a 20° circular orbit (altitude 865 km) that enabled up to five observations per day at the same location. CNES provided the SAPHIR and the SCARAB instruments.

Despite the loss of the MADRAS in January 2013 (which was in part mitigated by SAPHIR), Megha-Tropiques has provided relevant data for cyclone and monsoon prediction as well as for understanding our Earth's climate because of the impact of tropical processes on the global climate. Additionally, Megha-Tropiques forms part of a ten-satellite constellation of the Global Precipitation Measurement (GPM) project being put up by NASA, JAXA, ISRO and CNES. In view of the increasing demand for Megha-Tropiques' data, in May 2014 the mission was extended for a few more years.

SARAL

Launched aboard an ISRO's PSLV 16 months after the Megha-Tropiques mission, SARAL (Satellite for Argos and Altika, also meaning "easy" in Sanskrit) is a joint ISRO–CNES mission for operational oceanography development. It is designed to study the circulation of ocean currents and to measure ocean surface topography, by providing global measurements of the sea surface height with unprecedented precision (in the order of one millimetre).

As defined by the agreement, ISRO provided the satellite bus and the PSLV launch to a heliosynchronous circular orbit (altitude of 865 km), while CNES provided an integrated payload module including the instruments and a part of the ground system. The SARAL mission is two-pronged in that there are two independent payloads: the Altika radar altimeter operating at high frequency in the Ka band, and hence giving the measurements increased accuracy in terms of vertical and spatial resolution, time decorrelation of echoes, and range noise; and the Argos Data Collection System, which collects valuable and precise data from thousands of transmitters at sea and on land about various oceanographic developments and phenomena.

The data collected by SARAL have proved particularly valuable and with a quality generally better than Jason-2's products. They have enabled better observation of ice, rain coastal zones, lakes, rivers and wave heights, which has led to important breakthroughs in the field of hydrology (Bouhey-Klapisz 2017). Over 20 Indian and 18 French institutions have been involved in the scientific activities of SARAL. In addition, SARAL data have contributed to the Global Ocean Data Assimilation Experiment (GODAE) and have been integrated into both EUMETSAT and French oceanography programmes. Likewise, the mission has also played an instrumental role for the SWAT altimeter mission of CNES and NASA.

Fig. 6.3 ISRO–CNES
joint Post Initiative (Credit:
CNES)

two sides organised a number of space-related events, including a joint ISRO–CNES exhibition and the release of joint stamps carrying images of Megha-Tropiques and SARAL (see Fig. 6.1). Even more importantly, they also signed a new agreement covering new cooperative ventures and detailing a "Programme for Reinforced ISRO–CNES Cooperation in Space Activities" (Fig. 6.3).

As underlined by CNES, the purpose of the agreement is to "strengthen the cooperation in place in remote sensing, telecommunication and weather satellites, space science and planetary exploration, data collection and location, operation on ground receiving stations, management of space missions, research and applications" (Centre National D'Etudes Spatiales 2015). More specifically, the new agreement puts the spotlight on three major cooperation projects, for which implementing arrangements were signed and cooperation structures put in place in January of 2016, on the occasion of President Hollande's state visit to India:

- A joint Thermal Infrared Satellite (TIR) mission for global heat exchange mapping
- A joint Argos-4 mission to be embarked aboard the 2018 Oceansat-3 satellite for the collection of climate and wildlife data
- A cooperative undertaking in the field of planetary exploration

Both the TIR and Oceansat-3–Argos missions are a consistent continuation of the fruitful Indo-French cooperation in the fields of climate- and weather-related Earth observation through Megha-Tropiques and SARAL. Primarily designed to map heat exchanges on the Earth's surface, the TIR satellite will provide remote sensing data

> in the thermal infrared and a visible part of the electromagnetic spectrum, at a high spatial scale (resolution of a few tens of metres) coupled with a high temporal scale (revisit of one to three days) with two main objectives: ecosystem stress and water use monitoring; and coastal zone monitoring and management (Blamont 2017).

Similarly, the Oceansat-3 mission, scheduled for launch in 2018, will incorporate CNES' latest-generation Argos-4 instrument for data collection and global location with the objective of supporting the observation of ocean parameters such as currents, temperature and colour and the collection of environmental and wildlife data such as ocean temperature profiles, river levels and animal migrations.

More broadly, the Oceansat-3–Argos mission and the joint TIR satellite confirm CNES and ISRO's commitment to maintaining continuous monitoring of climatology parameters through space and represent concrete contributions to the fight against climate change that was endorsed by the world's space agencies with the "Declaration of New Delhi". Strongly supported by both ISRO Chairman Kiran Kumar and CNES President Jean-Yves Le Gall and adopted on 3 April 2016 by more than 60 nations, the Declaration of New Delhi expresses the will of all the major spacefaring nations to work together towards the establishment of a global system of satellites dedicated to measuring and monitoring climate change indicators. The declaration thus affirms the strong commitment to the COP21 Paris Agreement of December 2015, which, in fact, highlighted the need for an independent measuring system to tackle greenhouse gas emission and the key role of satellites in studying and preserving the Earth's climate (Centre National D'Etudes Spatiales 2016).[31] The Oceansat-3 and the TIR satellite missions hence represent important contributions in this respect.

In addition to these two missions, the 2016 ISRO–CNES agreement also envisaged the creation of a joint space science working group tasked with the definition of France's possible participation in India's future interplanetary missions. While details on how exactly this involvement will materialised remain unknown as of January 2017, there are indications of a probable contribution to the Venus mission ISRO is planning for the early 2020s.

As a final confirmation of the renewed dynamism in this decade-old partnership in space, three additional developments deserve to be mentioned. The first is CNES' nascent interest and involvement in the activities of India's New Space ecosystem. In line with the recent emergence of India as one of the most promising hubs for New Space activities, CNES and Axiom Research Lab/Team Indus, an Indian start-up company participating in the Google Lunar X Prize,[32] reached an agreement in January 2017 for the provision of two latest-generation CASPEX microcameras, developed by CNES in partnership with French firm 3DPlus.[33] Equipped with

---

[31] The Declaration of New Delhi, which came into effect on 16 May 2016, more specifically calls for the maturation of "space-based operational tools combining in-situ measurements and increased computing resources. To this end, space agencies will need to develop new technologies and encourage their research community to contribute actively with new models. Success will depend above all on cooperation to cross-calibrate instruments and validate their measurements. Certain satellites like GOSAT for JAXA and OCO-2 for NASA are already paving the way, and in the near future TANSAT for China, the Copernicus programme's Sentinel series and of course MERLIN for CNES and DLR, and MicroCarb for CNES" (Centre National D'Etudes Spatiales 2016).

[32] The Google Lunar X Prize is a competition that awards a cumulative $30 million in a competition to land an unmanned spacecraft on the moon by the end of 2017 and record a high definition video of at least half a kilometre of travel on the celestial body before returning back to Earth.

[33] CASPEX (Colour Cmos-Based Microcamera for Space Exploration) are microcameras using patented technology that reduces the size of an optical imaging instrument by a factor of ten.

CNES' cameras, Team Indus' lunar rover is scheduled to be launched aboard the Indian PSLV on 20 December 2017. This interesting participation of CNES in the activities of a non-institutional actor "reflects a diversification of its partnerships in India and its desire to work closely with new players of the New Space movement" (Consulate General of France in Bangalore 2017).

A second important development in Indo-French space relations was the establishment, in early 2017, of a joint working group dedicated to rocket technology. Following the organisation of a dedicated workshop in Paris in October 2016, on 9 January 2017 CNES President Jean-Yves Le Gall and ISRO Chairman Kiran Kumar signed an agreement covering engineering exchanges and the setting-up of a joint working group to study future launcher technologies, including those related to full or partial rocket reusability. Experienced ISRO engineers will now receive training at CNES, Arianespace and ASL before the technical working group starts to "gauge synergies between ongoing developments and look at proposed future concepts, especially in the domain of reusable launch vehicles" (Centre National D'Etudes Spatiales 2017; de Selding 2017).

A third and perhaps even more audacious subject area of future Indo-French space cooperation that is now under consideration relates to the possible sharing of defence-related space products for SIGINT and maritime security. While the zero-threat dimension and the bipartisan political support that the relationship enjoys in both countries could certainly allow for more strategic cooperation including the sharing of technologies and intelligence exchange (Observer Research Foundation 2013), it remains to be seen how far space-based defence cooperation can go before clashing with sovereignty issues.

6.3.4   Evaluating India–Europe Space Relations

Over the past 50 years, Europe has played a major role as cooperation partner for India, be it at national or pan-European level. The amount of cooperative ventures that has taken place years is remarkable, and in many cases, such cooperation has proved of utmost importance for the emergence of India in the international space arena. In spite of this importance – or precisely because of it – the relationship does not seem to have been sufficiently leveraged by the European partners.

ESA–ISRO cooperation has been fragmented – not to say erratic, with very little going on after the cooperation experience on Chrandrayaan-1 almost 10 years ago. EU–India space relations have been likewise devoid of an overarching policy purpose and commitment, but have proved even more displeasing in terms of actionable policy measures and deliveries. As for EUMETSAT–ISRO cooperation – which appears today reliable, mutually beneficial and enduring – it is still far from being

---

CASPEX is reprogrammable and radiation-tolerant, making it suited to a range of space missions. After their maiden flight for the Team Indus lunar mission, CASPEX cameras will next equip the NASA's Mars 2020 Rover Mission (Consulate General of France in Bangalore 2017).

ground-breaking when compared, for instance, to the much deeper relations EUMETSAT has with NOAA.

On a more positive side, there is a need to acknowledge the special relation that CNES has with ISRO. But also in this case – as Jacques Blamont highlighted – many

> ingredients have been wanting in the course of India-French cooperation in space, kept at relatively low level by lack of motivation and general sluggishness. If the ISRO-CNES balloons and satellites have provided good scientific results, these operations have happened on a sporadic, haphazard, roadmap without any strategic view. Balloons studies of monsoon or pollution were interrupted when LMD scientists obtained their DSc. […] The Megha-Tropiques programme was virtually abandoned after many years of hopeless discussions between engineers of Toulouse and Bangalore and the project was saved in extremis by ISRO Chairman. The programmatic management of the cooperation has not been up to par with the excellence of the instrumentation and to the potential of Saphir and Altika (Blamont 2017).

An analogous line of discourse could certainly be made for other space cooperation experiences at the level of other national space agencies in Europe as well as for industry, which has generally left much cooperation potential unexplored, despite both Airbus and Thales being major industrial partner for India in the broader aerospace and defence sector.

From an overall perspective, what appears to have thwarted the deepening of Indo-European space cooperation ties is both the presence of a plethora of objective barriers as well as the dearth of a strong political vision and commitment to make cooperation happening. A more detailed reflection on these programmatic, industrial and political hurdles will be provided in Chap. 8, which more specifically aims to elaborate on the possible future evolution of India–Europe space relations.

References

Agenzia Spaziale Italiana. (2014). *Piano Triennale delle Attività 2015–2017*. Rome: Agenzia Spaziale Italiana.
Airbus. (2017). *Airbus in India*. Retrieved 2 February, 2017, from Airbus: http://www.airbusgroup.com/int/en/group-vision/global-presence/india.html
Al-Ekabi, C. (2016). *The future of European commercial spacecraft manufacturing*. Vienna: European Space Policy Institute.
Aliberti, M. (2013, April). Regionalisation of space activities in Asia? *ESPI Perspectives 66*.
Aliberti, M. (2015). *When China goes to the Moon….* Vienna: Springer.
Antrix Corporation. (2015). *Satellite platform*. Retrieved 22 August, 2016, from Antrix: http://www.antrix.gov.in/business/satellite-platforms
Bigot, J.-C. (2017). 40 Years of Cooperation between ESA and ISRO. In*India in space: The forward look to international cooperation*. Vienna: European Space Policy Institute.
Blamont, J. (2017). Cooperation in space between India and France. In R. P. Rajagopalan & N. Prasad (Eds.), *Space India 2.0 commerce, policy, security and governance perspectives* (pp. 215–233). New Delhi: Observer Research Foundation.
Bommakanti, K. (2009). Satellite integration and multiple independently retargetable reentry vehicles technology: Indian–United States Civilian Space Cooperation. *Astropolitics: The International Journal of Space Politics & Policy, 7*(1), 7–31.
Bouhey-Klapisz, C. (2017). Space cooperation between India and France. In*India in space: The forward look to international cooperation*. Vienna: European Space Policy Institute.

Brown, P. J. (2010, January 22). India's targets China's satellites. Retrieved November 18, 2016, from Asia Times: Brown, Peter J "India's targets China's satellites. *Asia Times* http://www.atimes.com/atimes/South_Asia/LA22Df01.html

Busvine, D., & Pinchuk, D. (2016, October 15). India, Russia agree to missile sales, joint venture for helicopters. Retrieved May 20, 2017, from *Reuters*: https://www.reuters.com/article/us-india-russia-helicopters/india-russia-agree-to-missile-sales-joint-venture-for-helicopters-idUSKBN12F058

Centre National D'Etudes Spatiales. (2015, April 10). *CNES and ISRO sign new cooperation agreement at the Elysée Palace.* Retrieved 20 October, 2016, from CNES: https://presse.cnes.fr/en/cp-9834

Centre National D'Etudes Spatiales. (2016, May 18). *New Delhi declaration comes into effect. World's Space Agencies working to tackles climate change.* Retrieved 11 September, 2016, from CNES: https://presse.cnes.fr/en/new-delhi-declaration-comes-effect-worlds-space-agencies-working-tackle-climate-change

Centre National D'Etudes Spatiales. (2017, January 9). *France-India Space Cooperation. Agreement signed on future launchers and lunar exploration.* Retrieved 18 February, 2017, from CNES: https://presse.cnes.fr/sites/default/files/drupal/201701/default/cp003-2017_-_cooperation_spatiale_entre_la_france_et_linde_va.pdf

Consulate General of France in Bangalore. (2017, April 10). *Signing of ISRO-CNES and CNES-TeamIndus agreements.* Retrieved 11 April, 2017, from Consulate General of France in Bangalore: https://in.ambafrance.org/Signing-of-ISRO-CNES-and-CNES-TeamIndus-agreements-14402

Correll, R. R. (2006). U.S.-India space partnership: The jewel in the crown. *Astropolitics: The International Journal of Space Politics & Policy, 4*(2), 159–177.

Council of the European Union. (2005, September 7). *The India-EU strategic partnership joint action plan.* Retrieved 2017, from Council of the European Union: http://www.consilium.europa.eu/ueDocs/cms_Data/docs/pressData/en/er/86130.pdf

Council of the European Union. (2016, March 30). *EU-India Agenda for Action 2020.* Retrieved 10 April, 2016, from Council of the European Union: http://www.consilium.europa.eu/en/meetings/international-summit/2016/03/30/

De Cosmo, V. (2006). *La partecipazione ASI alla missione ISRO OCEANSAT-2.* Roma: Agenzia Spaziale Italiana.

de Selding, P. (2011, October 12). ISRO launches joint French-Indian satellite to study tropical monsoons. *Space News, 2016* (September), 10.

de Selding, P. (2017, January 10). *France and India to study future rocket designs, including reusable vehicles.* Retrieved 11 February, 2017, from Space Intel Report: https://www.spaceintel-report.com/france-india-to-study-future-rocket-designs/

Defence Aerospace. (2002, January 9). *India joins international charter on space and disaster.* Retrieved 10 November, 2016, from Defence Aerospace: http://www.defense-aerospace.com/articles-view/release/3/8175/india-signs-two-agreements-on-space-(jan.-10).html

Der Spiegel. (2015, June 19). *We could build all kinds of things with Moon concrete.* Retrieved 20 November, 2016, from Spiegel Online: We Could Build All Kinds of Things with Moon Concrete.

Deshpande, R. (2006, October 16). *India may quit EU-led GPS project.* Retrieved 20 November, 2006, from Times of India: http://timesofindia.indiatimes.com/india/India-may-quit-EU-led-GPS-project/articleshow/2172710.cms

Diplomacy & Foreign Affairs. (2013, November 12). *The EADS Group in India.* Retrieved 7 December, 2016, from Diplomacy & Foreign Affairs: http://diplomacyandforeignaffairs.com/the-eads-group-in-india/

DLR. (2016, January 25). *A model for energy providers in India – the DLR test and qualification laboratory for solar power plants.* Retrieved 14 February, 2017, from DLR: http://www.dlr.de/dlr/en/desktopdefault.aspx/tabid-10081/151_read-16498/#/gallery/21772

Embassy of India. (2016, November 30). *India Germany foreign office consultations.* Retrieved 1 February, 2017, from Embassy of India: https://www.indianembassy.de/pages.php?id=241

EUMETSAT. (2014, February). *International cooperation: Benefits for global users.* Retrieved 30 March, 2017, from EUMETSAT: http://www.eumetsat.int/website/home/InSight/ NewsUpdates/DAT_2169186.html?lang=EN

EUMETSAT. (2016, April 1). *EUMETSAT discusses Asia-Pacific Cooperation.* Retrieved 3 April, 2017, from EUMETSAT: http://www.eumetsat.int/website/home/News/DAT_2994861. html?lang=EN

EUMETSAT. (2017a). *EUMETCAST.* Retrieved 20 March, 2017, from EUMETSAT: http://www. eumetsat.int/website/home/Data/DataDelivery/EUMETCast/index.html

EUMETSAT. (2017b). *Indian Ocean data coverage.* Retrieved 30 March, 2017, from EUMETSAT: http://www.eumetsat.int/website/home/Data/MeteosatServices/IndianOceanDataCoverage/ index.html

EUMETSAT. (2017c, February 1). *Meteosat-8 Now Prime IODC Satellite.* Retrieved 31 March, 2017, from EUMETSAT: http://www.eumetsat.int/website/home/News/DAT_3363066. html?lang=EN

European Commission. (2005, September 7). *The GALILEO family is further expanding: EU and India seal their agreement.* Retrieved 12 January, 2017, from European Commission.

European Global Navigation Satellite Systems Agency. (2014, February 19). *Boost for EU-India Cooperation.* Retrieved 5 January, 2017, from European Global Navigation Satellite Systems Agency: https://www.gsa.europa.eu/news/boost-eu-india-cooperation

European Global Navigation Satellite Systems Agency. (2015, June 26). *GNSS in Asia – Support on international activities.* Retrieved 1 February, 2017, from European Global Navigation Satellite Systems Agency: https://www.gsa.europa.eu/gnss-asia-support-international-activities

European Space Agency. (2005). *Opportunity for collaboration with India on Chandryaan-1 Mission to the Moon.* Retrieved 10 November, 2016, from ESA: http://sci.esa.int/science-e/ www/object/doc.cfm?fobjectid=37906

European Space Agency. (2006). *Summary. Europe's first Lunar adventure.* Retrieved 16 February, 2017, from European Space Agency: http://sci.esa.int/smart-1/31407-summary/

European Space Agency. (2007, September 25). *ESA and India tighten relations at.* Retrieved 8 January, 2017, from ESA: http://www.esa.int/About_Us/Welcome_to_ESA/ ESA_and_India_tighten_relations_at_IAC_2007/(print)

European Union External Action Service. (2008, September 29). *The EU-India Joint Action Plan (JAP).* Retrieved 7 March, 2017, from European Union External Action Service: http://eeas. europa.eu/archives/docs/india/sum09_08/joint_action_plan_2008_en.pdf

EUTELSAT. (2017). *Eutelsat Asia, expanding reach across the Asia Pacific.* Retrieved 1 March, 2017, from EUTELSAT Asia: http://www.eutelsatasia.com/home/eutelsat-asia.html

Global Times. (2014, February 18). India uncertain as Abe looks for anti-China Alliance. *Global Times.*

GNSS Asia. (2016). *Country profile – India.* Retrieved 3 January, 2017, from GNSS Asia: http:// www.gnss.asia/sites/default/files/Countryprofiles_INDIA_2016_v4.pdf

GNSS Asia. (2017). *Opportunity & partnering.* Retrieved 10 February, 2017, from GNSS Asia: http://www.gnss.asia/node/27

Harvey, B., Smid, H. H., & Pirard, T. (2010). *Emerging space powers.* Chichester: Springer – Praxis.

Horikawa, Y. (2017). Space security, sustainability, and global governance: India-Japan collaboration in outer space. In R. P. Rajagopalan & N. Prasad (Eds.), *Space India 2.0. commerce, policy, security and governance perspectives.* New Delhi: Observer Research Foundation.

Indian Space Research Organisation. (2002, January 9). *ISRO signs cooperative agreement with European Space Agency.* Retrieved 1 March, 2017, from Wikram Sarabhai Space Centre: http://www.vssc.gov.in/VSSC_V4/index. php/2002/79-press-release-articles/789-isro-cooperative-agreement-esa

Indian Space Research Organisation. (2006, February 20). *EADS Astrium-ISRO alliance sealed. First contract with Eutelsat for W2M satellite.* Retrieved 12 December, 2016, from ISRO: http://www.isro.gov.in/update/20-feb-2006/ eads-astrium-isro-alliance-sealed-first-contract-with-eutelsat-w2m-satellite

Indian Space Research Organisation. (2008, September 30). *India signs agreements with France on cooperation in space.* Retrieved 1 October, 2016, from ISRO.

Indian Space Research Organisation. (2012, February). *Global applications of Oceansat-2.* Retrieved 31 March, 2017, from United Nations Office for Outer Space Affairs: http://www.unoosa.org/pdf/pres/stsc2012/tech-49E.pdf

Indian Space Research Organisation. (2016, September 4). *Bengaluru Space Expo (BSX)-2016.* Retrieved 5 October, 2016, from ISRO: http://www.isro.gov.in/bengaluru-space-expo-bsx-2016

Indian Space Research Organisation. (2017a, February 15). *PSLV-C37 successfully launches 104 satellites in a single flight.* Retrieved from ISRO: http://www.isro.gov.in/update/15-feb-2017/pslv-c37-successfully-launches-104-satellites-single-flight

Indian Space Research Organisation. (2017b). *Government of India – Department of Space – Indian Space Research Organisation.* Retrieved from List of Communication Satellites: http://www.isro.gov.in/spacecraft/list-of-communication-satellites

Indo-German Science & Technology Centre. (2017). *About IGSTC.* Retrieved 3 February, 2017, from IGSTC: http://www.igstc.org/about_us.html

Japan Aerospace Exploration Agency. (2005, March 31). *JAXA Vision – JAXA 2025.* Retrieved 20 March, 2016, from JAXA: http://www.docstoc.com/docs/87745747/JAXA-Vision

Jayaraman, K. (2006, November 14). *ISRO seeks government approval for manned spaceflight program.* Retrieved 13 December, 2016, from Space News: http://spacenews.com/isro-seeks-government-approval-manned-spaceflight-program/

Korovkin, V. (2017). Evolution of India-Russia partnership. In R. P. Rajagopalan & N. Prasad (Eds.), *Space India 2.0; commerce, policy, security and governance perspectives* (pp. 246–262). New Delhi: Observer Research Foundation.

Kyodo News Service. (2008, September 25). Japanese official calls for space cooperation with China. *Kyodo News Service.* Retrieved from Space Daily: http://www.spacedaily.com/reports/Shenzhou_7_Astronauts_Brace_For_Space_Walk_999.html

Lele, A. (2013). *Asian Space race: Rethoric of reality?* New Delhi: Springer.

Lele, A. (2014). *Mission Mars: India's quest for the red planet.* New Delhi: Springer.

Lele, A. (2015, September). Space collaboration between India and France – Towards a new era. *AsieVisions, 78.*

Maass, R. (2017, March 14). *India test fires BrahMos extended range missile.* Retrieved from Space Daily: http://www.spacedaily.com/reports/India_test_fires_BrahMos_Extended_Range_missile_999.html

Malone, D. M. (2011). *Does the elephant dance? Contemporary India foreign policy.* New Delhi: Oxford University Press.

Mathieu, C. (2008). *Assessing Russia's Space cooperation with China and India.* Vienna: European Space Policy Institute.

Missile Technology Control Regime. (n.d.). Retrieved from http://mtcr.info/

Miller, J. B. (2013, May 29). The Indian piece of Abe's security diamond. Retrieved May 30, 2017, from *The Diplomat*: https://thediplomat.com/2013/05/the-indian-piece-of-abes-security-diamond/

Moltz, J. C. (2012). *Asia's space race. National motivations, regional rivalries, and international risks.* New York: Columbia University Press.

National Aeronautics Space Administration. (2016). *NASA-ISRO SAR Mission (NISAR).* Retrieved 8 February, 2017, from NASA-JPL: https://nisar.jpl.nasa.gov/

NDTV. (2013, August 16). *Human space flight mission off ISRO priority list.* Retrieved 28 September, 2016, from NDTV: http://www.ndtv.com/article/india/human-space-flight-mission-off-isro-priority-list-406551

Observer Research Foundation. (2013, 1 May). *Expert policy dialogue on India-France relations.* Retrieved 10 September, 2016, from ORF: http://www.orfonline.org/research/experts-policy-dialogue-on-india-france-relations/

OECD. (2014). *The space economy at a glance 2014.* Paris: OECD Publishing.

Peter, N. (2008). Developments in space policies programmes and technologies throughout the world and Europe. In K.-U. Schrogl, C. Mathieu, & N. Peter (Eds.), *ESPI yearbook 2006/2007: A new Impetus for Europe* (p. 96). Vienna: Springer.

Peter, N. (2009). Developments in space policies programmes and technologies throughout the world and Europe. In K.-U. Schrogl, C. Mathieu, & N. Peter (Eds.), *ESPI yearbook on space policy 2007/2008: From policies to programmes* (p. 81). Vienna: Springer.

Prasad, M. (2015). ISRO and international cooperation. In M. Rao (Ed.), *From fishing hamlet to red planet: India's space journey*. New Delhi: Harper Collins.

Reuters. (2015, June 13). *European space chief suggests making room for India, China on station*. Retrieved 15 February, 2017, from Reuters: http://in.reuters.com/article/space-station-china-idINKBN0OT0J520150613

Samson, V. (2017). India-US: New dynamism in old partnership. In R. P. Rajagopalan & N. Prasad (Eds.), *Space India 2.0 commerce, policy, security and governance perspectives* (pp. 235–244). New Delhi: Observer Research Foundation.

SES. (2017). *Networks in Asia*. Retrieved 1 March, 2017, from SES: https://www.ses.com/asia/asia

Sheehan, M. (2007). *The international politics of space*. New York: Routledge.

Smith, J. M. (2014, February 10). India and China: The end of cold peace? *The National Interest*.

Somasekhar, M. (2007, September 27). Italian space agency, ISRO to expand ties. *The Hindu Business Line*.

Sourbès-Verger, I. (2016, December 30). EU-India cooperation on space and security (*IAI Working Papers 16|38*).

Sputnik News. (2017, March 9). *BrahMos Aerospace to Be Indian DRDO's commercial wing abroad*. Retrieved from Space Daily: http://www.spacedaily.com/reports/BrahMos_Aerospace_to_Be_Indian_DRDOs_Commercial_Wing_Abroad_999.html

Suzuki, K. (2013). The contest for leadership in East Asia: Japanese and Chinese approaches to outer space. *Space Policy, 29*(2), 99–105.

Suzuki, K. (2017). An Asian space partnership with Japan? In R. R. Pillai & N. Prasad (Eds.), *Space India 2.0. commerce policy, security and governance perspectives*. New Delhi: Observer Research Foundation.

Television Post. (2013, December 4). *Dish TV to add transponder capacity, SES' new satellite launches successfully*. Retrieved 1 March, 2017, from Television Post: http://www.television-post.com/dth/dish-tv-to-add-transponder-capacity-ses-new-satellite-launches-successfully/

Times of India. (2004, November 8). *India, EU close to agreement on Galileo project*. Retrieved 12 January, 2017, from Silicon India News: http://www.siliconindia.com/shownews/India_EU_close_to_agreement_on_Galileo_project-nid-25955-cid-3.html

Track Dish. (2015, August 10). *38 Indian channels started on Eutelsat 70B*. Retrieved 1 March, 2017, from Track Dish: http://www.trackdish.com/38-indian-tv-channel-started-on-eutelsat-70b/

University Corporation for Atmospheric Research. (2017). *INDOEX: Indian Ocean Experiment*. Retrieved 31 March, 2017, from Earth Observing Laboratory Data: https://data.eol.ucar.edu/project?INDOEX

Vasagam, R. (2015). APPLE in retrospect. In M. P. Rao (Ed.), *From fishing hamlet to red planet. India's space journey*. New Delhi: Harper Collins.

Verma, R. (2016, February 25). *"Bringing U.S.-India space cooperation to the edge of the universe" Special Address by U.S. Ambassador to India Richard Verma at the ORF Kalpana Chawl*. Retrieved from US Embassy and Consulates in India: https://in.usembassy.gov/bringing-u-s-india-space-cooperation-to-the-edge-of-the-universe-special-address-by-u-s-ambassador-to-india-richard-verma-at-the-orf-kalpana-chawl/

Vicziany, M. (2015). EU-India security issues: Fundamental incompatibilities. In P. Winand, M. Wicziany, & P. Datar (Eds.), *The European Union and India: Rethoric or meaningful partnership*. Cheltenham: Edward Elgar Publishing.

Weiss, M. (2014, October). France-India: A fraternal relation. *CNES Mag* (pp. 54–57).

Wülbers, S. A. (2011). *The paradox of EU-India relations: Missed opportunities in politics, economics, development, cooperation, and culture*. Plymouth: Lexington Books.

Chapter 7
India's Path Forward in Space

This chapter provides a comprehensive reflection on India's long-term pathway in space. The aim is to identify and examine long-term factors that will likely drive the evolution of the Indian space programme and to describe a most probable scenario for the future of space activities in this country.

Considering that Indian policy-makers have not detailed a precise vision of the strategic interests, needs and priorities guiding the future evolution of their space programme, such reflection proves a particularly relevant – yet challenging – exercise. In fact, whereas short-term directions can be assumed by assessing the challenges that the Indian space sector, in particular ISRO, currently faces or prepares to face in the near future, it appears manifest that future space activities cannot be predicted in isolation from the wider national social, economic, political and technological context. In order to examine anticipated developments of Indian space activities, it is therefore essential to consider India's future path in space closely with broader issues for India.

Accordingly, this chapter will focus first on the country's future outlook and subsequently provide a reflection on the evolution of its civil, military and commercial space efforts. Considerations on how the space ecosystem will be adapted to address these evolutions will also be provided.

7.1 Visualising India's Long-Term Civil, Military and Commercial Space Efforts

This section offers a reflection on the possible long-term direction of India's civil, military and commercial space efforts. Given that any space programme evolves within a broader socio-economic, political and technological context, assessing what these activities could be and what posture India could adopt in the future requires to first elaborate on the country's long-term prospects.

© Springer International Publishing AG 2018
M. Aliberti, *India in Space: Between Utility and Geopolitics*, Studies in Space
Policy 14, https://doi.org/10.1007/978-3-319-71652-7_7

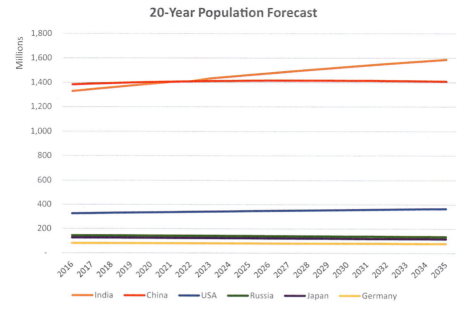

Fig. 7.1   India's population forecast (*Source*: United Nations)

7.1.1   India's Long-Term Prospects

Back in the early 1960s, when the space programme was initiated, India was a young nation inhabited by 468 million people and with a GDP of nearly $43 billion (current USD). In 2015, just over half a century later, India had a population of about 1,25 billion – making roughly every 6th human on Earth an Indian – and a GDP of $2,1 trillion (current USD): the 7th highest GDP worldwide and the 3rd highest in Asia behind China and Japan. Even more interesting than that is that the population and economy are expected to continually grow in the future, generating deep impacts at national, regional and global level.

Demographic Trends
According to the United Nations, India's population is set to outnumber that of China – the world's most populous country – in as little as 5 years from now, as graphically displayed in Fig. 7.1.
The weight and potential of the two Asian nations become obvious when observing the gap between the population forecasts of India and China in comparison to Japan, European nations like Germany or even world GDP leader the United States. Unlike Japan, Germany or Russia – all of whom are bound to see a significant decline in population until 2030 and an even steeper decline forecasted until the year 2100 – India is projected to grow steadily for the next five decades, at which point it will

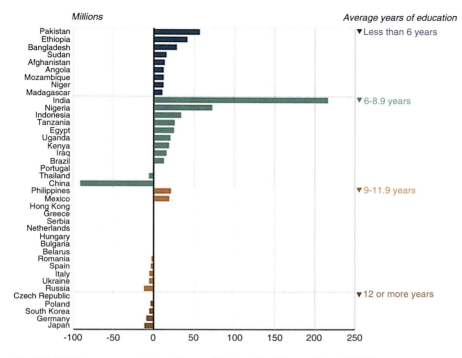

Fig. 7.2 Working age populations (*Source*: National Intelligence Council 2017)

have reached a peak population of around 1,754 billion. The slope declines slowly from there, guiding India to 1,65 billion inhabitants by 2100. India will have roughly 655 million, or 65%, more inhabitants than current world leader China by the turn of the century. That is approximately today's population of the United States, Russia, Germany, France and the United Kingdom combined (Our World in Data 2017).

A general trend typically associated with worldwide demographic patterns is ageing. While this is true for all OECD countries and even China, it will not be the case in India, where the number of young people is in fact increasing. By 2020, demographic growth will turn it into the youngest country on Earth, with roughly two-thirds of its population in working age. Between 2017 and 2035, India will be the country with the largest increase in working age population: more than 200 million people during that period. In contrast, during the same period, the populations of Europe and China will witness the largest decrease in working age populations, as illustrated in Fig. 7.2.

Having such a young population will entail both challenges and opportunities. Some studies have highlighted the negative drawbacks, such as the risk of a demographic disaster if not enough jobs are created. The National Intelligence Council,

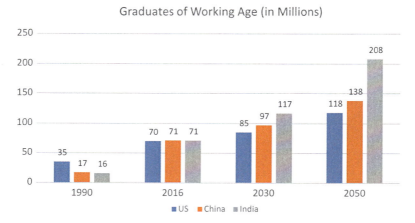

Fig. 7.3 India's demographic dividend (*Source*: IndoGenius 2016b)

for instance, has pointed out that growth in working age population in countries with low education levels such as India will put them

> at disadvantage in the evolving global economy, which will favour higher-skilled workers. […] Worldwide, low-value-added manufacturing – historically the steppingstone to economic development for poor countries, and a pathway to prosperity for aspiring workers – will tend toward needing ever-fewer unskilled workers as automation, artificial intelligence, and other manufacturing advances take effect. (National Intelligence Council 2017)

Other studies have, on the contrary, predicted a demographic dividend for India, with a copiousness of college-educated – and hence economically productive – young people. In this respect, some comparative figures on the number of graduates of working age prove particularly enlightening. As documented by IndoGenius, in 1990, the United States could boast the largest number of graduates of working age in the world, roughly 35 million people, while China had just 17 million and India 16 million. In 2016, there were about the same number of students enrolled in American, Chinese and Indian higher education systems. However, by 2030, India is projected to have 117 million graduates at working age, China 97 million and the United States 85 million, while 20 years later (by 2050), the United States will have 118 million graduates of working age, China 138 million and India an impressive 208 million (see Fig. 7.3). This "means that India is going to have nearly double the number of graduates of the U.S., and 70 million more than China. Young Indians, therefore, will represent the most important human capital on the planet" (IndoGenius 2016b). One of the key questions that needs to be asked, however, is whether the Indian government will be able to seize this potential and offer young Indians the tools they will need to thrive.

Urbanisation Prospects

Another key trend closely associated with demographic growth is urbanisation. With only a third of Indians living in the urban areas of the country, India has, today, one of the lowest urbanisation densities in the world. India ranks an underwhelming

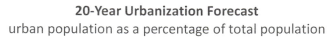

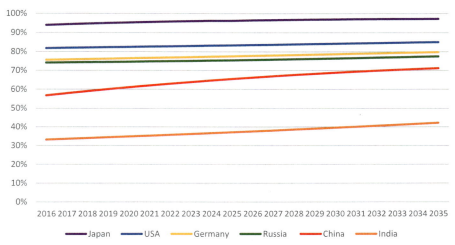

Fig. 7.4 Urbanisation forecast (*Source*: United Nations 2014)

156th of 193 nations in urbanisation – near the 20th percentile of all United Nations member states. In absolute terms, however, India already has the second largest urban population in the world (410 million people) after China (710 million) but well ahead of the United States (263 million), Brazil (173 million), Indonesia (134 million), Japan (118 million) and the Russian Federation (105 million). Taken together, these countries account for more than half of the world's urban population (United Nations 2014). Figure Fig. 7.4 projects the future increases in urban population of various nations.

While its high number of rural inhabitants puts India far behind several nations in comparison, the population flow towards cities is expected to be substantial leading to an average annual growth of urban population of 9.4% until 2035: second only to China's astonishing 15.5% of urban inhabitants added but significantly ahead of Germany's 4.4% or the United States' 3.7%. Given the sheer size of the Indian population, this projected increase in India's urban population will be impressive by any measure. The UN estimates that India's urban population will reach 600 million by 2031, roughly 218 million more than today, and by 2050, India is projected to add 404 million urban dwellers (United Nations 2014).

As a consequence of this massive urbanisation, Indian cities are expected to reach massive sizes and to face dramatic challenges. From this standpoint, it suffices to recall that 53 cities in India already have populations greater than one million. Eight of these cities have populations of more than five million people, and four of those cities (New Delhi, Mumbai, Kolkata and Bangalore) have huge populations above ten million. New Delhi, according to the UN, is likely to become the largest urban agglomeration in the world, overtaking Tokyo sometime in the early 2030s

(Ibid., p. 13). Four of India's present cities with five to ten million inhabitants (Ahmadabad, Bangalore, Chennai and Hyderabad) are projected to become megacities in the coming years (for a total of seven megacities projected in the country by 2030). India's urban growth, however, is not going to happen just in the big cities. In fact, it will also take place in the so-called "peri-urban" areas and in thousands of towns and villages across the subcontinent.

While economic growth is expected to contribute to poverty reduction with a parallel consolidation of India's middle class, the country's demographic and urbanisation prospects are also expected to lead to formidable challenges. In fact, all of these cities, towns and villages will have ever-increasing demands for education, healthcare, retail and of course infrastructure, among other things.[1] India's public and private institutions and organisations will then need to adopt ambitious strategies to tackle the diversity of urban transformation and also find ways to solve the problem of the sheer number of locations involved if they are to successfully deliver the products and services that Indian citizens will need to prosper.

Natural Resources
Another important component of India's socio-economic projections is the increasing pressure on key resources such as food, water and energy, as well as their growing relationship with the anticipated effects of climate change on the country. Principally owing to the increase of the Indian population and the expansion of its middle class, the demand for these resources is projected to substantially increase over the next decades.

India's demand for food is expected to rise considerably, in part because of its growing population and in part because of increasing food consumption per capita.[2] Food consumption growth will likely be accompanied by changes in composition. The FAO estimates that patterns of food consumption in India will shift towards higher quality and more expensive food such as dairy products, meat, sugar and dairy products (Alexandratos and Bruinsma 2012).[3] Future demand for meat and dairy products should be met easily, in particular because India's food demand is expected to remain below the levels consumed by other developed and developing countries. Actually, challenges will arise from the consequences of increased production on agricultural land and water resources.[4]

---

[1] According to some estimates, India will have to build each year 700–900 million m² commercial and residential spaces, i.e. the equivalent of a city like Chicago! (Khandekar, Towards an EU-India Partnership on Urbanisation, 2015)

[2] The projection of calories in 2050 is expected to rise to 2825 kcal/person/day, up from 2300 in 2007. This projection is per se remarkable, considering that thus far "food demand in India has not been growing at anything near the rates one would expect from the high economic growth and the high prevalence of unsatisfied food needs" (Alexandratos and Bruinsma 2012).

[3] A rapid increase will take place in meat (the per capita consumption of which will increase to 9 kg in 2030 and 18 kg in 2050, up to 3.1 kg in in 2007), dairy products (which will grow from 67 kg to 110 kg per person), sugar (which will grow from 18.8 kg to 29 kg) and vegetable oils (rising to 18 kg, up from 8 kg in 2007).

[4] The amount of water required for livestock farming is far more than required to produce an equivalent amount of grain or vegetables. As noted by the National Intelligence Council "given

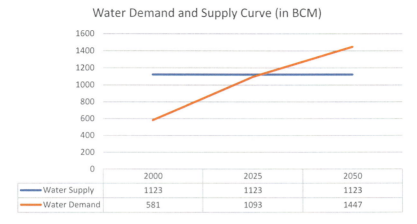

Fig. 7.5   Towards Water Scarcity? (*Source:* KPMG 2010)

The growth in annual water requirements is projected to reach 1,093 billion cubic metres (BCM) by 2025 and 1447 by 2050, well above the current utilisable water supply level of 1123 BCM estimated by the Ministry of Water Resources (FAO, UNICEF, SaciWATERs 2013).[5] The projected demand-supply gap – which is already affecting some parts of the country – is expected to seriously impact the availability of water per capita, which will decline by almost 36% in 2025 and 60% in 2050 as compared to 2001 levels (KPMG 2010). Efficient water management through technological innovation will thus be essential to avoid water scarcity and to meet the burgeoning requirements. Specifically, given that some 80% of water usage in India is used for agricultural irrigation, improving irrigation efficiency will be of utmost importance to preventing water shortage while ensuring long-term food security (Fig. 7.5).

Propelled by the projected demographic growth and economic expansion (see below), the demand for energy will also increase almost exponentially over the next 20 years, passing from 701 Mtoe in 2015 to 1603 Mtoe in 2035, accounting for 25% of global energy consumption growth (International Energy Agency 2015).[6] The International Energy Agency (IEA) projects that India will become the fourth largest consumer of energy by 2035 (after China, the United States and Europe) and its share of global energy demand will rise to 9% (see Fig. 7.6).[7] Demand for gas will

---

that agriculture uses 70% of global freshwater resources and livestock farming uses a disproportionate share of this, water management will become critical to long-term food security" (National Intelligence Council 2012).

[5] According to the Ministry of Water Resources, of the 1869 BCM of waters theoretically available, only 1123 BCM can be exploited due to topographic constraints and distributions effects.

[6] This demand growth (+129%) is more than double the non-OECD average of 52% and also outpaces each of the BRIC countries: China (+47%), Brazil (+41%) and Russia (+2) (BP 2017).

[7] Even if India is set to contribute more than any other country to the projected rise in global energy demand, its per capita energy demand in 2040 will still be 40% below the world average. Energy production is expected to grow by 122%, but imports will rise by 138%; hence, energy production as a share of consumption will decline marginally from 58% today to 56% in 2035 (BP 2017).

Fig. 7.6 Primary energy demand in 2035 (in Mtoe) (*Source*: International Energy Agency 2015)

grow by 162%, followed by oil (+120%) and coal (+105%). Renewables are expected to rise by almost 700%, nuclear by 317% and hydro by 97% (BP 2017).

Specific estimates apart, what needs to be underlined is the ever-stronger linkage between India's demand for resources such as energy, food and water and the problems stemming from the country's already grim environmental conditions. While increasing resource demand will most probably be met, the combination of growing resource demand with the anticipated effects of climate change will possibly generate costly trade-offs. In other words, "tackling problems pertaining to one commodity [will not] be possible without affecting supply and demand for the others" and without impacting on the climate (National Intelligence Council 2012). Absent positive synergies, the scope for such negative trade-offs is inevitably bound to increase. To echo the warning of the UK scientific adviser John Beddington, "the point is that we have to deal with increased demand for energy, increased demand for food, and increased demand for energy *together*, while mitigating and adapting to climate change" (Population Institute 2009).

India is one of the most vulnerable countries in the world to projected climate change impacts and is already "experiencing changes in climate and the impacts of climate change, including water stress, heat waves and drought, severe storms and flooding, and associated negative consequences on health and livelihoods" (Joint Global Change Research Institute; Battelle 2009). Because of its fast-growing population and the increased stress on natural resources, India will probably be even more severely impacted in the future. Global climate projections indicate several changes in India's future climate such as extreme heat, changing rainfall patterns triggering more frequent droughts as well as greater flooding in large parts of the country, glacier melt over the Himalayas and sea level rise (The World Bank 2013).

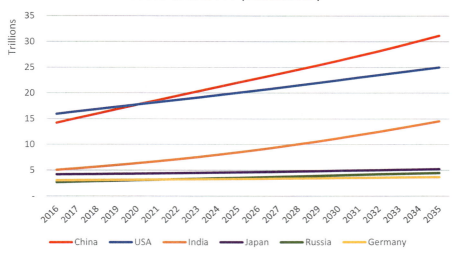

Fig. 7.7 Towards a new economic triad?

Economic Prospects

On a more positive side, although India's demographic and urbanisation trends are expected to lead to increasing pressure on production and management of food, water and energy, they will also offer important contributions to stimulating economic growth at national level. In this respect, India seems to have a bright future.

Several institutions and organisations, including the UN, have worked on predicting the future of the Indian economy. Forecasts consistently project that India's GDP growth rate will consolidate over the next years. More specifically, the International Monetary Fund (IMF) forecasts that the 7.6% annual increase we saw in 2015 will slowly ascend to an 8.6% growth rate increase by 2021 (International Monetary Fund 2017).

Looking further into the future, the OECD's long-term forecast gives insight on projections up to 2060, showing that India's economic growth will slow down gradually and in small fraction percentage increments after 2022, with 3.15% annual growth being the last projected value in 2060 based on calculations at 2005 purchasing power parity (PPP) (OECD 2017).

As far as the total value of GDP is concerned, Fig. 7.7 – using OECD-projected long-term data – shows Indian GDP forecasts compared to those of other nations over the next two decades.

According to these projections, China is set to dethrone the United States as the world's GDP leader by early 2020, while India will maintain its third place and march steadily towards the $15 trillion mark, leaving behind stagnating European countries, Russia and Japan. What remains uncertain is how potential international

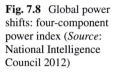

Fig. 7.8 Global power shifts: four-component power index (*Source*: National Intelligence Council 2012)

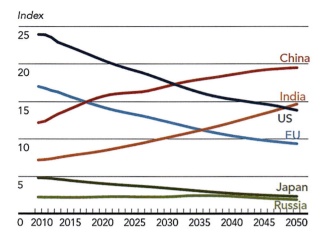

conflicts, technological advancements and market innovations or other external factors and "black swans" will directly or indirectly influence India's economy.

Be this as it may, there appears to be a wide consensus that world of the 2030s will witness a pronounced shift in economic power from the West to non-OECD Asia. Barring a major backlash, China alone is projected to have the world's largest economy, outstripping that of the United States some time before 2030, while India will probably become the rising economic superpower that China is today.[8] In the meantime, the economies of Europe, Japan and Russia will likely witness a relative decline, primarily because of their ageing and shrinking populations.

Power Shifts

This shift in the geo-economic barycentre of international affairs will be more importantly accompanied by a parallel shift in political power. Indeed, India's economic growth is likely to propel it to new levels of global influence, thus inching closer to the potential dethroning of many of the dominant world powers today. While from an overall perspective China, the United States and Europe will ostensibly remain the major geo-economic "hubs" of the world, accounting for slightly more than one half of global output and exercising the most influence on global issues, India will be able to take a more leading position, with a stronger influence on the regional and global scene (see Fig. 7.8). According to the National Intelligence Council, over the next three decades, India will reach the level of the United States and Europe in terms of national power based upon GDP, population size, military spending and technological investment (National Intelligence Council 2012).[9]

---

[8]During the next decade, China's rate of economic growth will continue to slow down, while India's will continue to rise. These forecasts have been largely built upon the demographic patterns of the two countries and in particular on China's progressive decline in working age population and India's continuous increase (National Intelligence Council 2012).

[9]The four-component power index is based upon GPD, population size, military spending and technological investment. The multi-component power index also includes health, education and governance (National Intelligence Council 2012).

This transformation may well turn India into an indispensable superpower exercising major influence and perhaps even vetoes over many international decisions. Issues such as global security and terrorism, climate change, environmental sustainability, development and international trade will also witness the exercise of a stronger influence by India (Sachdeva 2015).

It appears already evident that India – as pointed out by Henry Kissinger – "will be a fulcrum of twenty-first century order: an indispensable element, based on its geography, resources and tradition of sophisticated leadership, in the strategic and ideological evolution of the regions and the concepts of order at whose intersection it stands" (Kissinger 2014).

Concluding Remarks: The Need for a Different Path

India certainly has a bright future to look upon – *achhe din,* to echo Prime Minister Narendra Modi. India's development will also bring its share of challenges in the coming decades, which will likely dictate the need to pursue a different path and make different developmental choices from those made by the West.

This need is already visible in a wide range of issue areas. Taking mobility, or urban development more generally as an example, the rate of car ownership in India is today very small – just 6%, compared to 88% in the United States and 85% in Germany (Statista 2017). But even so, India's cities are already confronted with huge traffic problems that spawn major quality of life and public health issues, for instance, by aggravating the already high pollution levels or by creating enormous commuting times in Indian cities (IndoGenius 2016a). As poignantly asked by Nick Booker, what will these cities look like if Indian rates of car ownership continue to rise as per capita incomes rise? And what will be the effect on local, regional and even global pollution levels, if India were to adopt the same approach as Western countries? (IndoGenius 2016a)

To India's credit, the country's policy-makers seem fully aware of the need to pursue a different path from that of the West and are, in fact, trying to adopt leapfrog developments that will enable the country to seize most of its growth potential. With respect to the urbanisation issue, for instance, India's Ministry of Urban Development has put forward an ambitious plan for transforming 100 cities into smart cities, hence providing a key leapfrog opportunity to rethink future urban development by fostering deployment of new promising energy, transport, water management, sanitation and housing technologies to create sustainable cities able to meet both growth and sustainability objectives.

The country's need for a different path is likewise visible in many other issue areas such as education, healthcare or financial inclusion for the lower strata of Indian society, which will all require devising innovative solutions. For each of these grand challenges, there is an invaluable leapfrog opportunity that the Indian policy-makers are looking to seize. As mentioned in earlier chapters, the Indian government has already launched a number of long-term flagship initiatives that can boost the transformation of India, including *Digital India*, which focuses on digital empowerment of citizens through better provision of government services through ICT; *Skill India*, which aims to train over 400 million citizens by 2022 in key areas such as agricultural and industrial training; *Startup India*, which seeks to promote innovation and foster entrepreneurship, with the ultimate goal of transforming India into a nation of job creators rather than job seekers; *Make in India*, which seeks to position India as a global manufacturing hub as well as the top destination globally for Foreign Direct Investments (FDIs), generating economic growth and employment opportunities in the process; and *Clean India*, which is a cleanliness campaign aiming to enhance the quality of Indian citizens, by cleaning up the streets and infrastructure of India's cities, towns and rural areas and making India free of open defecation by 2019.

What is important to highlight is that for many of India's future grand challenges and prospective ways forward, there is an invaluable role that can be played by space, which can hence be expected to become an even more crucial backbone in supporting India's path towards economic prosperity as well as an increased geopolitical weight on the world scene. More specifically, the potential of space – which has been widely acknowledged by Indian policy-makers since the very beginning of the space age – will be most certainly enlisted to:

(a) Address economic and societal challenges through the development of novel applications and services.
(b) Support India's military build-up and soft power projection through the development of robust Command, Control, Communications, Computing, Intelligence, Surveillance and Reconnaissance (C4ISR) and SSA capabilities as well as bolder efforts in the area of space exploration and human spaceflight.
(c) Boost Indian industry's competitiveness and innovation as a means to further India's position in upper-end industrial sectors (including spin offs for defence, aviation and other sectors) and international (space) economy.

Specific considerations for each of these fields will be offered in the next sections.

7.1.2 Future Civil Space Efforts: Space Applications for India's Grand Challenges

When trying to visualise the future direction of India's civil space undertakings, a first consideration is that space technologies and associated services will certainly continue to be used to offer innovative solutions to India's societal grand challenges and fully seize its future growth potential. In this respect, India's socio-economic prospects that have been described in the previous section clearly shed valuable light, because they pinpoint key areas in which the role of space will likely become part of the solution. Below are some of the most important cases in point.

Urbanisation and Mobility
The growing urbanisation prospects highlighted in Sect. 7.1.1 will have a number of far-reaching impacts for India. As engines of productivity, urban centres will

stimulate economic growth and innovation, but at the same time, they will put serious pressures on the environment and on key resources such as food, water and energy. Because of the massive urbanisation expected in the next two decades, there will be an increasing need for enhancing and optimising urban planning, governance, physical and communications infrastructure, resource management, transportation systems and security services, among others. In order to meet these needs, Indian cities will have to become smart. The government of India has already enacted key long-term initiatives to support this goal – flagship initiatives being promoted by the Indian Government include the 100 Smart City Mission and the ARMUT (Atal Mission for Rejuvenation and Urban Transformation) programme.[10]

In the context of future Indian smart cities, there is a strong role that space will almost certainly come to play. It can be expected that space assets will, inter alia, support the operations of future city management by complementing and augmenting terrestrial technologies with EO, GNSS and telecommunications capabilities, enhancing the quality of the situational awareness of cities and allowing better urban planning and management (e.g. monitoring traffic flow, making possible real-time congestion planning, providing automated warnings at unmanned level crossings[11]). This will likely be accompanied by efforts to create synergies with other technologies and applications (e.g. in the area of resource technologies, automation systems, logistics and mobility). It can be also expected that in order to maximise citizens' quality of life and increase economic productivity, special attention will be devoted to the creation of people-centred applications (e.g. smart transportation applications and geo-intel) in which the role of navigation, communication and Earth observation satellites will be essential.

Education and the Digitally Empowered Society

Another issue area calling for innovative solutions and in which the role of space can be empowering is education and more broadly the advancement of a knowledge-based digital society. As mentioned earlier, young Indians potentially represent the most significant human capital on Earth, provided that the government will able to

---

[10] Smart cities will have the primary goal of minimising resource consumption and environmental degradation, while maximising citizens' quality of life and increasing economic productivity and competitiveness. Within the context of future smart cities, ICT will play the pivotal role in many areas of applications. From households to industry and institutions, more and more networks (electricity, gas, heat, water, transport) and systems (e.g. buildings) will make increased use of sensors, converters and actuators and progressively become integrated within a system of systems. This will support the end users in a more sustainable fashion and ensure an optimal utilisation of different utilities and services while providing city managers with a real-time comprehensive situational awareness of the state of their cities. So-called city dashboard solutions will offer invaluable inputs for modelling and simulation activities that will help cities to "work" and grow more efficiently.

[11] Pilot studies have been already conducted to create a "geospatial database containing the locations (the geographical coordinates of unmanned level crossings) and have GAGAN-enabled devices mounted on train engines. A train mounted with such a device will know the location of unmanned level crossings and train's hooter will automatically be activated when the train approaches an unmanned crossing" (Dasgupta 2017).

fully nurture this potential. Ensuring access to quality education to isolated and remote populations has been and will remain a Sisyphean challenge for India. As reported by IndoGenius co-founder, Nick Booker, any calculation of the number of young people in India and the current ability of India's education system – large as it may be – to successfully educate that population is difficult. By some estimates, India would need to create a new university every single week for the next 10 years: it simply will not be able to meet the mounting demand! (IndoGenius 2016a). Because "the brick and mortar approach" prevalent in the West is not practical in the case of India, New Delhi's government is now looking at new leapfrogs opportunities, including digital technology as a mandatory part of the educational offering.

> India's interest in blended learning is probably higher than in most Western countries. This puts India and Indian students at the forefront of educational trends. One can imagine the impact that personalized learning, adaptive learning, virtual reality, virtual labs, and other future techniques are going to have on improving the power and effectiveness of digital education (IndoGenius 2016a).

But tapping into this potential will clearly require meeting the challenging task of connecting to the Internet the remaining 70% of the country's population, the majority of which resides in semi-urban and rural areas. Considering India's specificities (including India's sheer size in terms of landmass, the still poor terrestrial infrastructure, the high share of remote and isolated populations and India's high susceptibility to natural disasters), the standard approaches using underground optical cables are not the best options to bridge this gap. Satellites, on the contrary, provide a quick, cost-effective and resilient solution particularly adapted to India's situation. It has been estimated that the cost of space solutions to cover a square km varies between $3 to $6 as compared to terrestrial solutions that "incur at least $3000 to cover the same area using cables and towers" (Satak et al. 2017). Areas where ground infrastructure is difficult and cost-intensive to penetrate can be easily covered by satellite infrastructure (Ibid).[12] Since bandwidth depends only on the satellite, the location of the user does not really matter: "for a given cost, consumers in either a big town or a tiny Himalayan hamlet will get the same bandwidth" (Ibid).

Whereas just a few high throughput satellites would be in principle sufficient to cover a country as vast as India and provide high bandwidth Internet, thought has also been given to the possible development of constellations in LEO, since they would have lower latency as compared to GEO satellites (Rajgopalan and Jayasimha 2016).

In a country such as India, satellite infrastructure will also free ground infrastructure from important burdens and reduce congestion in already overloaded networks. More importantly, it could ideally speed up the fulfilment of the government's long-term goal of connecting 250,000 villages (gram panchayats) in the country and hence the possibility of fully harvesting India's demographic dividend.

---

[12] Even the ITU/UNESCO Broadband Commission for Broadband Development in its report The State of the Broadband has observed that "Satellite systems should be given full consideration as solutions for next-generation broadband network deployments in rural and remote areas, as well as in diverse environments and deployment scenarios" (ITU/UNESCO Broadband Commission 2016).

Indeed, reaching out to these rural areas will not only enable offering education services but more broadly empowering the deployment of communications networks that are a vital enabler of social inclusion, digitally empowered society and economic growth. An ORF study has shown that 100% Internet connectivity by 2020 could add an extra $1 trillion to India's GDP (Rajgopalan and Jayasimha 2016). These economic paybacks will be accrued "from dramatic changes in the way people connect with each other, the government, and businesses. With the advent of satellite Internet, one can expect dramatic transformations in the way some of the essential services are delivered" (Satak et al. 2017). In this respect, it suffices to recall the financial inclusion mechanisms (e.g. the direct electronic transfer of subsidies to the bank accounts and the Unified Payment Interface) that were recently introduced by Modi's government to improve governance and push towards building a digital economy by encouraging cashless transactions (see *Box 4.1: The India Stack* for an overview of these initiatives). With the eventual introduction of satellite-based Internet, this and other highly valuable services contemplated in the *Digital India* flagship programme have the potential to reach every corner of the subcontinent. In the process, the Internet will also open up a myriad of space commerce opportunities for start-ups to develop new citizen-centred products and services.

Healthcare

A similar line of reasoning on the empowering effect of space assets can be made for healthcare. India faces and may continue to face in the future a huge healthcare challenge. As in education with schools, the "brick and mortar approach" (in this case, building hospitals in every town) will prove challenging when considering India's demographic prospects. India will clearly require an innovative healthcare approach that is able to expand reach while also improving quality and reducing cost (IndoGenius 2016a). The identified leapfrog opportunity in India's healthcare is to reduce the cost of health and medical care by cross-fertilising ICT technologies with the coming revolution in biosciences. From this standpoint, India has already become a pioneer in telemedicine, shifting the point of care from hospitals to patient homes (Bonnefoy 2014; IndoGenius 2016a). And also in this respect space is clearly part of the enabling technologies that, according to McKinsey, could transform healthcare in India and enable an additional 400 million people to access healthcare by 2025 (McKinsey Global Institute 2014).

In the future, innovative e-health services are expected to be required at an increasing pace and will create many businesses around the e-health domain, such as remote healthcare, the fully digitalised and integrated medical profile, big data disease tracking or even personalised medicine, not just for treatment but also for probabilistic prevention as a function of environmental and genetic parameters. Personalised medicine promises to deliver the right treatment or prevention strategy for a specific person at the right time, capitalising on advances in health research and IT. People will be able to manage their health with highly customised lifestyle plans (e.g. diet, exercise and relaxation), as well as deal with acute health problems with precision.

The new concept of personalised medicine will require a combination of technologies to be in place in a fully integrated manner, such as innovative sensors to monitor the health condition of people, mobile applications that will collect, process and distribute the data to the Cloud and artificial intelligence systems to provide personal medical solutions based on a digital patient record. There will be an increased need to make use of satellite assets, such as telecommunications and EO satellites and navigation services, to support this new and dedicated personalised medical care for Indian citizens. Space assets-enabled services will more specifically include remote healthcare services (e.g. tele-diagnostics, tele-radiology, tele-prevention and tele-consultations); bid data disease tracking (for the detection and mapping of disease outbreaks, prevention and containment of epidemics, planning treatment capacities, etc.); and automated in-patient care (e-learning and simulated learning to shrink training time for nurses; smart ICU systems for reminders, protocols, alarms, and recording patient data) (McKinsey Global Institute 2014).

Environment and Resources

As noted above, India is one of the most vulnerable countries in the world to projected climate change, and with its fast-growing population and a rising demand for food, water and energy, it probably will be severely impacted by such changes. The country is already experiencing the impacts of climate change, including water stress, heat waves and drought, severe storms and flooding and associated negative consequences on health and livelihoods (The World Bank 2013). Hence, it will be essential for India to tackle these issues. From this perspective, space has once again an important role to play. And India, is already taking the lead in the use of space assets for monitoring climate, as evidenced by the Declaration of New Delhi of 2016, which highlighted the world's space agencies' commitment to exploiting satellites' unique role in observing climate, understanding its variations and mitigating its effects by supporting the implementation of precise actions (Committee on Earth Observation Satellites 2016).[13]

Given that climate change already heavily affects living conditions in India, it can be expected that India will progressively plan and implement innovative programmes to monitor and predict the effects of climate changes while limiting the consequences of natural disasters. It is likely that in the future, satellites will be used to offer their aid ahead of such events through improved ability to provide early warning. It is also foreseeable that dedicated programmes for the development of next-generation meteorological satellite observation systems will be boosted in the future, since they

---

[13] In fact, out of the 50 essential climate variables (ECVs) defined by the Global Climate Observing System (GCOS), 26 can only be monitored from space (e.g. soil moisture, fire, land cover, glacier, Antarctica, Greenland, ocean colour, sea level, sea surface, temperature, sea ice, GHG, ozone, aerosol, clouds). Satellites acquire the data needed to develop climate models and guarantee global, precise and multi-parameter observations, allowing to measure sea level rise and global warming of the atmosphere, two of the most serious consequences of climate change. And space-based Earth observation enables detailed measurements of greenhouse gases such as carbon dioxide and methane. Space is also a precious ally in predicting and managing natural disasters that are another major consequence of climate change (Aliberti et al. 2016).

provide valuable modelling and forecasting capacities. Consistently with the 2016 New Delhi Declaration, it can be also expected that international coordination and cooperation will be an integral part of India's future efforts in this field.

However, the role of space assets will probably not be limited to monitoring the effects of climate change in India but also to actively ensuring better exploitation and creation of synergies in the use of natural resources such as food, water and energy. GIS- and PNT-based systems will likely be developed to support food security and water management efforts through the provision of precision farming solutions (guide planting and irrigation, yield monitoring, variable rate technology to fine-tune inputs, improve yields, water and fertiliser efficiency); the creation of so-called technology-enabled supply chains[14] to reduce wastage and ensure quality throughout the agricultural supply chain; or the amplification of programmes for effective water management such as the National Mission for Clean Ganga (NMCG), the Accelerated Irrigation Benefit Programme (AIBP)[15] and the Integrated Watershed Management Project (IWMP) already supported by ISRO.[16]

Similarly, in the field of energy production and distribution, space-based GIS will be possibly leveraged to exploit the full potential of renewable energy, for instance, by mapping the optimal areas for the deployment of solar energy infrastructure and smart power grids for optimised energy distribution and consumption, thus reducing the carbon footprint of human activity. Looking a step further, a proposed contribution of space to India's energy (and climate) challenge might be the construction of an orbiting solar photovoltaic plant which can benefit from about ten times higher solar irradiance than a terrestrial plant (Prasad and Basu 2015).

All in all, there is an abundance of fields in which the role of space applications will continue to be – or become – of crucial importance in addressing India's future societal grand challenges and accompanying the country's economic resurgence.

---

[14] Technology-enabled supply chain refers to the use of information technology (in particular advanced GIS/PNT technologies) to track food across the distribution system in order to improve demand and supply matching across markets and locations and also reduce waste while helping farmers capture more value from the sale of their output. Such applications are particularly relevant for India, because "one of the biggest challenges in the Indian agriculture and food sector is post-harvest crop losses. Some 30% of fruits and vegetables (40 million tonnes per year) are lost to spoilage on the way through the distribution system. This adds up to losses of about $4 billion per year in the Indian food distribution system and contributes to food scarcity. Building warehouses, cold-storage facilities, roads, and rail links in remote areas could cut losses in Indian food distribution to just 4–7%. Using Internet of Things tagging and tracking technologies, Indian farmers and food distributors could monitor the progress of crops from farm to market and cut down on the amount of food spoiled in transit because of avoidable delays" (McKinsey Global Institute 2014).

[15] AIBP is a project of the Ministry of Water Resources for "generating an inventory of existing irrigation infrastructure using high resolution satellite data and comparing it with the planned irrigation infrastructure to make an assessment of the critical gaps and additional irrigation potential created" (Dasgupta 2017).

[16] IWMP aims to ensure optimal soil and water conservation and assured water resources harvesting in order to provide sustainable livelihoods to the people residing in the watershed area. Already, the IWMP is being carried out by ISRO in collaboration with the Ministry of Rural Development to monitor 52,000 micro-watersheds in 50 identified districts across the country.

What is important to highlight here is that enabling all these advancements will require considerably expanding and improving Indian space infrastructure and supply capacity, in particular with respect to HTS for Internet connections and telecommunications, space imagery satellites and PNT services. The feats will be not only financial but also technological and organisational. The specific requirements to fully enable applications development will be considered in Sect. 7.2.

7.1.3 Human Spaceflight on the Horizon?

For a fast-rising power like India, there are many indications that space will be used not only to meet future societal grand challenges and fully seize the country's socio-economic potential but also to accompany and substantiate India's rise on the international scene, particularly through reverberating undertakings such as space exploration and human space spaceflight.

Since the onset of the space age, human spaceflight activities have been perceived as a "marker" of space powers. This is not only true for the leading spacefaring nations such as the United States, Russia and China but also for India. New Delhi's high evaluation of human spaceflight activities as a geopolitical tool was demonstrated by its tentative forays in this arena as a reaction to the glaring light of China's *Shenzhou* missions (see Chaps. 2 and 6). Although the ambition of launching its own "vyomanaut" (from Sanskrit vyoman, "sky" + -naut) into space had to be dropped out of ISRO priorities – primarily because of the much delayed development of the GSLV Mk-III and of failed cooperation with Russia leading to the inability to keep pace with China's advancements in the short term – it appears unlikely that, with India's international stature growing further in the future, India's political leadership will not opt to acquire this "marker" to put the country at the forefront in the international space arena and contribute to its image as a major player.

For one thing, this line of reasoning is based on the fact that techno-nationalist considerations have remained important drivers of India's posture in space, as demonstrated inter alia by the fact ISRO has been looking at alternative paths to gaining international prestige, including the previously discussed reorientation of its space exploration programme towards Mars, which allowed the country to achieve a "space first" in comparison to China.

In addition, when announcing the decision to put human spaceflight on the backburner, former ISRO Chairman Koppillil Radhakrishnan was keen to emphasise that concrete efforts may well be taken later (stating "We are not going to see the Human spaceflight as a programme in the 12th Five-Year Plan. *We will see maybe later*") and that the country would not start from scratch should New Delhi's government decide to take up ISRO's plans as a fully fledged programme (NDTV 2013). Indeed, what is perhaps more important to highlight is that ISRO has continued to fund and develop a number of critical technologies that could speed up the implementation of the human spaceflight programme (HSP) when governmental approval comes (Vikram Sarabhai Space Centre 2017). As more specifically articulated by Unnikrishnan Nair, Project Director of the HSP initiative, at present, "ISRO's focus is on four major lines of activity:

(a) Development of new technologies required for HSP in the areas of

- Crew Module systems including re-entry and recovery elements,
- Environmental control and life support systems and flight suits, and
- Crew escape system.

(b) Undertaking unmanned flight-testing of the Crew Module systems in LVM-3-X flight.

(c) Demonstrating the performance of the Crew Escape System through Pad Abort Tests.

(d) Development of Environmental Control and Life Support Systems (ECLSS) and carrying out an integrated test of ECLSS with the Crew Module on the ground.

The successful completion of this phase of critical technology developments will enable ISRO to prove its capability leading to the full-fledged HSP" (Nair 2015). The concept of the venture around which ISRO is working on remains the same as that proposed in 2007 and envisages "the development of a fully autonomous orbital vehicle carrying two (three) crew members to about 300 km low earth orbit and their safe return. The spaceship would be launched by the GSLV Mk-III launcher" and would comprise a service module and a crew module with a docking port in anticipation of future Indian programme development or potential international cooperation (Vikram Sarabhai Space Centre 2017) (Fig. 7.9).

As reported by ISRO,

a demonstration of the atmospheric re-entry flight of the proto crew module (CM) was successfully carried out on 18 December 2014 as part of the first experimental flight of GSLV Mk-III. This was used as a platform for testing the atmospheric re-entry technologies envisaged for a crew module and also for validating the performance of end to end parachute based deceleration system (Vikram Sarabhai Space Centre 2017).[17]

In parallel to this, the preliminary design of the Environmental Control and Life Support System (ECLSS) has been completed. A number of

critical components like air–liquid heat exchanger, liquid–liquid heat exchanger, space radiator and LiOH (lithium hydroxide)-based canisters have been realised and detailed characterisation tests are in progress [...] Design and development of flight suit have also crossed many milestones and a few development models have been realised and tested to prove the concept. ISRO is now looking to integrate the ECLSS subsystems and carry out an integrated test with one of the CM models as part of the critical technology development (Nair 2015).

---

[17] In the Crew Module Atmospheric Re-entry Experiment (CARE), "the crew module got separated from the launch vehicle at an altitude of 126 km and re-entered the Earth's atmosphere at an altitude of 80 km and further descended in the ballistic mode. Three axis control was provided for damping out the rates and for ensuring zero degree angle of attack during re-entry. The crew module made a safe splash down at the predicted impact point about 600 km away from Port Blair approximately 1243 s after lift-off. Subsequently the module was recovered by Indian Coast Guard and brought to Ennore port Chennai on 21 December 2014" (Vikram Sarabhai Space Centre 2017).

Fig. 7.9 Artist's view of
India's human spaceflight
technologies (*Credit:
Vikram Sarabhai Space
Centre 2017*)

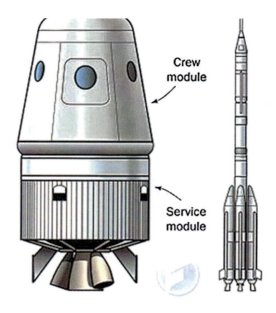

Crew
module

Service
module

The most challenging technological feat to go forward in the direction of an HSP
remains, however, the development of a man-rated heavy-lift launcher. The GSLV
Mk-III has already been identified as the launch vehicle for the Indian HSP because
thanks to its payload capacity, long-duration human spaceflight with a habitat module could be ideally configured in the future. In this regard, it can be expected that
once the GSLV Mk-III enters full exploitation stage in 2018/2019, dedicated funding to human rate, this vehicle will be allocated thanks to the savings realised
through the reduced dependence on foreign launch services.

Based on this technological growth trajectory, there are clear indications that India
may proceed with a formal decision to undertake the programme by the early 2020s,
with a first flight around 2024, at a time, however, when India's alleged rival, China,
should have already completed the assembly of its *Tiangong-1* space station.

Looking further into the future, possible directions of the human spaceflight programme may include the development of a habitat module in LEO, as already indicated by several Indian commentators and scientists. However, it is also quite
possible that India's political leadership may decide to skip the step of building its
own space station to join the next wave of future human space exploration in the
post-ISS period as a leading agency. Space exploration and human spaceflight are
spearheads for international collaboration, and stakes are high for India when it
comes to future possible partnerships. Indeed, pursuing a cooperative path in its
exploration and human spaceflight programme would not only allow catching up to
China's achievements – which is per se already an important consideration for New
Delhi's leadership – but more broadly enable the country to become more integrated
in the international space arena and participate in possible ground-breaking international programmes that would otherwise be out of its reach.

7.1.4   Future Military and Security Efforts

Whereas the very likely involvement of India in human spaceflight initiatives will be primarily driven by the need to acquire new tools of soft power projection and bolster India's stature on the international space scene, the actual geo-strategic weight India will exercise globally will be primarily supported by its military build-up. From this standpoint, space assets – as a force multiplier providing unique capabilities for secure and resilient communications, military intelligence (early warning, signal interception and active observation) and augmentation of terrestrial technologies such as missile guidance or UAV piloting and data transfer – can be anticipated to be an indispensable component of India's future security and defence strategy.

Major space powers such as the United States, Russia and China have already moved beyond these considerations and are increasingly contemplating means to protect their space assets from threats (be they intentional or not) or to incapacitate foreign assets. Irrespective of whether New Delhi adopts a proactive or a reactive strategy, it appears that the creation of a strong national SSA and C4ISR capability is a minimum requirement for an aspiring military power such as India. Indeed, given the global trends and developments where the use of space is linked to both nuclear and conventional war, no country that aspires to take a strategic position on the global scene can afford to ignore the power that comes from a robust SSA and a Command, Control, Communications, Computing, Intelligence, Surveillance and Reconnaissance capability. Once SSA and C4ISR capabilities will be mastered, India will have to possibly use them as a force multiplier for either a proactive (offensive) or reactive (defensive) strategy.

Looking at the use of space for security purposes, it can firstly be anticipated that, to support India's military build-up, the next two decades will witness more robust investment in its military space, with the launch of both dual-use and dedicated military satellites for the Navy, Army and Air Force to support C4ISR functions. According to a study published by NIAS (Chandrashekar 2015), the basic requirements for establishing a fully fledged C4ISR capability in India encompass:

- A four-satellite constellation of advanced communications satellites in GSO that use ion propulsion for carrying out vital C4 functions
- A constellation of 40 satellites in LEO that provide Internet services for the military
- Three clusters of three satellites each for performing the ELINT functions
- A constellation of 12 EO and SAR satellites in appropriate Sun-synchronous orbits (SSO) for meeting ISR needs
- A constellation of 24 small satellites in LEO for meeting ISR needs during times of crises
- Three TDRSS satellites in GSO for performing the tracking and data relay functions needed for a C4ISR capability
- Two operational satellites in GSO along with three orbiting satellites in an 800 km SSO for meeting operational weather requirements

- Seven satellites in geostationary and geosynchronous orbits for meeting core navigation functions
- A 24-satellite constellation in MEO, established over a 10-year period, for providing an indigenous navigation solution
- A constellation of 100 small satellites in the 5–100 kg class.

To meet the needs listed above, India will have to launch 17 advanced satellites into various GSO, SSO and other orbits every year. In addition, 28 smaller satellites will also have to be launched as part of constellations that cater to various C4ISR needs. To meet this requirement, India will need 15 PSLV, 7 GSLV and two modified Agni 5 launchers every year. As for the promotion of the small satellite initiative, it would require one dedicated PSLV launch along with five launches of the Agni 5 per year. The details of both the satellite and launcher requirements for establishing a fully fledged C4ISR capability in India – as determined by the NIAS – are summarised in Table 7.1.

Given these requirements,

> a significant augmentation of satellite and launcher capacities will be clearly needed, as India can currently build and launch about four advanced satellites a year. To go from the current capacity of four to 17 satellites along with their launch requirements implies a four-fold increase in capacity. If the small-sat constellations are added, the satellite and launch capacities need to be multiplied at least five times to meet the C4ISR of an aspiring space power (Chandrashekar 2015).

To achieve these objectives, a transformation of the national ecosystem in terms of both hardware capacities and organisational setup for space activities is clearly of utmost importance for India. Specific considerations on these transformations will be discussed in Chap. 7.2; for now, suffice it to highlight that achieving these capabilities will also entail major financial efforts. While no precise indication on the required budgetary resources can be offered at this stage, in the best-case scenario, it is anticipated that it will take India a period of approximately 10 years of major efforts to achieve robust C4IRS competences that could complement its technological prowess in civilian space and make it a premier military space power in the geopolitical landscape. Although India's dual or dedicated space assets will almost certainly remain smaller than those of such major powers as the United States and China, they will play an important role in positioning India as a major powerhouse exercising increased strategic influence in the international (space) arena. Hence, at the very least, these assets will need to be protected against the direct and indirect consequences of actions taken by the other space powers. This brings us to the second key pillar of India's future military and security efforts: the securitisation of its space assets, a pillar that will be primarily linked to the expansion of the country's SSA capabilities and deterrence options.

Given that satellites are vulnerable to a broad range of natural and man-made threats (including space weather, space debris, collisions with other satellites, cyber-attacks and ASAT attacks), characterising the space environment through sound SSA capabilities remains essential for any established or aspiring space power. Even if one were to ignore the military relevance and implications of what is hap-

Table 7.1 Dual-use satellite and launcher requirements

Function	Satellite mass (kg)	Orbit (km)	No per year	Launchers			Comment
				PSLV	GSLV	Agni 5	
C4 functions							
4 satellite C4 System Military	2000–4000	GSO	2		2		Use of Ion Propulsion
Civilian Advanced Satellites*	2000–4000	GSO	2		2		Use of Ion Propulsion
40 satellite LEO Military Internet	150–200	LEO	20	5		1	Use of Agni 5 Launcher
ISR functions							
3 × 3 ELI NT Clusters Military	300–400	63° 1100 km	3		1		One cluster single GSLV launch
6 EO satellites (civilian + military)	500	500–1200 SSO	2	2			Small satellite capacity available
6 SAR Satellites (civilian + military)	1500	500 km SSO	2	2			Dedicated launch may be needed
24 satellite LEO Military	100–150	500 km SSO	8	2		1	Inclined orbit regional coverage
Space support services							
3 satellite TDRSS system	2000–4000	GSO	1		1		Civilian & Military needs
2 Satellite GSO weather service	2000–3000	GSO	Small		Small		Civilian & Military needs
3 Satellite LEO weather service	1500	800 km SSO	1	1			Civilian & Military needs
7 Satellite GSO Navigation service	2000	GSO	1		1		Civilian & Military needs
24 Satellite MEO Nav service	1500	20,000 km	3	3			Indian Navigation system 10 years
Small Satellite Initiative (R&D)	5–100	LEO	100	1		5	Multiple satellites – Many Builders
Total			*17 + 128*	*16*	*7*	*7*	

Source: Chandrashekar (2015)

pening in space, the great economic stakes associated with space assets dictate protecting these assets from the risks posed by space weather events and the exponentially increasing population of dead satellites and space debris.

Because Indian current capabilities for tracking inactive satellites and space debris are extremely limited and because considerations of technological autonomy continue to drive the country's space efforts, it is not unreasonable to argue that, alongside a likely expansion of cooperation with the United States in the short term, India will in the medium-term place more efforts on the creation of a suitable network of long-range radars and optical sensors.[18] A primary objective would be to track space objects in order to provide collision avoidance assessments, independently.[19] In addition, considering that satellites form an important part of network-centric warfare doctrine envisaged in the Technology Perspective and Capability Roadmap (TPCR) (Headquarters Integrated Defence Staff – Ministry of Defence, 2013), it is also well possible that the country will eventually opt for the development of space-based systems and radar-independent tracking methods such as lasers that will not only track space objects but also determine the capabilities of various space systems in orbit, the intention of their owner and hence their implications for Indian national security.

For this, a dedicated space situational awareness initiative within the national security complex (the much discussed Aerospace Command or Defence Space Agency) might be established in the course of the next decade to gather space-based network-centric intelligence capabilities of adversaries and mapping out evolving ground-based counter-capabilities (Prasad and Rajagopalan 2017).

The establishment of robust SSA capacities is intimately connected to India's deterrence considerations and consequently to the possible development and use of space weapons, specifically BMD and ASAT, to support India's future defence strategy (Bommakanti 2017). As contended by Chandrashekar, to a large extent, this is because "one of the major requirements for a more advanced BMD and ASAT system is the ability to track missiles and satellites in space. The network of short-, medium-, and long-range radar systems complemented by suitable optical and laser ranging stations could also be used in BMD or ASAT tests" (Chandrashekar 2015).

Although India's official position has always been against active military uses of space, it is also fully aware that strategies for coping with and managing future conflicts (be they conventional or nuclear) will necessarily involve the development and deployment of weapon systems that can neutralise space assets (Headquarters Integrated Defence Staff – Ministry of Defence, 2013). Towards this, India has conducted experiments with some components of a BMD system. Limited intercept tests just outside the atmosphere have been carried out as a demonstration of capa-

---

[18] As Gopalaswamy already noted, "Indian policymakers believe that an Indian space situational awareness would pose great challenges in terms of resources needed to provide extensive geographical coverage. In the short run, therefore external cooperation would be the best strategy. But in the long-term they are of opinion that India should develop its own assets to perform space situational awareness functions" (Gopalaswamy 2013)

[19] In addition to its multi-object tracking radar (MTOR), ISRO is already installing two 1-m optical telescopes for tracking debris at Mont Abu and Ponmudi, respectively.

bility. The Indian missile arsenal is also considered adequate to demonstrate an ASAT capability to LEO if needed (Chandrashekar 2016).[20]

Indian space officials have, however, made it clear that "there are no plans for offensive space capabilities", because India's position on the weaponisation of space is similar in spirit to its nuclear "no-first-use doctrine". They have also explained that India's focus will remain on the development of retaliatory capabilities but that such capabilities will not be tested unless geopolitical "circumstances" force India into demonstrating them. Ultimately, it must be underscored that any Indian ASAT test in the future will be a "response-based event" (Gopalaswamy 2013).

The question is thus whether future geopolitical reality will eventually push India to prove an operational weapon system. There are several indications that India may in fact proceed in that direction, particularly in view of China's continuous build-up of its military space power and disregard of India's security concerns, which render for many imperative (and inevitable) the development of an operational retaliatory capability.[21] Be this as it may, it appears for now safe to argue that preference will be in any case accorded to the development of technology demonstration for BMD and ASAT. Indeed, Indian policy-makers seem to prefer technology demonstration over the introduction of operational military assets. As Gopalaswamy puts it, "technology demonstration is less provocative externally, allows long lead times for technologies to mature, and is sensitive to the difficulties of building consensus within the Indian political system. In the past, India has adopted such approaches in developing nuclear weapons, ballistic missiles, and chemical weapons. It can hence be expected that a similar path will be followed for the development of deterrence space capabilities, possibly through ground tests or as part of the BMD" (Gopalaswamy 2013).

At present, "India's BMD and potential ASAT capabilities only cover the space over India. Extending these capabilities to cover midcourse and boost phase BMD and to be able to target satellites at higher altitudes will require significant augmentation of both the ground infrastructure as well as space borne assets" (Chandrashekar 2015). While the extension of BMD and ASAT capabilities outside Indian airspace may not be an immediate priority, India's military space strategy will certainly keep these prospects in mind for the future. Also for this reason, "the establishment of a sound SSA capability would go a long way in ensuring that some of these BMD and ASAT options are preserved should India need to embark on them at some future date" (Chandrashekar 2015).

---

[20] This is mainly because both ASAT and BMD weapons draw upon the same domains of technology and knowledge; hence, capabilities in one also necessarily mean that capabilities exist for doing the other.

[21] As, for instance, pointed out by Bommakanti, a direct ascent ASAT "is invaluable in the context of the Sino-Indian conflict dyad. [...] For deterrence and compellence to have any credibility in the India-China space dyad, New Delhi will require a potent retaliatory capability [...which] will denude opportunities for China to threaten India's space infrastructure and gives New Delhi the opportunity to limit the means and the ends that Beijing pursues in conflict against India. Otherwise, space deterrence that is inextricably linked to terrestrial deterrence between India and China will remain fragile" (Bommakanti 2017).

In conclusion, while technology options on space weapons will certainly be kept open, operational decisions on these will probably be deferred by a few years till a robust SSA and C4ISR capability has been built up in the country.

7.1.5   India's Footprint in the Future Space Economy

In what seems to be a "great acceleration" of space activities, much attention has been paid to the New Space paradigm emerging in Western countries as a result of public strategies to foster the growth of a space private sector both in upstream and downstream portions of the value chain. Little consideration has been paid to India and to the footprint it may eventually have on the future space economy. While this is to a large extent understandable, given that India's engagement in the international commercial space markets has so far been very limited, many indications suggest that New Delhi will seek a larger share and role in the global space market and that such quest may eventually be successful. There are a number of factors underpinning this projection.

Inherent Potential and Private Dynamism
For one thing, one should not overlook that Indian demographics and economic projections highlight high potential for the development of a space-based services private economy in India. This projection is closely related to India's already large and ever-growing market, the specificities of India's geography and state of development for the adoption of space-based services, and the existence of an industrial base including technologies, competences and infrastructure to develop, deploy and operate space assets.

In addition, it is important to highlight the structural and policy reforms that India is currently introducing and the impact that these reforms will have on the growth trajectory of India's traditional space industry, both in local and international markets. As discussed in previous chapters, with the increase in demand for space-based services in the country, ISRO is moving towards a two-pronged strategy for the future – on one side, it is focusing on supporting the development of advanced space technologies through public investments, and on the other, it is carving a clearly larger role for the Indian private sector that will take over a substantial share of activities related to manufacturing and servicing operational space assets and downstream application services (Murthi and Rao 2015).

Towards this, industry consortia are currently under implementation for the production of both the PSLV and the satellite systems, in particular the IRNSS. This approach is expected not only to allow ISRO to focus completely on novel technology development over a longer horizon period but also to

> bring along a unique opportunity in the Indian industry ecosystem to build up systemic capacity to be able to deliver end-to-end space systems for the first time in the country. [...] The effect of such increased trust in Indian industry by ISRO shall provide further confidence for the industry to invest into infrastructure that can deliver to ISRO's needs. The

satellite AIT industry consortium has already accounted commitments of Rs. 100–150 crore in investments to set up facilities that can support the operations required. In the current growth trajectory, Indian industry will be equipped to deliver rockets and satellites by 2020. (Prasad 2017c)

Once these structural reforms of organisational procedures enter into force, their impact will not be limited to the Indian commercial space ecosystem but will also be probably felt in global commercial markets. Taking launch services as an example, the success of Indian rockets on the commercial market has been so far limited by the fact that ISRO has only been able to utilise the limited opportunities offered by excess capacity on its launchers. With industry taking over both launcher manufacturing and exploitation (through privatisation operations), it can be anticipated that the ensuing increase in production rates will not only enable meeting domestic demand but also offering competitive launch solutions in the global launch service market, with the future competitiveness of Indian rocket also backed by the country's burgeoning internal demand, which will allow generation of economies of scale and keep competitive prices.

While these developments in traditional space industry are certainly encouraging from an Indian commercial perspective, it should not be overlooked that India is also seeing a growing New Space phenomenon. An increasing number of companies are now pursuing the development of solutions independently of public programmes and hold "the potential of creating a multiplier effect on the country's space economy unlike the circulation of tax money that normally happens within traditional space industry approaches" (Prasad 2017a, b, c).[22]

The country's start-up ecosystem is already the second-fastest growing and third-largest in the world, and New Space is not remaining oblivious to this growth story, with many Indian entrepreneurs kick-starting space ventures to provide innovative end-to-end solutions in both the upstream and downstream segments. Companies such as Earth2Orbit, Bellatrix Aerospace, SatSure and Astrome Technologies, among several others, are now progressively positioning their offerings within the national market and aim not only to disrupt the way space activities are conducted in India but also to integrate in the global supply chain and compete internationally.

While the emergence of these New Space initiatives in India is still at its infancy and, if compared to the United States or Europe, much slower and more cumbersome, their growth potential is real, as they can take full advantage of the unique opportunities available in the country. Among the most relevant, one needs to acknowledge India's surging market demand for space products and services, its remarkable assets on the supply side and the growing number of positive externalities (Aliberti et al. 2017).

---

[22] As addressed in Chap. 4, they typically operate either on the basis of an integrated approach with respect to satellite development and the provision of related commercial services (e.g. Astrome) or through the provision of new services based on existing satellite capacities (e.g. Earth2Orbit).

Fertile Demand and Supply Conditions of the Indian Market

One of the most relevant factors that could possibly boost and ensure the success of private space undertakings in India is related to the conditions of the country's market demand.

As highlighted in the previous section, because of its size and scale, India already has a very large market for space-based products that highlight high potential for the development of a space-based services private economy in the country. What is perhaps more remarkable than this already huge demand is that the projections associated with Indian demographic and economic growth – both of which are expected to generate a large-scale middle-income population of working age – highlight an even higher potential for growth in demand for satellite capacity and for new downstream services. In this respect, it should be reiterated that "space-based services in India have reached a stage where demand is outpacing supply and is creating a unique opportunity for developing space industry" (Murthi and Rao 2016). It is here sufficient to highlight that only 25% of India's transponder capacity is being served by Indian satellites, while the rest is leased from foreign satellites via 3-year deals made by Antrix Corporation. By the same token, "the market for satellite remote sensing data is also inadequately covered by domestic systems and industry, particularly in the high-resolution data segment" (Murthi and Rao 2016), while the market for PNT applications remains yet to be fully exploited.[23] Clearly, there is not only an opportunity but also a necessity for more private sector engagement.

When looking at demand conditions, it is also important to stress that India proves a fertile ecosystem for space entrepreneurship also because the specificities of the country and state of development make space solutions particularly apposite. More specifically, India's still-present developmental challenges (e.g. in terms of access to education, healthcare and government services), the concentration of its population in rural areas and the sheer number of locations involved given the size of this country all underscore a strong case for the development of private space-based solutions. The connectivity issue and the solution proposed by Indian start-up Astrome Technologies – a Bangalore-based start-up aiming to provide Internet connectivity to the remaining 70% of the country's unconnected population – is a clear case in point. Considering the relative advantage offered by space solutions and that ISRO is still tackling a large supply gap in its telecommunications satellites, there is immense scope for private space companies to step up with their solutions. Towards this, Astrome Technologies intends to launch 150 satellites into space by 2020 providing high-speed affordable Internet to remote locations across the country (Satak, Putty, & Bhat, 2017).

An analogous line of reasoning can be certainly made for meeting the requirements of other upstream systems or downstream services. For instance, there is a clear need to further develop GIS- and PNT-based systems to support food security and water management efforts through the provision of precision farming solutions and through the creation of so-called technology-enabled supply chains to reduce

---

[23] It can be for instance highlighted that vehicle navigation systems still have a very poor presence in India with less than 40 cities having been mapped (GNSS.Asia 2016).

wastage and ensure quality throughout the agricultural supply chain (McKinsey Global Institute 2014). Towards meeting these requirements, a New Space company like SatSure, for instance, has been developing analytics tools and predictive models that make use of Earth observation satellites, big data, cloud computing, machine learning and IoT technology to deliver insurance companies, pesticide and seed companies, etc. with clear decision solutions in the field of crop damage assessment, crop health monitoring and precision farming (SatSure 2017).

Overall, there appears to be important market requirements for the development of a private space industry in India. Arguably, such requirements can not only support the economic viability of several Indian New Space companies but may also likely empower small businesses to achieve economies of scale that may eventually position their products/services offerings into the global space markets.

Together with the favourable conditions of India's market demand, an equally important factor that can potentially boost the future growth of New Space in India is given by the valuable assets and talent the country can boast on the supply side. As inter alia noted by the Observer Research Foundation (ORF), India has "inherent advantages such as the availability of skilled workforce, a stable and business-friendly government, positive investor climate, and low cost of labour and operations" (Observer Research Foundation 2016). In addition, as already underlined in Chap. 4, India has not only proven to be a hotbed of so-called frugal (*jugaad*) and reverse engineering; it has also shown a growing ability to innovate in a variety of forms that often remain invisible to end consumers in the West – from groundbreaking B2B solutions to R&D outsourcing innovation and programme management improvements. There is no doubt that India's New Space entrepreneurs can leverage this talent pool to offer highly competitive industrial approaches or innovative market solution offerings that can be scaled up to international space markets.

Growing Domestic and Foreign Levers

Closely related to this aspect are the unfolding possibilities for India's private space companies to benefit from relevant levers on both the domestic and international front that can help this nascent ecosystem to move towards long-term growth.

On the domestic front, Indian astropreneurs see increasing opportunities to create synergies and cross-fertilisation with other industrial sectors such as information technology (IT), an area where India is clearly thriving. By the same token, there are also clear possibilities to link and team up with India's broader start-up ecosystem, in order only to access growing capital investment but more importantly to improve the overall conditions of the space industry ecosystem and move towards a viable growth.

Another domestic lever is given by the inherent possibly to exploit the already established capacity to deliver highly reliable technology products and services, including the infrastructure created over five decades of investments into the space sector by the government. For instance, Earth2Orbit has been using remote sensing data from ISRO's IRS constellation to provide its customers with "actionable intelligence" applications. As also argued by Dhruva space co-founder Narayan Prasad, "with the backbone technology know-how foundation in place mainly by the efforts of ISRO and its vendor base, there is immense scope for New Space enterprises to leverage these cluster-based externalities such as technologies, infrastructure and manpower to build space-based services" (Prasad 2017a, b, c).

On the international front, the most important lever is the possibility for Indian SMEs and New Space companies to attract attention for potential B2B solutions with international partners to realise end-to-end upstream and downstream systems offerings. For instance, in 2014, Indian start-up Dhruva Space has signed an MoU with the German start-up Berlin Space Technologies (a company specialised in small satellite systems and technology) for the joint development and manufacturing of Earth observation micro-satellite in India (Murali 2015).

Another possible involvement of foreign partners lies in investment opportunities through foreign direct investments (FDI), mergers and acquisitions (M&A) or joint ventures (JV). Indeed, with the recently introduced relaxation of the investment threshold in the aerospace sector to 100% FDI, as well as the above-mentioned surge in Indian demand, there are stronger incentives for international firms to invest in (and create B2B alliances with) Indian New Space companies. In addition, the same market conditions (e.g. the availability of skilled workforce, low cost of labour and operations, engineering services backbone, a positive investor climate, etc.) that MNCs have already exploited successfully in technology sectors such as aerospace and IT continue to make India an attractive destination for foreign companies looking for replicating outsourcing innovation paradigm in the space sector through innovative industrial approaches and/or disruptive market solutions for space manufacturing/applications (Prasad 2017a, b, c).

Emerging Institutional Support

The possibilities for India to attract foreign investment and to leverage the already present infrastructure base put in place by ISRO relates to a fourth type of externality that will prove of utmost importance in nurturing the growth potential of New Space initiatives in India: the increasing encouragement and support from the public sector. The importance of governments' role in supporting space industry's competitiveness has been widely articulated in a number of studies. Even for a country like the United States, which certainly has the most vibrant space industry in the world, the reality is that without the US government as either an anchor tenant or a major customer, very little of the still small but growing private activity that we now see would exist (OECD 2016). As inter alia pointed out by innovation economist Mariana Mazzucato, the current mythmaking about Silicon Valley's wildcatting entrepreneurs often forgets that many of the seeds were planted by public sector agencies through a series of institutional, financial and legislative tools (Mazzucato 2015). An invaluable role has been in particular played by such public institutions like NASA or DARPA, which have been key in backing the competitiveness of many private undertakings, including SpaceX. More broadly, public institutions remain key performers and customers of innovation in the space sector (OECD 2016); it can be expected that "the future of the New Space dynamic, although commercially driven, will be highly dependent on the success of implementation of new public strategies" (Vernile 2017).

In light of the burgeoning national demand for space services and of ISRO's purported inability to manage this demand, policy-makers in India are becoming aware of the need to support private industry in India to complement efforts of ISRO in the provision of new and innovative services and have come to realise that "there is immense scope for the bringing a change in the commercial space landscape and achieve economies of scale in space-based services if New Space is enabled. Parallels can be drawn to the rise of IT industry which enabled the creation of an environment that today is one of the pillars of export activity in the country" (Prasad 2017a, b, c).

As a demonstration of this increased awareness about the role of private space industry in India, it is important to remind the structural reforms IRSO is introducing in the organisation/execution of its space activities but also the fact that the government of India has already launched several key initiatives such as *Make in India* and *Digital India*, which provide Indian "traditional" and the "New Space" industry with important opportunities of engagement. Further, Prime Minister Narendra Modi's government has also launched a *Startup India* initiative in January 2016 which seeks to foster entrepreneurship and promote innovation in a variety of sectors, including space.[24]

Given the importance of a dynamic and strong private space industry for India, it can be expected that in the coming years, dedicated initiatives to support the rise and long-term competitiveness of Indian private industry more strongly will be launched. These initiatives may include the creation of a space start-up incubator similar to the ones successfully put in place by ESA (Prasad 2017a, b, c), as well as the setting up of a national fund for the promotion of space entrepreneurship, and more broadly policies that can encourage entrepreneurial activity through grants, procurement mechanisms and grants.

While these instruments are yet to be put in place, should the government start to back the private sector through a series of institutional, financial and legal tools along the line of US policy, Indian space industries have the potential to become very active players, exercising a great impact on Indian and the global space economies alike.

Whereas the exact positing of India's "traditional" and New Space companies in the future space markets are certainly hard to disentangle and no conclusive projections on what their growth will mean for the global space industry can be drawn at this point, it can be expected that their rise will possibly pose a rising threat for the historical US and European industry position as major competitors for both upstream space systems and downstream applications on the global space scene. At the same time, however, new cooperation scenarios in the global space industry can potentially emerge.

---

[24]The most important points of this flagship initiative include the freedom from tax in profits for 3 years, the freedom from capital gain tax for 3 years, the creation of a 10,000-crore fund of funds, a single window clearance system, a reduction in patent registration fee, self-certification compliance, an innovation hub under the Atal Innovation Mission and new schemes to provide IPR protection to start-ups and new firms (PM Jan Dhan Yojana 2016).

7.2 Enabling India's Space Efforts: Transforming the Ecosystem

As emerges from the previous section, space systems and related services will continue to be an integral part of the Indian geopolitical and socio-economic growth strategy. Increasing national motivation for boosting societal empowerment and economic growth as well as new geopolitical challenges will push India to make great strides forward in space, focusing on the above-mentioned development of space applications for coping with economic and societal grand challenges such as support to communication networks and healthcare, precise agriculture, smart power grids, smart cities, space industry entrepreneurship for boosting Indian competitiveness in upper-end industrial sectors and innovation, dedicated military space assets to enhance its international stature and geopolitical projection, human spaceflight programme and more ambitious exploration endeavours, among others.

Achieving all these efforts will certainly prove challenging and will require adaptation of the space ecosystem to overcome new major challenges associated with the prospected fast-paced advancements. These challenges are already apparent with respect to all the layers forming India's space ecosystem: the hardware, the orgware and the software.

In this categorisation, the term hardware refers to the infrastructure, material and financial resources and technological capabilities of a country. It basically comprises the national capacities in terms of space systems (e.g. launchers, satellites and ground facilities), the industrial base and the budgetary expenditures. Orgware, on the other hand, comprises the organisational structures set up to develop and run the hardware. The software of the space programme denotes the norms and rules applied to use the technological capabilities for specific purposes. These are generally embedded in national space policies and strategies.

7.2.1 The Hardware: Improving Technological Competences and Supply Capacity

From a hardware perspective, the potential for the successful delivery of space products and services in India is today primarily limited by the need to develop new technological competences, scale-up activities and subsequently increase the supply capacity. These requirements are evident in both the upstream and downstream sectors.

Looking at the upstream first, access to space remains a major hurdle that India's space programme will have to clear in the near future, by achieving a significant augmentation of its launchers capacities, in terms of both launch frequency and launch mass at lift-off. India is currently in the final stages of developing the GSLV Mk-III heavy launch vehicle, powerful enough to place 4 tons in GTO. Adding these capabilities is imperative for India to avoid dependence on foreign launch services (in particular Ariane) for its telecommunication and meteorological satellites weighing more than 2.5 tonnes. This is primarily an issue of political autonomy, but alongside such considerations, the unavailability of a heavy-lift GTO launcher has also proved resource consuming. In fact, as already discussed in Chap. 3, an important part of ISRO's budget has been regularly dedicated to procurement of foreign launch services for payloads based on I-3K and I-4K satellite buses, underlining the criticality of developing GSLV Mk III. Consequently, as GSLV Mk III nears entering the exploitation phase following the successful first orbital test flights carried out in June 2017, new available resources for other programmes as a result of reallocation is foreseeable. Once the new GSLV vehicle enters exploitation in the second half of 2018, two major developments can be anticipated, respectively, on the institutional and commercial fronts.

On the institutional front, India will be free to engage fully in more ambitious development plans that are currently being funded at very low levels. These plans include the possible development of a human spaceflight programme for which the first steps have already been taken with the first re-entry tests of a crew module prototype. They also include the development of critical technologies for advanced launch capacity such as the maturation of semi-cryogenic propulsion (critical for increasing GSLV payload capacity to GEO to 6 tonnes and enabling India to deploy larger communication satellites and/or human spaceflight habitat modules in LEO) and to develop technologies for a fully reusable two-stage-to-orbit launcher.[25,26] The development of these new technical capabilities will be a key factor in positioning ISRO as a leading space agency in the new global space arena. Nevertheless, these objectives will require more substantial and steady investment in the development of advanced launch capabilities beyond PSLV and GSLV.

Stakes are high for India in the launcher domain. At a time when all space powers are already engaged in the upgrade or introduction of new families of launchers at both sides of the mass spectrum including super-heavy lift in the United States, ISRO is still struggling with the establishment of its own heavy-lift GEO capacity. Hence, without a stronger commitment to technological innovation and launch development, India runs the risk of lagging behind international developments and continuing playing a catch-up game.

---

[25] The first test flight of the 1.5 t demonstrator was undertaken in December 2014 atop a PSLV, targeting an altitude of 70 km to test hypersonic aerodynamic properties. Four demonstrator flights have been planned in total.

[26] As also argued by Rao et al. (2015) in a thorough research conducted by the NIAS, "Looking ahead of the PSLV and GSLV, India will need to develop advanced launch capability – reduced number of stages to improve cost performance; semi-cryogenic propulsion improving safety and cost factors and also increase geo stationary payload capacity to 6t and 10t - thereby giving the nation a capability to embark upon more ambitious planetary missions as also an ability to build a space station/habitat module in Low Earth Orbit develop a reusable launch vehicle which can deliver men and materials to space and then return back to earth for refurbishment, refuelling to embark upon next mission; technology for sub- orbital trans-atmospheric transportation systems. Thinking for an aerospace plane may also be required as a technology development for long- term future".

On the commercial side, the introduction of GSLV Mk-III could enable India to offer increasingly reliable launch service solutions that will possibly allow India to address new segments of the global commercial launch market. However, this possibility remains closely related to another visible challenge faced by India's space hardware, namely, the need to substantially increase launch vehicle production rates in view of enhancing launch frequencies to better respond to the growing domestic demand and allow more commercial launches per year.[27] A prerequisite for this objective is the establishment of a new modus operandi in the organisation of Indian space activities (see next section), as ISRO alone is unable to fully cope with the required increase of launch frequencies. Until then, the production rate of GTO launch vehicles can be expected to continue to be sized to satisfy India's domestic demand with a continued limited presence on the commercial market for LEO. This would be assured by the very slightly increased PSLV launch rate, which will allow the occasional accommodation of foreign primary payloads.

In the upstream hardware segment, satellite systems and associated services constitute another area that will certainly see particular attention in the near future. Here, the most visible challenge ISRO will have to face is the growing demand for orbital bandwidth, which will require, in addition to the launch of new satellites, advancement and fortification of satellite communications technology. The general application of such technology must be extended into wider societal and commercial arenas to meet the rapidly growing demand for communications and broadcasting services of the ₹250 million-strong television and entertainment service Indian industry. Research shows a 165% increase in satellite transponders demand in India from 2012 to 2017 (more specifically from 104 in 2012 to 276 in 2017 as expressed in TPE), primarily owing to increased DTH and pay TV penetration, as well as growth in bandwidth occupied by TV channels in relation to HD, UHD and 3D channels (Rao et al. 2016; PwC 2016). So far, as discussed in Chap. 4, ISRO has been unable to tackle the demand-supply gap and has been forced to sublease capacity from foreign operators.

In the coming years, addressing this capacity crunch and keeping up with transponders' demand both logistically and technically will be crucial to avoid self-inflicted stunting of potential economic growth. According to the National Institute of Advanced Studies (NIAS), "bridging the present gap in transponder availability against demand requires more than doubling of capacity" (Rao et al. 2015). It hence appears evident that such challenges cannot be addressed only through the utilisation of traditional satellite communications technology but, to the contrary, require novel techniques and approaches. As also argued in a thorough research by Rao, Murthi and Raj:

> Satellite communications technology needs to expand in newer areas and with newer methods (like Hybrid Satellite-Terrestrial Network systems, Sensor Networks; High Capacity Satellite Links – Ka-band; large class Ku- satellites; advances in modulation techniques,

---

[27] While the commercial market represents an important driver for increasing launch frequencies of both the PSLV and GSLV, there is a more important need to tackle the ever-growing domestic demand for space services, particularly communications and broadcasting.

spot beam-based geosynchronous and medium earth orbit High Throughput Satellite (HTS) technologies and Internet applications; enhanced mobility services with aeronautical and maritime applications; Machine to Machine (M2M) satellite applications; emerging ultra HD technologies; and electric propulsion etc.) – for these newer areas need to be developed to meet the multi-carious social broadcasting needs across languages/states and also for commercial enterprises. Already delayed, this needs to be quickly done to fill the technology gap (Rao et al. 2015).

Apart from these new technologies, capacity crunch can – and will presumably – be also addressed by supporting the emergence of Indian private companies and commercial solutions on the market, as explicitly recommended by TRAI in May 2017 (Bhattacharjee 2017).[28] Again, this is an issue calling for changes in the orgware of Indian space activities.

In the field of satellite systems, another set of objectives calls for the improvement of Earth observation satellites, specifically in terms of their resolution and availability for a variety of users. Current IRS satellite imaging comes within 1-metre resolution accuracy, with ISRO planning to bring it under half a metre by 2019. However, average resolution of the major spacefaring nations' EO satellites is in many cases already under 0.3 metres, and by that time, the market is in addition projected to be inundated by almost instant imaging capability.

Major efforts can thus be expected also in this area, particularly in terms of fine-tuning GIS tools that are capable of near-real-time and ubiquitous imaging and smart data delivery systems to users while in parallel enabling operations by private players. Indeed, as provocatively highlighted by NRS analyst Prateep Basu,

> the Indian market is currently well primed for satellite EO services, with ISRO signing 170 MoUs with various public agencies for applications such as infrastructure monitoring, crop insurance, watershed development, asset management, and mapping. But the question ISRO and the government must ask is — can ISRO alone deliver to all the customers in a timely fashion? (Basu 2016).

The necessity of enabling private contributions appears evident and so does the need to alleviate policy restrictions related to access to such imagery.

With respect to PNT satellite systems, the immediate challenge for India will be in the operational exploitation of GAGAN and NAVIC and delivery of related services. Another immediate challenge is related to system evolution, particularly in terms of the accuracy of PNT services, to better serve India's growing needs for national security, governmental, commercial and citizen services.

In addition, while NAVIC was since its onset designed as an RNSS, a possible expansion of the regional system NAVIC into a fully fledged GNSS has been advanced by several Indian stakeholders. Any decision in this regard is directly connected also to questions of techno-political autonomy as well as to the place India wants to occupy among major spacefaring nations and military powers.

Closely related to this possible development, it is also important to highlight the above-mentioned need to build up more robust C4ISR capabilities through the

---

[28] The recommendation was put forward by on 20 May 2017, arguing that rising communications needs make it necessary for private players to venture into space.

launch of dedicated military satellites for the Navy, Army and Air Force, as well as the development of a suitable ground and space infrastructure for autonomous SSA.

All in all, to support long-term advancements in its space programme, in the near future, India will be pushed to make great strides forward in terms of its hardware capacity, focusing on the aforementioned communication satellite capacity, Earth observation improvements, enhanced access to space as well as possible moves towards a global satellite positioning system and human spaceflight programme.

In 2015, India's National Institute of Advanced Studies (NIAS) undertook a simulative exercise to determine the number of missions that India would have to carry years ahead of the 12th plan period (from 2017 to 2025) and estimated that the number of missions required to meet national needs would be near 150 missions (Rao et al. 2015). These missions include:

- Missions for advanced large bandwidth and bulk-transponder communications that will be required to meet large demand for societal and commercial needs
- Missions for EO – including high-resolution, ocean-monitoring, meteorological and earth science applications to meet national security, governmental, commercial and citizens' services
- Planetary and science missions to maintain a continued dove-tailed strategy;
- Enhanced positioning system missions to augment regional services into global high-precision services of positioning
- Technology tests and demonstration and early experimental missions for human spaceflight, including man-rating of launchers and a range of technological development missions for human spaceflight
- Access to space and advanced launch technology activities (e.g. for re-entry testing)
- Critical technology R&D and demonstration missions – including advanced communications technology, advanced EO technology, etc.

Realising all these missions will certainly require major financial efforts, which will hence require a substantial increase in India's space budget. In this respect, it may be relevant to try to estimate the evolution of India's space budget over the next decades. A projection can be made by directly linking the increase in the space budget to the forecast growth of India's economy, in particular to the estimated GDP growth rate (Table 7.2). As in previous budgetary estimates, these figures should be treated with caution as they neither account for inflation nor reflect possibly deeper political involvement in space activities but only display the pace of growth of India's space budget in relation to its forecast economic performance to 2035 assuming that the space budget will continue to represent 0.06% of India's GDP. Since labour and manufacturing costs as well as market prices remain rather low in India, it appears more useful to express both GDP and the associated space budget evolution in purchasing power parity (PPP) terms.

Notwithstanding the possible pitfalls, Table 7.2 gives quite a good indication of the likely pace of growth of India's space programme. Like its economic growth rate, the space budget can be expected to increase at a compound annual growth rate

Table 7.2 India's GDP and space budget projections

Year	GDP	Space budget (0.06& of GDP)
2017	5.340.142.000.000	3.204.085.200
2020	6.337.715.000.000	3.802.629.000
2025	8.437.521.000.000	5.062.512.600
2030	11.162.212.000.000	6.697.327.200
2035	14.504.379.000.000	8.702.627.400

India's GDP and budget estimates 2017–2035 *Measured in USD at 2010 purchasing power parities* https://data.oecd.org/gdp/gdp-long-term-forecast.htm

(CAGR) of 7.0 % between 2017 and 2035, when it will reach approximately $8.7 billion in PPP terms.

While it can be anticipated that such a growth rate will make India one of the largest space spenders in the world (after the United States, Europe and China), by the end of the next decade, this increase will possibly not suffice to cope with the level of India's possible space ambitions. Indeed, according to many commentators, the investment level required is so great that India may find it difficult to simultaneously deal with civilian, commercial, military and human spaceflight initiatives (Chandrashekar 2015). India may also have to make some hard choices on the trade-offs between the efforts required to put its *vyomanauts* in space and the requirements to address societal needs and build robust space-based C4ISR capabilities. Unless the human and financial resources currently available in the country are expanded in an even more significant manner, pursuing all the objectives simultaneously may not be feasible or may sensibly delay progress in the identified areas.

A utilitarian approach would involve according to priority societal needs while retaining technology development and a more cautious approach for military space activities and the human spaceflight programme. However, as both a developing country and developing power, by necessity India will have to maintain a coordinated approach to its space efforts, meaning that the different facets of the space programme should be advanced concurrently. Clearly, all this will imply the emergence of even more considerable strains on ISRO, which for this reason can be expected to introduce important changes in the very organisational and managerial aspects of the national space ecosystem.

7.2.2   The Orgware: Introducing Changes in the Organisation of Space Activities

The above outlined need to expand and improve India's hardware will clearly dictate changes in the *orgware*, that is, in the organisational structure of space activities in India and in particular the relations of ISRO with other major stakeholders such as industry, academia and the military establishments. The ISRO model has been very successful for the past 40 years, setting ground-breaking avenues for addressing societal pull through an end-to-end approach. However, as mentioned earlier, the fundamental nature of the Indian space ecosystem will have to be adapted in the coming years to meet new objectives and overcome current challenges. This is because ISRO is unlikely to have the resources and means to manage the explosive growth in demand for space assets on its own. After all, up to 150 missions are estimated to be scheduled for the next 5–10 years (Rao et al. 2015): a major increase as opposed to the 58 missions undertaken during the 12th (and last) Five-Year Plan. Assuming the accuracy of these prognostications and taking into consideration the fact that ISRO's workforce has remained fairly constant at 16,000 with little change as of recently leads to the assumption that – even with a more substantial budget increase – ISRO simply lacks the capacity to face these sizeable but needed increases in space activity. In short, there is a clear need to transform the orgware by in particular taking the necessary steps to accommodate the involvement of private actors.

An Increased Role for Industry

As primarily observed in the United States over the past few years, space activities involve the private sector now more than ever. What started as a space race between superpowers amidst global geopolitical tension – thus mainly serving political interests in the hands of government – has since evolved to include a commercial private-industry platform where incentives along the traditional profit motive of the free enterprise system have a chance to fundamentally change the space industry. India's future in space, as for any spacefaring nation nowadays, relies heavily on the emergence and development of a competitive and sustainable private industry, thus making it paramount to create a favourable environment for public-private partnerships and to enable entrepreneurship and development of private endeavours in the operational space activity of India.

Private actors' involvement can range from the manufacturing of space assets or anything in between through "contractorship" to the private ownership and exploitation of said assets. Not only would this fulfil the primary objective of meeting skyrocketing Indian demand faster and at a lower cost through commercial solutions, it would also contribute to the further development of Indian space capabilities and pave a lucrative path for Indian products and services in the global marketplace. India has already demonstrated its capacity to capture a share of international commercial demand with competitive solutions. PSLV, for instance, has proven superb reliability in several successful payload launches and has proven commercial applications as part of satellite launch services. As India expands beyond covering its own national demand for space services, increased international commercial transactions could help boost the domestic economy while simultaneously spreading a good reputation for Indian space products and services, which naturally increases demand for them and incentivises other space programmes or private entities to seek cooperation with India. Policy models such as the "transfer and buy-back" scheme (i.e. the procurement of products manufactured by industry as a result of technology transfer by public agencies) have been staples of the Indian mining, telecom and insurance industry's privatisation efforts and may thus also be a suitable option for India's space ecosystem (Prasad and Basu 2015).

An increasing contribution of private space industry does not necessarily imply a reduction of ISRO importance but rather suggests a refocus of the space agency's activity. In a new ecosystem, ISRO would take on the vital role of guiding and empowering industry with their vast array of experience, achievements and expertise. As is the case for any major shift in an industrial sector, it will ideally happen gradually and with good communication and mutual support among the involved players working towards complementary objectives. In the name of furthering the nation's joint interests in space, ISRO would thus assist private entities in reaching their individual roles within the grand scheme while assuring quality and safety across the board. In essence, ISRO would take leadership, hand in hand with private actors, to transform India's space industry, providing mutual and multilateral benefits through an appropriate space governance, industry and market organisation driven by Indian strategic objectives. Involvement with foreign partners – as part of a bilateral cooperation agreement, for instance – could provide this new Indian space ecosystem with the tools (e.g. technologies, business models, etc.) necessary for future success on a domestic and global level alike.

Certainly, one must not overlook the structural challenges that come with progressing and refining the Indian space industry. Consideration of changes in the larger governing structures must be taken into due account to enable the space landscape to evolve. As the Antrix/Devas fiasco shows, the current organisational structure inherently bears potentially adverse downsides, owing to the central role that the government continues to occupy in the market. The fact that the chairman of ISRO is simultaneously the chairman of Antrix and the secretary of the Space Commission makes ISRO/Antrix play overlapping and conflicting roles as a supplier, an intermediary, policy formulator and arbitrator. As PwC notes, "This results in direct conflicts of interest and runs counter to regulatory best practices adopted in other sectors and other geographies. [...Hence] there must be a segregation of the roles of ISRO/Antrix into a satellite operator, a research institute and an independent commercial entity. The roles of a policymaker and enforcer should be assigned to independent entities" (PwC 2016). This of course is primarily a legal issue for the government, which must clearly be considered in future regulatory moves.

Enhancing the Role of Academia

The above-mentioned separation of roles and enhancement of industry's role in launch vehicle and satellite manufacturing would allow ISRO to truly focus on technology R&D, which is its core mandate as a research and development organisation. In fact, as commented by many – including ISRO Chairman Kiran Kumar – for a number of reasons, ISRO has turned into the India Space Production Organisation (ISPO), while it should rather go back to a research and development organisation, with production handed over to industry.[29]

---

[29] See 2nd ORF Kalpana Chawla Annual Space Policy Dialogue (Observer Research Foundation 2016).

A stronger focus on R&D will be extremely important and beneficial for more ambitious space activities such as future space sciences and planetary missions. However, also in this respect, some steps to optimise this role will likely be made. This is because the stronger involvement of academia and the scientific community in these fields will be an all-important factor for the advancement of India's space sector at large. Despite tools such as RESPOND, ISRO is still far from streamlining the participation of research institutions in the ecosystem and deriving more substantial outcomes (Prasad and Basu 2015).

To this end, it is likely that ISRO will pursue active engagement of clustered activities involving academia-industry-agency, which are of significant importance for creating systemic changes in establishing a globally leading research output environment. One of the methods of moving away from an "Islands of Excellence" model to actively promoting interdependent engagement of academia-industry-agency is by setting up programme consortia (where each of these stakeholders have concrete involvement in deliverables and gain significant benefits having long-term ecosystem prospects of spin-offs). As also recommended by NIAS, it would be desirable to apply open public procurement based on free market principles to the R&D process so as to create the optimal climate for ground-breaking scientific breakthroughs to be made, which in turn would contribute to developing Indian sciences and enabling certain missions (Rao et al. 2015).

Orgware Improvements for Military Space Activities

In addition to civil and commercial space activities, the evolution of India's space orgware also comes to the fore when considering military space. As discussed in Chap. 5, with the recent proliferation of military space activity at both regional and international level, the Indian government has started to reconsider its traditional focus on civilian space and to contemplate development of both passive and active military space capabilities to protect its national interests and enhance its security.

As Narayan Prasad argued

> it has to be noted that the civilian nature of ISRO creates fundamental policy complexities in trying to realise a space security deterrent.... With the armed forces of India effectively trying to integrate space capabilities into their operations, the challenge for the government lies in creating an effective organisational architecture that can allow the space programme to effectively continue delivering civilian applications, while allowing enough room for the armed forces to expand their capabilities to utilise space assets in areas such as ELINT, COMINT, SIGINT, IMINT. An independent organisational structure for security-related utilisation of space assets has been realised in the area of the launch segment with the Defence Research and Development Organisation (DRDO) taking over the development of launchers for armed forces (Prasad 2016).

In this frame, a dedicated organisational structure – e.g., a fully fledged aerospace command or alternatively a military space agency (Rajagopalan 2013) – should clearly be established in the near future to deal with all security-related aspects of the programme (particularly if considering the possible development and testing of deterrence capabilities).

7.2.3 A New Software: Towards a National Space Policy

Introducing the changes outlined above in the modus operandi and organisational framework of space activities will have to be accompanied also by the implementation of a new, adapted, *software* (i.e. a comprehensive space legislation and regulatory system that clearly defines specific responsibilities and procedures). Together with the orgware evolution, the introduction of new software is indeed a key element to enhancing the *confidence* and the *level* of private investments by (local or foreign) industries as well as for defining the roles of academic institutions and other stakeholders involved in the space ecosystem.

The need for national space legislation comes clearly to the fore when considering the proposed privatisation of launch vehicle and satellite manufacturing by industrial JVs, as well as when considering the current restrictions for accessing and using EO data and authorising the launch and operations of satellite systems or when considering the lacunae in emerging areas such as PNT services. As succinctly stressed by Prasad and Basu:

> Enacting space legislations within the country to define regulatory, legal and procedural regimes with transparent timelines for pursuing space activities by the private space industry is currently at a nascent stage with no national legislations governing space activities, which however remains critically necessary. Enacting stand-alone policy packages for regulating a service/product as a retrospective measure on commercial interest can only be an initial step to the leap required in development of a holistic act for active promotion and encouragement of commercial space activities. The need of the hour is the development of time-bound, transparent procedural aspects of delivering authorisations, licences, frequency allocations, and others (Prasad and Basu 2015).

Given that the need for a comprehensive space law is becoming more pressing, it can be expected that its enactment will be carried out in the near future. A draft of India's national space law is already circulating among Indian stakeholders, though it is raising some criticism for having been defined within the framework of ISRO rather than being defined as an overarching policy vision by the political leadership. As inter alia argued by ORF senior fellow R.P. Rajagopalan:

> As India's power and influence grows, it is important that New Delhi spells out how India approaches outer space. Equally importantly, India's outer space policy must be driven and announced by the national political leadership rather than being left to the discretion of individual bureaucracies such as the Indian Space Research Organisation (ISRO) or Indian military services. Each organization will have its own interest, and these sometimes competing and even conflicting interests can only be shaped into a national policy by the political leadership (Rajagopalan R. P., 2017).

Such an overarching policy document would then form the foundation for lower-level government policies in specific sectors such as national security, civil or commercial space and for interested actors, such as ISRO, industry, academia or the military. For many Indian commentators, if a comprehensive policy vision is to be introduced, at least three elements should be incorporated:

- The overarching rationales and objectives with respect to (a) civil and commercial space activities, (b) military space and (c) international cooperation

- The general principles setting the foundation for the norms, rules and implementing procedures of sub-sectorial policies, including (a) new satellite communication policy, (b) a geospatial data policy, (c) a space transportation policy, (d) a space science policy, (e) a space applications policy, (f) an industrial policy and (j) a national security policy
- The long-term vision for Indian space activities in alignment with the overall vision of national development

7.2.4 Considerations on India's Future Space Diplomacy

With the forecast multiplication of civil, military and commercial ventures and increasing autonomy in the whole spectrum of space activities, it is important to address how Indian space diplomacy will look like in the future, by looking in particular at how India can be expected to choose its partners and what role it will likely play in future space governance.

Certainly, these projections cannot be drawn accurately, as they prove to be highly contingent on the broader evolution of the international space arena and broader political dynamics. Some considerations can nonetheless be offered also in this respect. For one thing, with India moving towards self-reliance in many space areas, it appears that its cooperation approach will likely be recalibrated in light of the lower level of support or mentorship that is needed from more experienced spacefaring nations. In the field of access to space, for instance, the end of dependency on Ariane launch services for India's large telecommunication satellites will necessarily lead to a readjustment in cooperation with France/ESA.

At the same time, however, the new level of technological prowess reached by India will open new possibilities for embarking upon more ground-breaking cooperation projects on an equal footing. In addition, it can be argued that cooperation on a bi- or multilateral level will continue to make sense for India in the pursuit of its space objectives. Whether it be for the utilitarian advantage of joint technological advancement, of gaining access to additional data and of increasing the marketability of their technologies and commercial services around the globe, or the political advantage of linking India with the activities of other space organisations and strengthening political ties, partnerships with other nations will continue to prove greatly beneficial to India.

So far India's participation in international partnerships has been rather limited. As it becomes increasingly capable and autonomous, India will have the possibility to choose between an open and an isolationist path in space. The Indian general policy posture would suggest that the path of opening its space programme to an increasing number of international partnerships is the most likely. This will entail new stakes for its space strategy. It is in fact evident that in order to become a key partner and ensure its place in the control cabin, India will have to demonstrate unique values on the international stage (e.g. specific technologies, resources, cost-effective solutions, etc.) that will likely affect its overall strategy in accordance with the position it will want to occupy on the global scene.

Moreover, there are dimensions to India's future challenges such as climate control, large-scale environmental degradation and natural resource depletion that Indian governments have always felt most appropriately addressed through collective action by the international community, rather than through specifically national policies (Sheehan 2007). Such issues will continue to be addressed through a multilateral rather than a purely national effort.

All in all, international cooperation will certainly remain an important dimension of the Indian space programme. While programmatically driven cooperation can be expected to sensibly reduce, if considering India's emerging uses of space, politically driven cooperation will certainly grow further over the next decades. For instance, it can be expected that New Delhi will increasingly use its space programme to offer South Asian countries a stake in India's economic and political growth, thus contributing to India's broader efforts to assert its role as the predominant power in the region. Likewise, it is likely that India will expand space cooperation with the United States, Japan and Australia with the aim of reinforcing strategic ties with these countries as a way to balance China's power.

Looking at the broader positioning of India in the future governance of space activities, the forecast multiplication of civil, military and commercial ventures and great stakes surrounding its programme will almost certainly make India more assertive as compared to today. Indeed, because of India's visions of a larger international role and its achieved recognition of its space power status (be it in the civil or military realm), it appears quite likely that India will want to express a greater voice in global space governance over the next decades. Indeed, one could even imagine India positioning itself as a fulcrum of the twenty-first-century space order and becoming the source of new proposals for advancing international regimes for outer space activities. What remains unknown, however, is whether this future role will be conducive to greater international synergies or will make the management of space issues even more difficult than today.

7.3 Conclusions: India's Future Path Between Internal Needs and External Projection

Thrust by its fast-pacing economic and demographic growth and its surging geopolitical weight on the global scene, India will certainly make great strides forward in space, focusing on an impressive quantity of efforts that span the entire spectrum of space activities: enhanced access to space, communications capacity, Earth observation satellites improvements, moves towards a global satellite positioning system and dedicated military space assets, planetary exploration and human spaceflight, to name the most important. It can be estimated that the number of Indian operational satellites and annual launches will have to be multiplied at least five times to meet

India's projected undertakings in space. To tackle this ambition, important transformations in the overall organisation and modus operandi of space activities in India will have to be introduced, with new governance schemes including in particular the integration of the private sector both as a lever for future public efforts in space and as an independent sector with high potential.

Turning attention to the future policy vision, the first pillar around which India's future pathway will be based will be meeting domestic needs and overcoming future societal grand challenges while fully seizing future Indian generations' potential through the leapfrog opportunities offered by space. The second pillar sustaining India's path forward in space will be the bolstering of its external power projection. In the course of the next decades, space, as the ultimate expression of what being a scientifically and technologically advanced nation means, will be used to support India's visions of a larger international role for itself. Ground-breaking endeavours such as launching Indian astronauts in space, increasing the Indian footprint in the global space economy as well as building a space military capacity and ensuring security of Indian space assets will all be instrumental towards achieving this objective.

Thanks to all its future undertakings, India will certainly gain a more prominent role in the international space hierarchy of the decades to come. However, the exact position India will occupy on the global space scene will eventually be determined by a number of key factors including:

(a) The level of political ambition and allocation of appropriate funding
(b) The capacity to develop a vibrant private space sector that can contribute to public space efforts while gaining ground in the global space economy
(c) The resolve to develop a comprehensive space capability, particularly in domains thus far addressed with limited financial and human resources (such as military space activities and human spaceflight)
(d) The capacity to trigger innovation and develop advanced technologies to gain weight on both the geopolitical and commercial chessboard of global space relations

Should India eventually fulfil these requirements, a New Space giant will certainly emerge.

References

Alexandratos, N., & Bruinsma, J. (2012). *World agriculture. Towards 2030/2050: The 2012 revision*. Rome: FAO.
Aliberti, M., Ferretti, S., Hulsroj, P., & Lahcen, A. (2016). *Europe in the future and the contributions of space*. Vienna: European Space Policy Institute.
Aliberti, M., Nader, D., & Vernile, A. (2017). Understanding India's new space potential: implications and prospects for Europe. In *68th international astronautical congress (IAC)*. Adelaide: International Astronautical Federation.

Basu, P. (2016, December 1). *New horizon for satellite Earth Observation.* Retrieved December 2, 2016, from Observer Research Foundation: http://www.orfonline.org/expert-speaks/new-horizons-satellite-earth-observation/

Bhattacharjee, S. (2017, May 20). TRAI bats for private players in satellites. *Business Standard.*

Bommakanti, K. (2017). The significance of an Indian direct ascent kinetic capability. *ORF Space Alert, 2*(2).

Bonnefoy, A. (2014). *Humanitarian telemedicine.* Vienna: ESPI.

BP. (2017). *BP energy outlook. Country and regional insights – India.* New Delhi: BP P.iC.

Chandrashekar, S. (2015). *Space, war and security. A strategy for India.* Bangalore: National Institute of Advanced Studies.

Chandrashekar, S. (2016). Space, war, and deterrence: A strategy for India. *Astropolitics: The International Journal of Space Politics and Policy, 14*(2–3), 153–157.

Committee on Earth Observation Satellites. (2016, April 3). *Heads of space agencies decide to join efforts in support of COP 21 decisions.* Retrieved May 10, 2017, from CEOS: http://ceos.org/document_management/Meetings/SIT/SIT-31/160408%20Declaration%20NewDelhi_V3.docx

Dasgupta, A. (2017). Unlocking the potential of geospatial data. In R. Rajagopalan & N. Prasad (Eds.), *Space India 2.0 commerce, policy, security and governance perspectives.* New Delhi: Observer Research Foundation.

FAO, UNICEF, SaciWATERs. (2013). *Water in India: Situation and prospects.* New Delhi: FAO/UNICEF.

GNSS.Asia. (2016). *GNSS opportunities in India.* Retrieved June 10, 2017 from GNSS.Asia: http://india.gnss.asia/gnss-opportunities-india

Gopalaswamy, B. (2013). India's perspective. In A. M. Tellis (Ed.), *Crux of Asia. China, India and the emerging global order.* Washington, DC: Canergie Endowment for International Peace.

IndoGenius. (2016a, August). Creative solutions to grand challenges. *The Importance of* India. IndoGenius (Online Lecture).

IndoGenius. (2016b, August). The youngest population on Earth. *The Importance of India.* (Online Lecture).

International Enegry Agency. (2015). *India Energy Outlook.* Paris: OECD-IEA.

International Monetary Fund. (2017, April). *World economic outlook.* Retrieved May 10, 2017, from International Monetary Fund: https://www.imf.org/external/pubs/ft/weo/2017/01/weodata/index.aspx

ITU/ UNESCO Broadband Commission. (2016). *The state of broadband 2016: Broadband catalyzing sustainable development.* Geneva: ITU/UNESCO Broadband Commission for Sustainable Development.

Joint Global Change Research Institute; Battelle. (2009). *India: The impact of climate change to 2030 A Commissioned Research Report.* Washington, DC: National Intelligence Council.

Kissinger, H. (2014). *World order.* New York: Penguin Books.

KPMG. (2010). *Water sector in India: Overview and focus areas for the future.* New Delhi: KPMG.

Mazzucato, M. (2015). The Innovative State. *Foreign Affairs.*

McKinsey Global Institute. (2014). *India's technology opportunity: Transforming work, empowering people.* New Delhi: McKinsey.

Murali, M. (2015, January 26). *Dhruva Space, German startup Berlin Space Technologies ink MoU on satellites.* Retrieved June 8, 2017, from Economic Times: http://economictimes.indiatimes.com/small-biz/startups/dhruva-space-german-startup-berlin-space-technologies-ink-mou-on-satellites/articleshow/46015590.cms

Murthi, S., & Rao, M. (2015). Future Indian (New) Space – Contours Of A National space policy that positions a new public-private regime. In *3rd Manfred Lachs international conference on newspace commercialisation and the law.* Montreal: NIAS.

Murthi, S., & Rao, M. (2016). Privatising space misssions: The critical route to boost India's space economy. *New Space Journal.*

Nair, S. U. (2015). Initiatives on India's human spaceflight. In P. M. Rao (Ed.), *From fishing hamlet to red planet: India's space journey*. New Delhi: Harper Collins.

National Intelligence Council. (2012). *Global trends 2030: Alternative worlds*. Washington, DC: National Intelligence Council.

National Intelligence Council. (2017). *Global trends. Paradox of progress*. Washington, DC: National Intelligence Council.

NDTV. (2013, August 16). *Human space flight mission off ISRO priority list*. Retrieved September 28, 2016, from NDTV: http://www.ndtv.com/article/india/human-space-flight-mission-off-isro-priority-list-406551

Observer Research Foundation. (2016). *2nd ORF Kalpana Chawla annual space policy dialogue*. New Delhi: ORF.

OECD. (2016). *Space and innovation*. Paris: OECD Publishing.

OECD. (2017). *GDP long-term forecast*. Retrieved April 29, 2017, from OECD Data: https://data.oecd.org/gdp/gdp-long-term-forecast.htm

Our World in Data. (2017). *Future world population growth*. Retrieved April 30, 2017, from Our World in Data: https://ourworldindata.org/future-world-population-growth/#un-population-projection-by-country-and-world-region-until-2100

PM Jan Dhan Yojana. (2016). *Startup India scheme*. Retrieved June 8, 2017, from PM Jan Dhan Yojana: http://pmjandhanyojana.co.in/start-up-india-stand-up-india-scheme/

Population Institute. (2009). *2030: The "Perfect Storm" scenario*. Retrieved December 9, 2016, from Population Institute: https://www.populationinstitute.org/external/files/reports/The_Perfect_Storm_Scenario_for_2030.pdf

Prasad, N. (2016). Diversification of the Indian space programme in the past decade: Perspectives on implications and challenges. *Space Policy, 36*, 38–45.

Prasad, N. (2017a). Developing a space startup incubator to build a new space ecosystem in India. In N. Prasad & R. P. Rajagopalan (Eds.), *Space India 2.0. commerce, policy, security and governance perspectives*. New Delhi: Observer Research Foundation.

Prasad, N. (2017b). Space 2.0 India – Leapfrogging Indian space commerce. In N. Prasad & R. P. Rajagopalan (Eds.), *Space India 2.0. commerce, policy, security, and governance perspectives*. New Delhi: Observer Research Foundation.

Prasad, N. (2017c). Traditional space and new space in industry in India: Current outlook and perspectives for the future. In N. Prasad & R. P. Rajagopalan (Eds.), *Space India 2.0 commerce, policy, security, and governance perspectives*. New Delhi: Observer Research Foundation.

Prasad, N., & Basu, P. (2015). *Space 2.0: Shaping India's leap into the final frontier*. New Delhi: Observer Research Foundation.

Prasad, N., & Rajagopalan, R. P. (2017). Creation of a defence space agency. A new chapter in exploring India's space security. In S. Singh & P. Das (Eds.), *Defence primer 2017. Today's capabilities, tomorrow's conflicts*. New Delhi: Observer Research Foundation.

PwC. (2016). *Capacity crunch continues assessment of satellite transponders' capacity for the Indian broadcast and broadband market*. Hong Kong: Casbaa.

Rajagopalan, R. P. (2013, October 13). Synergies in space: The case for an Indian aerospace command. *ORF Issue Brief, 59*.

Rajagopalan, R. P. (2017, January). The need for India's space policy is real. *ORF Space Alert, 5*(1), 7–8.

Rajgopalan, R., & Jayasimha, S. (2016, December 6). *Non-Geostationary Satellite Constellation for Digital India*. Retrieved from Observer Research Foundation: http://www.orfonline.org/expert-speaks/digital-india-non-geostationary-satellite-constellation/

Rao, M. K., Murthi, S., & Raj, B. (2015). *Indian space – Towards a "National Eco-System" for future space activities*. Jerusalem: 66th International Astronautical Federation.

Rao, M. K., Murthi, S. K., & Raj, B. (2016). Indian Space – Towards a "National Ecosystem" for Future Space Activities. *New Space, 4*(4), 228–236.

Sachdeva, G. (2015). *Evaluation of the EU-India strategic partnership and the potential for its revitalisation*. Brussels: European Union.

Satak, N., Putty, M., & Bhat, P. (2017). *Exploring the potential of satellite connectivity for digital India*. New Delhi.

SatSure. (2017). *Solutions*. Retrieved May 20, 2017, from SatSure: http://www.satsure.in/#solutions

Sheehan, M. (2007). *The international politics of space*. New York: Routledge.

Statista. (2017). *Percentage of households owning a car in selected countries in 2014, by country*. Retrieved June 8, 2017, from The Statistics Portal: https://www.statista.com/statistics/516280/share-of-households-that-own-a-passenger-vehicle-by-country/

The World Bank. (2013). *India: Climate change impacts*. Retrieved June 8, 2017, from World Bank: http://www.worldbank.org/en/news/feature/2013/06/19/india-climate-change-impactxandratos

United Nations. (2014). *World urbanisation prospects: The 2014 revision, highlights (ST/ESA/SER.A/352)*. New York: Department of Economic and Social Affairs, United Nations.

Vernile, A. (2017). *The rise of the private actor in the space sector*. Vienna: European Space Policy Institute.

Vikram Sarabhai Space Centre. (2017). *Crew module atmospheric re-entry experiment (CARE)*. Retrieved June 18, 2017, from Vikram Sarabhai Space Centre: http://www.vssc.gov.in/VSSC_V4/index.php/care

Chapter 8
Nurturing the India Opportunity: A European Perspective

The ongoing diversification and future ambitions set forth in India's space programme certainly provide the international community with an increasing array of opportunities for engaging and collaborating with this country. This final chapter provides a reflection on what the opportunities for Europe are to create mutually beneficial space cooperation with India, both at pan-European and national level. It also discusses how such cooperation would fit within broader political cooperation between Europe and India on general flagship economic and societal challenges. The chapter starts with an assessment of the current status of Europe–India relations in the political and space arena. It subsequently elaborates on the stakes, drivers and opportunities underpinning the case for closer Indo-European space cooperation and concludes with the identification of key domains for future cooperation.

8.1 Introduction: Unleveraged Potential?

India has embarked on an important project of self-renovation and re-emergence on the global scene urging the world to reconsider its importance. Strikingly, whereas countries such as the United States, Russia, Japan, Israel and Australia are already courting India to strengthen economic and political ties, using space as a central cooperation tool, on Europe's side, there has been very little interest from the top in deepening cooperation with New Delhi.

From a political standpoint, India occupies a rather marginal role in Europe's strategic thinking, and vice-versa. Even though some European countries such as France – and more recently Germany – have included India among their top foreign policy priorities, the overall strategic interplay between the two continents has remained negligible (a more detailed description of Europe–India economic and political relations is offered in Sect. 5 of Annex).

Admittedly, in 2004, India was acknowledged as a strategic partner of the EU. But despite the "initial enthusiasm, abundant affinities, and a genuine desire to

© Springer International Publishing AG 2018   289
M. Aliberti, *India in Space: Between Utility and Geopolitics*, Studies in Space
Policy 14, https://doi.org/10.1007/978-3-319-71652-7_8

interlink the already sound economic relations with geopolitical considerations, the partnership proved unable to match words with deeds, failing not only to develop a real strategic dimension (i.e. security and defence-related), but also to receive sufficient political attention from both sides" (Aliberti 2017).

Officials and political commentators from both sides have regularly pointed out that the EU–India strategic partnership had become "a charade without any strategic content" (Jaffrelot 2006), "a loveless arranged marriage" (Khandekar 2011), "not delivering on its ambitious goals" (Basin 2009). This may appear rather odd, considering that India and the EU share common values and beliefs that should make them natural partners and provide fertile soil for extensive cooperation. But the reality is that the two actors have been unable to transform shared values into shared interests and priorities because of "a big disconnect in world views, mind-sets and political agendas and because the two are at different levels of socio-economic development, come from different geopolitical milieus and have different geographical and geopolitical priorities" (Jain 2014).

Largely absorbed by their own internal problems and immediate neighbourhoods, the two actors have dismissed the potential of their political relations. While extensive policy dialogue and consultations at ministerial and expert level have been held on a variety of political issues (e.g. on disarmament, counter-terrorism and piracy, and cybersecurity), these "have been confined to bureaucratic exchanges, in which the counterparts engage without much commitment or therefore progress" (Benaglia 2016). Furthermore, EU's difficulties in affirming itself as a security actor and its ensuing lack of assertiveness have inevitably led to diminished credibility and interest in cooperation with the EU on defence and security matters. Trade and investments have remained the primary focus of EU–India relations. But here too, the relation seems to have become hostage to the never-ending negotiations on a Free Trade Agreement, despite the EU being India's largest trading and investment partner. As a result, much cooperation potential has remained untapped.

This story is well mirrored also in the space arena, an arena where, despite promising foundations, Europe has been unable to position itself as a key cooperation partner for India. Certainly, the cooperative ventures taking place over the past 50 years have in many cases proved of utmost importance for the emergence of India in the space arena. It suffices to remind the assistance provided by France in setting up India's sounding rocket programme in the 1960s, the transfer of Viking engine technology in 1972, which is still used nowadays to power the PSLV and GSLV launchers; the assistance offered by ESA in the development and launch of APPLE, India's first telecommunication satellite in 1980; or the decade-long utilisation of Ariane rockets for launching India's geostationary satellites. But it is exactly because Europe has played such a crucial role in the advancement of the Indian space programme that the relationship does not seem to have been sufficiently leveraged by the European partners. ESA-ISRO cooperation has been slow moving and fragmented – not to say erratic, with very little going on after the cooperation experience on Chrandrayaan-1 almost 10 years ago. EU–India space relations have been likewise devoid of an over-arching policy purpose and commitment but have proved even more displeasing in terms of actionable policy measures and deliveries. With the partial exceptions of EUMETSAT and CNES, an analogous line of reasoning could certainly be made for all the other space cooperation experiences at the level of national space agencies in Europe as well as for industry, despite both Airbus and Thales being major industrial partners for India in the broader aerospace and defence sector.

If it is undeniable that for European stakeholders there has been a plethora of hurdles standing in the way of closer Indo-European space ties, from an overall perspective, the underlying reason needs to be found in the lack of a deep conviction by the heads of space agencies and, more importantly, in the dearth of a strong political commitment that makes things happening.

This situation, however, is progressively changing. It is hence pleasing to see that a re-emerging convergence of interest towards closer interaction between Europe and India has recently come to the fore. Such convergence was duly reflected in the last EU–India summit of 30 March 2016, which was held after a remarkable break of 4 years. On that occasion the president of the European Commission and of the European Council and India's prime minister strongly committed to giving new momentum to their bilateral ties by endorsing a vision document to jointly guide and strengthen the India-EU Strategic Partnership in the period 2016–2020 (European Commission 2016). The EU-India 2020 Agenda for Action sets out a comprehensive roadmap for injecting new impetus to the India–EU Strategic Partnership through broad and concrete engagement in the implementation of poli-cies and projects of common interest, including in space (see Sect. 5.1 of Annex). As explained in the next sections, an expanded level of cooperation in this field could arguably become an effective tool to achieving the ambitious goal set for the future EU–India relations.

8.2 Outlining the India Opportunity: The Case for Closer Ties

Before assessing the potential for future European cooperation with India in key issue areas, it is essential to disentangle the general drivers and opportunities that could be seized through closer engagement. This section provides general consider-ations underpinning the case for closer Europe–India space ties, while taking into account also possible constrains and drawbacks. Given the variety of European stakeholders possibly involved (ESA, EUMETSAT, the EU, national space agencies and European space industries) and the different types of objectives they pursue, such considerations have been grouped under three different headings, namely, pro-grammatic, industrial and political.

8.2.1 Programmatic Drivers and Barriers

India's space activities are undergoing an impressive acceleration. The Indian space programme is gaining momentum with an ambitious roadmap of more than 70 satellites to be launched within the next 5 years and a target to increase launch frequency to 12–18 annually during this period. These programmatic objectives, however, will be met with great difficulties without an opening towards international cooperation, mainly because the closed structure of ISRO has prevented Indian private industry from reaching the maturity achieved by American and European counterparts (Blamont 2017). Structural reforms have been initiated to empower the domestic space industry, but, at least in the short term, India is not in a position to meet the rising demand by relying completely on its own capacity. It needs to find solid partners to meet its growing ambitions.[1] This offers unique opportunities for national space agencies in Europe and for ESA to deepen their space relations with India through various forms of collaboration. It is important to highlight, however, that windows of opportunities are not everlasting, as more spacefaring nations, including the United States, Russia, Japan and Israel, are already expanding their reach into India's space programme.

Equally important is that by pursuing cooperation, European stakeholders would not only be able to seize short-term opportunities but also plant important seeds for the future. India is poised to become a major space power, and there is an opportunity – not to say a necessity – to prepare the ground for what will certainly become a promising important partner in future space endeavours. Indeed, with India moving towards increased levels of technological capabilities and a more ambitious profile in the whole spectrum of space activities, closer ties will open up possibilities for Europe to embark upon more ambitious and enduring partnerships (e.g. space science missions, robotic exploration to celestial bodies, human spaceflight, etc.). Securing European participation in large and complex cooperative undertakings with India could not only offset economic costs but also enable ESA to take advantage of Indian contributions to expand ESA programmatic deliveries.

At the same time, one should also recognise inherent difficulties, including the possible mismatch in programmatic priorities, the different level of technological development, and the application of different standards for project management as well as the development and testing of hardware and software.[2] In addition, European stakeholders have often lamented continuous delays on delivery and lack of transparency and efficient communication,[3] as well as complex management and

---

[1] See Observer Research Foundation (2016) and Rao et al. (2015).

[2] These are essential elements for any cooperative venture as it can compromise the functional integrity and compatibility of the various elements of the project or at very least cause additional overhead costs that would increase the overall cost of any international cooperative project.

[3] The problems experienced on Chandrayaan-1 are a case in point, not only highlighting weaknesses in Indian space technology but also transparency. Indeed, on that occasion, ISRO refused to share information with its foreign partners regarding a problem with the overheating of the Chandrayaan-1 spacecraft until July 2009 (almost a year after the spacecraft's launch), leaving

policy-making processes that are perceived to slow progress. Moreover, India has continued to seek technology transfer from its cooperative missions, which has become somewhat of a hindrance in its relations with European partners. In this respect, it is useful to raise "the sensitive issue of technology transfer inherent to dual-use systems, an area in which Europe is largely bound to U.S. decisions" (Sourbès-Verger 2016). In the past, for instance, the launch of ASI's ROSA instrument aboard Oceansat and the development of ISRO-CNES' Megha-Tropique satellite have encountered a certain opposition from the US government. While these missions eventually took place, it is clear that Europe has not been fully free to engage in cooperation undertakings. However, ITAR restrictions on collaboration with India have been loosened substantially over the last few years, thus opening up new opportunities for furthering Indo-European cooperation in space.

There are a number of important factors that can now make India a partner of choice for Europe. To begin with, ISRO has a well-demonstrated ability to innovate frugally and to operate its space programme in a cost-effective manner, which is certainly a highly relevant asset for Europe, particularly when considering its plateauing financial possibilities. In addition, India's attractiveness also stems from the fact that, once approved, its space programmes and funding have demonstrated – as compared to those of the United States and Russia – a high degree of stability, which is ultimately an important guarantee for international partners.

More broadly, cooperation with India would allow a diversification of the portfolio of European partnerships. Such diversification is particularly important, as it would reduce European critical dependency on its historical partners (the United States and Russia), while at the same time providing more back-up opportunities for the implementation of programmes. As the experience of ExoMars clearly shows, potential back-up opportunities are essential, as they might ultimately prevent the collapse of a mission in case of withdrawal by another partner. China is certainly an option Europe is looking at, but cooperation with India will not be hindered by the same level of political and regulatory hurdles (above all, export control restrictions) as with China.

By the same token, associating with India could allow Europe to gain access to additional space-related infrastructure (e.g. ground stations for TT&C, deep space antennas, laser ranging stations, optical telescopes, etc.), thus contributing to securing stronger operational support to ESA missions.

Equally importantly, cooperating with India would be likely to optimise ESA missions' exploitation and data access via dedicated agreements to secure ESA access to non-ESA missions, while ensuring better exploitation and expansion of ESA missions' data to meet global needs if defined as of interest to ESA programmes (e.g. climate change, sustainable development, etc.). This is equally significant for ESA as it is for national space agencies and EUMETSAT. All European actors hence appear to have considerable interest in being able to access Indian ground and space infrastructures in the future; Indo-European cooperation would consolidate the prospect of achieving these objectives.

---

other space programmes frustrated, particularly after the spacecraft experienced altitude-control problems and crashed into the Moon's surface only a month later.

8.2.2 *Industrial Drivers and Concerns*

Alongside programmatic considerations relevant for national and pan-European public space actors, there is also a wealth of drivers and opportunities to be seized by the European space industry.

For one thing, India is an expanding market with largely untapped potential when it comes to space products and services.[4] Whereas this market has typically been rather restrictive for private actors' involvement, the Indian government has adopted business-friendly policies under its recent flagship initiatives that can ease the doing of business for foreign space industries. These include a relaxation of the FDI threshold in the aerospace sector from 49% to 100%, as part of a broader initiative (*Make in India*) that seeks to position India as a global manufacturing hub.

With India now willing to attract FDI in the space sector, there is a clear case for European firms to either outsource production (e.g. low-cost components) or gain market entry into India by positioning their offerings independently or by integrating their market strategies via partnerships with Indian SMEs and start-ups to manufacture satellite components and subsystems or even realise end-to-end upstream and downstream systems offerings (Prasad 2017).

Indian demand for space-related products and services is expected to increase dramatically over the coming years in several domains, thus creating unprecedented opportunities that could be best seized by European stakeholders if closer cooperation is pursued, for instance, through the creation of B2B alliances such as joint ventures (JVs), merger and acquisitions (M&As) or special-purpose entities (SPEs). While regulatory mechanisms for private sector activities in both the upstream and downstream value chain are still affected by some considerable issues (e.g. in terms of licensing, intellectual property rights, liability, etc.) that inhibit predictability, transparency and responsiveness,[5] an alliance with India's state-run and private industry might enable European firms to further access this growing institutional market.

In addition to the size of the market, India's market conditions, including low labour costs, talent pool and engineering services backbone, particularly in terms of

---

[4] Additional positive side aspects include the stability of the government, positive investor climate, eased banking interest rates, low cost of operations, improving logistics, etc.

[5] Indeed, India's general business and regulatory ecosystem have certainly played a crucial role in inhibiting cooperation. India's government mantra is "ease of doing business", but actually India still ranks 155th out of 189 countries in the World Bank's Doing Business Report. Long-standing problems such as an unpredictable bureaucracy, corruption, unclear procurement and licensing procedures, cap on FDI or IPR issues among others have jeopardised possible industrial cooperation. European industries that have a presence in the broader Indian aerospace and defence sector (like Airbus and Thales) have often "found themselves frustrated by opaque bureaucratic procurement processes, onerous domestic offset, work share requirements, and seemingly endless delays. They consider that the preference among stakeholders for a protectionist approach hinders cooperation. The win-win approach is not understood in its full potential, and therefore seldom applied" (Benaglia 2016). These barriers are certainly evident in the telecommunication sector and in the market for GIS solutions but are even more burdensome in the upstream segment.

information technology enabled services and software developments, create a strong case for European firms to invest in India. Thanks to its large educated English-speaking population and wealth of skilled low-cost labour, cooperation with Indian industry could be ideal in pursuing a cost leadership competitive advantage, where European manufacturers trade market access for low-cost labour (Al-Ekabi 2016).

India's creativity and capacity to innovate are likewise important drivers that the European private sector could size. Today, India is the fastest-growing start-up ecosystem in the world. According to the 2015 NASSCOM Industry Report, India now has over 4200 start-ups (generating over 80,000 jobs) and is well on its way to reaching its goal of 10,000 by 2020. This makes India the third largest home for start-ups, behind the United States and the United Kingdom, but ahead of countries such as Israel and China (NASSCOM 2016). Importantly, the Indian technology landscape has seen tremendous growth towards creation of innovative start-ups also in the space sector. While many of them still exist only on paper, the overall space ecosystem is progressively taking off.

What is also noticeable is that Indian companies have not only proven to be a hotbed of so-called frugal (*jugaad*) engineering and grassroot innovation, but they have also shown a great ability to convert and build on new concepts based on existing knowledge and technologies by using new processes and business models[6] (Radjou and Prabhu 2010; Zinnov 2015). As also argued by *Dhruva Space's* founder Narayan Prasad: "this makes a case for international firms to utilise India as a hub for 'reverse innovation' before 'trickling up' to either allied markets in emerging economies or developed countries themselves" (Prasad 2017). European firms could capitalise on these local market conditions

> to replicate the cost-competitive IP creation and co-creation in the country that has been a construct that global firms have already used successfully in technology sectors such as aerospace and IT […] Based on the long-term strategy, global firms can typically benefit from exploiting India as a foundation for devising growth strategy under competitive pressures by developing expansion plans, decreasing time to market, improving service levels, business process redesign, adopting an industry practice, testing differentiation strategy, access to new markets, enhancing system redundancy (Prasad 2017).

To sum up, there are several drivers and opportunities for European industries to engage India's. What is important to note, however, is that closer B2B ties would not only profit Europe's space industries but rather nurture mutually beneficial, win–win solutions with India's, since operations by European companies could likely spur the growth of an investment ecosystem in India and, in turn, also contribute to addressing India's increasing demand for space products and services.

Indeed, one of the most critical challenges for most of the country's space start-ups and SMEs is the possibility to access sufficient capital investment that can allow them to scale up their offerings and move from the proof-of-the-concept phase to

---

[6]For example, as already explained in Chap. 4, General Electric's ultra-low-cost ECG machines were led by development teams in India, initially for use in these countries, but with inputs from local and foreign subsidiaries.

the actual provision of products and services. Considering that India has a rather undeveloped private investment landscape in the space sector, Indian SMEs and New Space companies would greatly benefit from possible investments, acquisitions or creation of joint ventures with international partners (Prasad 2017; Aliberti et al. 2017).

In addition, it is also clear that the resources Indian partner would tap into through cooperation would not just be financial, but also technical and technological. By contributing to the marketability of European technologies, cooperation with India could also embed the use of related standards, which is essential for Europe's industry, given that it would strengthen the positioning of the European space sector in the global market for space technologies and services. For Indian partners, achieving "collaborative partnerships between Indian and international firms can lead to SMEs in the country to learn from international partners to work with new standards (e.g. European Cooperation for Space Standardization). Such an exercise can spread the capacity within the industry to replicate the success achieved in the auto-industry within the country" (Prasad 2017).

More broadly, by enabling an expansion of Indian industry's capacity to realise and deliver end-to-end upstream and downstream systems, Indo-European industrial cooperation would, in turn, also contribute to addressing India's growing demand for space products and services. At the same time, it is undeniable that, over time, "India's space industry might also gain a considerable benefit from the know-how provided by European partners, thus allowing its industry to enhance its own market position at Europe's expense" (Al-Ekabi 2016).

8.2.3 Political Drivers

Besides the set of programmatic, technological and industrial opportunities, cooperating with India can be key to also advancing diplomatic and broader political goals, which are a key driver of the EU still nascent space diplomacy.

At a strategic and global level, India is for Europe a partner that simply cannot be dismissed "given its geography, demographics, economy, military and nuclear power status" (Benaglia 2016). In addition, its path towards prosperity, the already strong Indo-European economic interdependence and their sharing of democratic values invite Europe to deepen its cooperative relations with New Delhi. Concrete policy measures in key areas of mutual interest are however needed if the EU is to give new momentum to its relations with India. Space cooperation could hence become a very symbolic – yet tangible – tool in a broader initiative aimed at building a more effective strategic partnership, as well as furthering the global impact of the two actors on the international system.

While space is only one of the issue areas where the EU and India have pledged to re-energise cooperation, it can nonetheless serve as a catalyst for achieving a plethora of shared policy objectives (Aliberti 2017). Indeed, for each "sector policy cooperation" identified by the Agenda 2020 – namely, Climate Change, Urban

Development, Research & Innovation, Information and Communication Technology, Energy, Environment, Transport and the 2030 Sustainable Development Goals – there is an evident role for space to play in achieving those objectives.

Thanks to its pervasive utility, space also provides a wide range of opportunities for the establishment of more tangible cooperation schemes with respect to foreign policy and security issues such as terrorism and maritime security, among others. This opportunity should not be dismissed lightly, because by establishing good relations in a prominent field such as space with a strategic partner such as India, the EU would also advance its foreign policy to a more mature stage and hence enhance its credibility as an effective international actor.

Equally important, cooperating with India would allow Europe to emphasise and actively promote its core values within the Indian space community with regard, for instance, to the sustainability and governance of outer space activities. Europe and India have expressed a number of shared views and concerns about the safety and stability of the space environment, and being both Tier-2 space powers, it should be in their mutual interest to team up by strengthening their policy dialogue. Teaming up with India through bilateral level consultations and in multilateral fora could prove key to advancing new normative solutions on the international stage that are in the interests of both.

For instance, cooperation could be conducive to more active Indian engagement in revitalising the International Code of Conduct for Outer Space Activities, implementing transparency and confidence-building measures (TCBMs), as well as more appropriate policy measures with regard to the broader governance of space activities. Indeed, cooperation can be considered an important TCBM in itself, as it leads to a higher degree of mutual trust and understanding between partners, reducing possible misunderstandings.

Cooperation can also go a long way in diluting the ambiguities in posture that arise from India's lack of a declared space policy,[7] as it would more specifically allow Europe to better understand New Delhi's underlying intentions and concerns as well as to learn more about India's interests, needs and priorities for the future. And this, in turn, would further help European partners to better identify the scope for further policy dialogue, data and information exchange and joint activities in areas of mutual interest.

In conclusion, whether looked at from a programmatic, industrial or political perspective, there proves to be a wealth of reasons underpinning the deepening of Europe's space cooperation with India. Whereas the several barriers and potential hurdles described in the previous section still dictate cautiousness – space cooperation cannot be overhauled in a matter of few bilateral consultations – there are several fields that might be ripe for a progressive expansion of ties. How and in what issue area this expansion could be achieved will be assessed in the following section.

---

[7] Indeed, in several occasions, Europe has indeed been left with ambiguous impressions about India's underlying goals (think, for instance, New Delhi's reluctance for a more stringent regime against the weaponisation of space).

8.3   Seizing the Opportunity: Cooperation Options
in Selected Areas

This section proposes four key issue areas in which cooperation between European stakeholders and India could be developed, namely, scientific research in the field of climatology, downstream space applications, space exploration and human space-flight and space security. This particular selection has been primarily made on the basis of targeted interviews with European and Indian stakeholders. In addition, the four domains are among the most substantial elements in the programme of any spacefaring nation and are loaded with important connotations that can benefit Europe, India and the broader international community. Finally, in all four cases, there is clear room for added value generated by cooperative activities that a unilateral approach could not bring.

8.3.1   Scientific Research: Climatology

Scientific research, particularly in the field of Earth Sciences such as climatology, oceanography and meteorology, is an optimal candidate to gradually bolster the spectrum of Europe–India cooperation. Good bases already exist in this area, and there is a clear convergence of interests between India and European stakeholders (including the EU, ESA, EUMETSAT as well as national space agencies) to strengthen activities in this field. Furthermore, this is an area that can leverage substantially cooperative actions.

European and Indian commitment to contribute to a better understanding of planet Earth and to its preservation has been well highlighted by their current policy dialogue on the topic (European Commission 2017) as well as by their active support for the 21st Conference of the Parties (COP21) and subsequent 2016 *Declaration of New Delhi*, which stressed the will of some of the major space agencies to play a major role in climatology research and to maintain continuous global monitoring of climatology parameters. As underlined by the Committee on Earth Observation Satellites (CEOS), the ultimate long-term goal is to have operational LEO and GEO constellations measuring greenhouse gases in the atmosphere (CEOS 2016).

CNES and ISRO are already paving the way in this direction, with the joint development of an infrared climate-monitoring satellite designed to map heat exchange on the Earth surface. ESA and EUMETSAT have also shown interest in cooperating with India towards the development of science-based climate services in synergies with weather and ocean services. Similarly, at the EU level, a regular dialogue on environment and climate change has been initiated, leading to the establishment of a Joint Working Group on Environment, an EU-India Environment Forum and a EU-India Initiative on Clean Development and Climate Change,[8]

---

[8]For an overview of the ongoing EU-India projects, see European Commission (2017).

which are two joint commitments aimed at protecting the environment and curbing climate change. The Joint Action Plan forms the backbone for enhancing cooperation on the environment and climate change. And the EU-India 2020 Agenda of Cooperation has listed both climate change and Earth observation as key issue areas for cooperation between India and the EU. Hence, it is in the interest of both India and European stakeholders to explore the cooperation potential in this area.

Ideally this cooperation would be based on an incremental approach. The primary step could be focused on satellite data and product exchange of respective Earth observation satellites in support of climate monitoring. This should enable data access and redistribution rights to European satellite programmes (including the Copernicus' Sentinels and contributing missions) by the Indian user community and vice versa. The overarching goal would not just be to gain access to more data but rather to optimise the exploitation of existing capabilities and data through international synergies.

In addition, given that exploiting the full potential of satellite data for climate products requires dedicated international efforts for recalibrating and reprocessing data, extracting climate records and making them available to downstream applications and scientists –[9]such cooperation should be extended to cover also calibration/validation, data processing and reprocessing methodology and satellite data applications, including numerical weather prediction. Towards this – and building on the experience already matured by EUMETSAT – an institutionally entrenched mechanism for cooperation (e.g. a joint Indo-European cooperation committee) could be established to deal with these tasks and to elaborate on joint research projects in climatology.

A next step in this cooperation would be the possibility of hosting payloads within respective Earth observation missions (e.g. ESA's future Earth Explorer missions and ISRO's continuity missions of its Oceansat series) or by contemplating the possibility of flying a common set of instruments on their respective platforms so as to provide the user communities with more homogeneous and consistent data as a result of similar technologies and better options for cross validation.[10]

A final step in this cooperation trajectory would clearly be joint planning, development and operations of future satellite missions for climatology research. While ambitious, this cooperation format is already a reality in ISRO-CNES cooperation

---

[9] Re-calibration and cross-calibration are an essential prerequisite to arrive at a homogenous time series of measurements across successive satellites that are useable for climate studies. This is because "each instrument or satellite has its own characteristic such as sensitivity to Earth signals, evolution of performance over time, or orbit stability. In addition, calibration and processing algorithms are continuously improved over the course of a mission. Therefore, a simple concatenation of data in time would show jumps when a satellite is changed or artificial trends for some satellites in a series, and would not be useful for climate analysis". Therefore, in order to achieve an international, independent system for estimating the global emissions based on internationally accepted data, cooperation to cross-calibrate instruments and cross-validate their measurements is quintessential.

[10] A similar example of cooperation has been already offered by EUMETSAT-NOA collaboration on their Initial Joint Polar System for meteorology. The synergies created by this *modus operandi* have proved to be highly beneficial for both organisations. See Lahcen (2013).

and has already been tested successfully by EUMETSAT in its cooperation with NOAA. Future cooperation with India could also follow this path. What is also important to highlight is that such cooperation would greatly contribute to ongoing international efforts (e.g. the World Meteorological Organization (WMO) Global Observing System, the Global Climate Observing System (GCOS), the United Nations Environmental Programme (UNEP) and other related programmes). More importantly, it would be a concrete milestone towards the possible development of climate services within the Global Framework for Climate Services.

8.3.2 Downstream Service Applications

Another promising area for Europe–India cooperation is linked to downstream service applications for navigation, Earth observation and satellite telecommunications.

For one thing, applications are a key focus area for both the European and the Indian space programmes. In addition, this is a domain where not only space agencies but also the EU and European industries might become directly involved. In fact, joint activities in the development of downstream services could be primarily embedded within the framework of EU–India "Sectorial policy cooperation", where they could be linked to the issue areas the EU and India have committed to working on together. Here the drivers are both political, with the progress of the EU–India cooperation framework, and economic. The objective would be to build bridges between the two regions that would benefit business development of private actors.

The EU is an actor with a number of interrelated tools to promote such activities and provide opportunities for engagement with a variety of stakeholders, including the private sector. For instance, Horizon 2020 (H2020), the EU's latest Framework Programme for Research and Innovation, has already been providing funding opportunities in the development of downstream GNSS applications in cooperation with Indian stakeholders through the GNSS.Asia project. India is considered as a very important market for GNSS applications,[11] and international cooperation would hence be essential to increase the adoption of Galileo services outside EU borders. As already mentioned in Chapter 6.3, the GNSS.Asia project, led in India by the European Business Group India (EBGI), aims to develop and implement GNSS industrial cooperation activities between the GNSS industry in Europe and India, focusing on the downstream sector (applications and receivers).[12]

Over the past few years, the EU and India have been collaborating well together on ICT, by working on synergies between the EU's Digital Single Market and the Indian

---

[11] See the GSA's GNSS Market Report at https://www.gsa.europa.eu/galileo/international-co-operation#asia

[12] There is an inherent potential for cooperation here, since the technology base for GNSS product segments is underdeveloped and because India has shown great interest in benefitting from European know-how in both the level of certification and in SBAS market uptake in non-aviation sectors.

flagship "Digital India" (European Commission 2016). This cooperation framework could also be used to explore possible ventures in the field of space applications.

When looking at the possible areas of cooperation with India in terms of GNSS- or EO-enabled applications, particular attention could be devoted to sectors having synergies with topics of societal challenges for both the EU and India, such as transportation, urbanisation, agriculture, energy and climate change) and in which the EU-India Agenda for Action 2020 already provides a framework for joint activities.

Case: Integrated Applications for Smart Cities
Taking the field of urban development as a key example, the EU and India have been working for several years towards the creation of an EU–India partnership on urbanisation. The EU–Mumbai partnership was established in 2013 to build a long-term cooperation platform to address challenges of urbanisation in mega-cities.[13] In India, these challenges relate both to the daunting need to improve urban services – including sanitation, clean water, transport, healthcare, etc. – which are still a below-optimum levels, and the "Sisyphean task of having to build each year, according to McKinsey Global Institute estimates, 700–900 million m^2 commercial and residential spaces" (Khandekar 2015). Optimal urban planning will hence become crucial in the coming years.

In this respect, the EU and India could contemplate the creation of a joint task force dedicated to exploring the potential contributions offered by space assets in the creation of "smart cities dashboard solutions" for enhancing the quality of urban planning and situational awareness through the integration of Earth Observation, Navigation and Telecommunication services. Initiatives could also cover the provision of dynamic flow data, traffic messages, real-time information on public transport, parking information, etc., also in view of the fact that vehicle navigation systems still have a very poor presence in India with less than 40 cities having been mapped (GNSS.Asia 2016).

Irrespective of the variety of specific projects that could be implemented, what is important to underline here is that embedding space applications' cooperation in the overall cooperation framework of urbanisation would accrue multiple benefits. First, it would provide innovative solutions to address future urbanisation challenges. In addition, it would contribute to better partnering in the implementation of flagship initiatives being promoted by the Indian government such as the 100 Smart City Mission and the ARMUT programme. More importantly, it would provide the EU and India with actionable policy measures to meet the joint policy objectives set (for both space and urban development) by the EU-India 2020 Agenda for Action

---

[13] Other initiatives include the ECOCITIES project, through which the EU has been cooperating with five Indian cities to support bringing low-carbon strategies into urban development, and the World Cities project, through which the EU and India have established partnerships between four Indian and four European cities. In addition, with its development cooperation, the EU has assisted more than 20 Indian cities in infrastructure development. In 2016, the EU also launched the International Urban Cooperation Initiative; India aimed at twinning on sustainable urban development between up to 15 Indian and European cities and working with Indian cities on sustainable energy, climate adaptation and mitigation (European External Action Service 2016).

and enhance the level of cooperation. In addition, such cooperation would arguably have positive spill-over effects on other policy areas, opening up valuable opportunities for cooperation on research and innovation (for instance, within the INNO INDIGO framework for cooperation on the development and integration of European Research)[14] and for supporting business solutions as well as new start-up companies within such frameworks as the "Start-up Europe India Network" (Start-Up Europe 2017). Since a large part of the innovation in the application segment is expected to surge in a bottom-up manner, the launch of dedicated initiatives for supporting possible B2B solutions between European and Indian space start-ups should be taken in due account, as they could become major enablers of change by cooperation. Indeed, some B2B solutions for the provision of space-based smart city services have been already proposed by Indian and European start-ups, which would hence greatly benefit from dedicated EU–India initiatives in this field.[15]

More broadly, when considering the role of space in meeting the objectives set forth in Agenda 2020, an analogous line of reasoning could certainly apply to most of the issue areas the EU and India have committed to working on together, including the recently launched EU-India Water Partnership, the Resource Efficiency Initiative and the cooperation framework for rural development, transport and so on. Also, the programmes and initiatives launched by Modi's government (e.g. Digital India, Clean India, Skill India, etc.) may provide new and complementary entry point for novel cooperation formats.

In short, space applications offer numerous opportunities to unfold mutually beneficial cooperation between Europe and India and could actually become a backbone of the current cooperation framework as they provide multiple opportunities for valuable undertakings that in turn would bolster the overarching objective of deepening further the relationship and realising the full potential of India-EU Strategic Partnership.

8.3.3 Space Exploration and Human Spaceflight

India has rightly taken great pride in its space exploration missions to the Moon and Mars, and it is now progressively raising the level of ambition, by contemplating bolder robotic and space sciences missions such as Chndrayaan-2, Aditya L1, XPoSAT as well as forays into the area of human spaceflight (Annadurai 2017).

Europe, on its side, is an established space actor in space exploration, boasting long-term experience and invaluable contributions to space science, robotic explo-

---

[14] For more information, see https://indigoprojects.eu

[15] For instance, Climate City, a start-up company based in Toulouse aiming to become the world's premier supplier of local climate data to governments, municipalities, smart cities, their economies and populations, is cooperating with India's start-up company Earth2Orbit for the provision of "actionable intelligence" data on cities' air and water quality based on Earth observation big data. Financial support to both start-ups has, however, remained limited so far.

ration and human spaceflight. Over the years, it has reached a state-of-the-art technological level and acquired a solid set of critical capabilities, often making it a partner of choice for international cooperation. Furthermore, Europe has demonstrated the ability to provide essential elements to large scientific missions (e.g. Cassini-Huygens) and to the ISS infrastructure (e.g. the Columbus Orbital laboratory and the Automated Transfer Vehicles (ATVs) among others).

Despite having been the core of European cooperation and success in space since the early 1960s, ESA's space sciences and exploration programmes are currently undergoing financial burdens that are inevitably forcing the Agency to make important trade-offs in the selection of programmes and missions.[16] Leveraging international cooperation is therefore essential for embarking on more ambitious space sciences and exploration programmes that expand ESA programmatic deliveries. Similarly, not having mastered critical technologies for full autonomy in the field of human spaceflight, today the debate in Europe has to focus on how to promote exploration and human spaceflight through a scheme of "interdependence and partnership".[17] This position is duly reflected in the orientations that emerged from the 2010 Brussels and 2011 Lucca summits, with the communications issued by the EU and, more recently, with the decisions of the 2016 ESA Ministerial Council. In other words, it appears clear that, for Europe, any future space exploration projects, particularly those involving astronauts, must be achieved through international cooperation.

In light of India's growing footprint in the area of space sciences and exploration, and inevitable forays into human spaceflight, consideration should be given as to how to leverage Europe's assets to benefit from cooperation potentials with India. Europe is already exploring avenues for more ground-breaking cooperation with China. In addition, it is cooperating with NASA in the development of the service module of the Orion Multi-Purpose Crew Vehicle (MPCV) that is slated to launch aboard the SLS in 2019. However, the extent of European contribution to US plans remains today largely undetermined, as does the feasibility of a strategic partnership with China. Therefore, other options in the crafting of partnership configurations for future exploration and human spaceflight endeavours must be considered by ESA (and the EU), particularly those that will not force them to choose mutually exclusive sides or imply a rather minor role for ESA in space exploration. This is *why* ESA has

---

[16] Indicative of this is that at the 2016 ESA Ministerial Conference, it was decided to use part of the funding of the ESA space science programme to help Exomars 2020 to be realised. As a result, future space science missions like PLATO or ATHENA now risk being delayed or even cancelled.

[17] The 2013 EC working document "A role for Europe within a Global space exploration endeavour" emphasises the importance of an integrated approach at international level in the field of space exploration and proposes building the long-term European roadmap consistent with international plans. As for the 2016 ESA/MC resolution, it remarkably stressed as first objective for space exploration that the Agency shall "pursue a consistent, strong and forward-looking space exploration strategy designed to strengthen the Agency's exploration efforts over the coming years, thus attracting even more interest and support from international partners".

been trying to adopt a novel approach to space exploration with its "Moon Village" concept. And this is also *where* a future Indian contribution might come into play.

The international lunar base concept proposed by ESA Director General J.D. Woerner is an initiative to keep space exploration in focus and to unify the major spacefaring nations including their industry and academia for a joint goal, while keeping Europe "a world-class actor in space and a partner of choice on the international scene" (European Space Agency 2016).[18] ESA sees its role as "enabler of cooperation" or more properly as a "hub" to conduct different initiatives with multiple actors, starting from traditional ISS partners such as the United States (which could ideally offer Europe the forthcoming Deep Space Gateway (DSP) as a springboard for a manned lunar exploration, not being interested in taking the lead on a human lunar mission as primary project), Russia, Canada and Japan but necessarily extending also to China and India. Indeed, as stressed by ESA director general, any future large exploration scenario cannot afford to exclude the two Asian giants.

The ISS experience showed that complex international cooperation across divergent – and even conflicting – political borderlines is possible. However, with new additional partners and private companies invited to participate, the negotiations will be more difficult than for the ISS. In addition, the credibility of ESA in taking a leading role might be also questioned, not having developed critical technologies for HSF. If ESA (and the EU) is to ensure a degree of viability and international convergence on this bold cooperation framework, winning the support of new potential partners through proactive and tangible engagement is quintessential to providing greater thrust to its role and its vision. An established player such as China is certainly important, but India is perhaps even more promising to increase the appeal of the vision, being a less politically cumbersome partner for countries such as the United States, Canada, Japan and Australia.

Europe–India Cooperation as an Enabler

In order to achieve this objective, in the near term, ESA and ISRO could explore new angles of bilateral cooperation with respect to scientific missions and robotic exploration of the Moon.

As a first enabling step, mechanisms for an increased level of sharing of scientific discoveries among European and Indian scientists could be contemplated. The overarching aim would be to achieve an expanded quality and availability of, and accessibility to, their respective space science data as well as optimise exploitation of data. In light of the unfolding global shifts in the very way of doing scientific research in the twenty-first century,[19] there not only appears to be much scope to

---

[18] It is important to acknowledge that – to quote J.D. Woerner – "the Moon Village is not a single project, nor a fixed plan with a defined time table. It is a vision for an open architecture and an international community initiative". But it is precisely the open nature of the concept that could allow a broader cooperation around one ambitious goal to take place while allowing to purse individual interests.

[19] As inter alia evidenced by the progressive opening of the research processes and globalisation of the scientific community, the surge of producers and consumers of scientific data and, ultimately, by the emergence of a so-called *Space 2.0* phenomenon (European Commission 2016).

harvest this objective but also a necessity to enhance open-source cooperation between international partners by inter alia making use of centralised analytics that can expand the exploitation of scientific data by the different user communities and lead to the faster advancement of knowledge.[20]

As a next step, the possibility of hosting payloads on respective science and exploration missions could be taken into account, particularly when considering ESA's future medium- and large-class missions such as PLATO, and ATHENA, since their funding remains uncertain following the 2016 Ministerial Council.[21] On the Indian side, there is clear interest in participating in ESA's future science missions as well as in benefiting from Europe's expertise by hosting instruments aboard its future science missions. The instruments provided for the Chandrayaan-1 have already brought excellent scientific results from the analysis of the data collected. Complementary roles are hence offered, with ESA (and its Member States) providing instruments and equipment to ISRO's missions, while bringing in contributions from ISRO that expand ESA programmatic deliveries. ESA and India could, for example, envisage the joint development of a Chandrayaan-3 sample return mission (where many key technologies for furthering the Moon Village concept can be tested). Moreover, active exchange of information and coordination of architectural studies for future human space exploration could be pursued. One lesson from ISS agreement negotiations is that the earlier infrastructural interfaces and standards can be agreed, the less costly will be the adaptations during the various implementation phases. For the best results, a debate needs, however, to be stimulated also at the high policy level concerning the prospects of expanded cooperation on a politically relevant project such as human space exploration.

With respect to the long-term activities revolving around the Moon Village concept, a variety of cooperation options could be considered, including cooperation on a joint return manned vehicle. ESA and Airbus have already conducted a feasibility study concerning the development of a re-entry capsule for the ATV, an ATV Return Vehicle (ARV). It was envisaged as the basis for developing either a cargo return capacity or a manned version of the ATV but never materialised. Nevertheless, ESA has been developing important technologies by collaborating with NASA on the ATV-derived service module for Orion. As for India, in 2007 it began to study plans for a manned space vehicle and has already conducted tests for return capsules, with ISRO's Space Capsule Recovery Experiment (SRE-1) and CARE missions of 2007

---

[20] In this context, see, for example, the Open Universe Initiative launched by ASI and UNOOSA (United Nations 2016).

[21] PLATO – PLAnetary Transits and Oscillations – of stars is the third M-class mission in ESA's Cosmic Vision programme. Its objective is to find and study a large number of extrasolar planetary systems, with emphasis on the properties of terrestrial planets in the habitable zone around solar-like stars. PLATO has also been designed to investigate seismic activity in stars, enabling the precise characterisation of the planet host star, including its age. ATHENA – Advanced Telescope for High-Energy Astrophysics – aims to combine an X-ray telescope designed with state-of-the-art scientific instruments to address how does ordinary matter assemble into the large-scale structures and how do black holes grow and shape the universe. The mission has now entered the study phase; once the mission design and costing have been completed, it will eventually be proposed for "adoption" around 2019 (European Space Agency 2017).

and 2014, respectively.[22] An improved version of the Crew Module will eventually serve as the platform for a manned Indian spacecraft. Such vehicles, however, are costly and complicated and require a clear strategic goal and political vision concerning their benefits. Given the success of the ATV and India's more recent ideas to build a return vehicle by ISRO, both Europe and India could consider the prospect of collaborating on a joint return man-rated spacecraft. One such activity would clearly have multiple benefits for both India and Europe.

First, it would enable the joint development of critical technologies for space exploration, speed up India's forays into human spaceflight and possibly have positive spillovers also in other areas (e.g. space transpiration concepts for space tourism). In addition, it would greatly contribute to enriching Europe's partnership portfolio with its pool of scientific and technological capabilities as well as enhancing the quality and level of Europe–India relations. More importantly, it would provide a crucial stepping-stone for the eventual realisation of the Moon Village concept, a framework which would in turn reinforce the role of Europe as the centre of gravity in international (space) affairs and promote cooperative international relations between currently isolated players (for instance, India and China and the United States and China), while undermining confrontational stances in the emerging global order.

8.3.4   The Ultimate Cooperation Nexus: Space and Security

When considering the role of space in boosting Indo-European relations, opportunities for space cooperation should be examined with respect to the many issue areas the EU and India have committed to working on together, including efforts to support international stability and security (both on orbit and on Earth). Indeed, while most of the cooperation between Europe and India will understandably focus on civilian ventures, it is important to acknowledge that space might also act as an enabler for the establishment of more tangible cooperation schemes in the field of security (from – and for – space).

Cooperation on Security from Space
The 2016 European Union Global Strategy (EUGS) emphasises the need to create a stronger union on security and defence that will be able to tackle today's threats and challenges more effectively and to engage more proficiently with other world powers. Engagement is indeed a central tenet of EUGS (Leffler 2017). Consistent with this goal, in the joint statement of 30 March 2016, the EU expressed its resolve to deepen security and defence cooperation with India and work towards tangible outcomes with respect to counter-terrorism, counter-piracy, maritime security,

---

[22] While the RLV-TD flight of 2016 has nothing to do with human missions, the experiment has great utility, and hence the landing tests will add to competencies of ISRO.

non-proliferation and disarmament – all areas of increased interest to both actors and their respective space programmes. Accordingly, the two unions could ideally include an agenda item within their consultation mechanisms to discuss the possible sharing of space-based information for security purposes. This could be an ideal instrument to address shared security objectives and expand the scope of the India-EU strategic partnership, especially in today's digital and data-centric world, where the flow and depth of available intelligence have become more crucial than ever (Aliberti 2017). As also suggested by Isabelle Sourbès-Verger, with the EU and India

> facing a growing number of security issues identified in the Joint Agenda 2020, the multiplication of EO systems of increasing performance and the enhancement of Europe's and India's own competencies should allow the promotion of an effective cooperation scheme in which political mechanisms, abiding by the principle of sovereignty, would complete each other's capabilities and would allow both countries to benefit from a network of multi-sensor systems offering a tremendous flow of images and reduced delays in data delivery (Sourbès-Verger 2016).

Even though cooperation on security issues remains inherently complicated, particularly when it comes to the EU, such cooperation is not far-fetched, provided there is sufficient leadership and support from both sides. The International Chart on Space and Major Disasters initiated by CNES and ESA already offers a good example of cooperation for providing sensitive data in some situations (Blamont 2017). In addition, the fact that the two actors have different geographical priorities but no conflicting interests really has the potential to make such cooperation viable.

A specific issue area that could be ripe for cooperation is using space for maritime domain awareness (MDA) or sea surveillance. In the section "foreign policy and security cooperation", the EU-India Agenda for Action 2020 indeed highlighted the need to expand existing cooperation "in other areas mentioned in the EU-India Joint Action Plan, including promoting maritime security and freedom of navigation in accordance with international law (UNCLOS)" (European Commission 2016). Space for MDA would be a natural candidate for pursuing this objective, particularly given the increasingly important tools space assets offer for control and localisation of illegitimate vessels through the Vessel Detection System (VDS). Because both Europe and India have long shorelines demanding continued vigilance and are investing in relevant space infrastructure to provide information about possible threats/disaster, space-based MDA should be of interest to both India and the EU, also considering that their areas of strategic concern are neither overlapping nor conflicting.

Besides the material benefits, more tangible cooperation on space for security could also generate other important payoffs. As mentioned earlier, the EU's weaknesses in affirming itself as a fully fledge foreign policy and security actor have played a primary role in disrupting the strategic dialogue with India, which has understandably questioned the EU's legitimacy to intervene on such issues. Hence, by engaging with concrete security-related space cooperation efforts with India, the

EU would increasingly prove its ability/credibility to deal with security issues and also advance its foreign policy to a more mature stage.[23]

Business opportunities for the European defence industry could be also envisaged as a result of deepening political cooperation. Already, in March 2016 Thales Alenia Space and Airbus Defence and Space "jointly presented to the Indian Ministry of Defence an offer for the joint development of a constellation of reconnaissance and intelligence satellites (SAR imagery, ELINT, SIGINT) with technology transfer from France via indigenous companies" (Blamont 2017). This type of European business operations would be best supported in the future, should a more robust policy dialogue and cooperation on space-related defence issues between the EU and India take place. As noted by EU–India relation analyst Stefania Benaglia,

> defence cooperation involves a strong political dimension and is an indication of a close mutual understanding of the geopolitical environment. By stepping up EU–India collaboration and creating a conducive political space, the EU could also facilitate European defence cooperation with India's government and industrial sector. Inspired by the U.S., the EU could therefore assume a role, through which it uses its political weight to back shortlisted EU firms when negotiating with the Indian government, contributing to a positive outcome (Benaglia 2016).

All in all, adding an agenda item on space cooperation for security could be an important instrument for moving forward the EU–India security dialogue with actionable policy that would greatly enhance the overall level and quality of cooperation between India and the EU.

Cooperation on Security for Space

An analogous line of reasoning can also be applied to issues revolving around the security of space, which would also be well primed to become an area of fruitful policy dialogue and cooperation.

Europe and India have expressed a number of shared views and concerns about the safety and stability of the space environment. Issues related to SSA, space debris mitigation/remediation and responsible space behaviour are topics of growing interest for both actors, which, in fact, are now in the process of establishing a clear vision for a comprehensive space security policy. Considering the increasing relevance that space assets have for both India and the EU, and the fact that the two actors are Tier-2 powers within the international space pecking order, it appears to be in the interests of both to add an agenda item on the most pressing space security challenges in their new framework of "Foreign Policy and Security Consultations". Such a policy dialogue would not only demonstrate heightened awareness and priority accorded to this rapidly growing issue area but also provide a channel for sharing views on respective strategic interests, concerns and priorities. Given the

---

[23] As more broadly stressed by the Deputy Secretary General of the EEAS, "the creation of a stronger Union on security and defence should be of interest to India and directly respond to criticism that the EU is not a serious international security actor. The EU and India could consult closer as the EU deepens defence cooperation, enhances civil-military synergies in CSDP Missions and operations and develops, in the long run, a European hub for strategic information, early warning and analysis. We would welcome India's cooperation in these developments" (Leffler 2017).

orientations emerging in India's defence community with respect to the possible development of offensive space capabilities, institutionally entrenched mechanisms of policy exchanges would be of crucial importance for Europe. Understanding the character and context of each other's space efforts will aid the advancement of broad Europe–India cooperation.

Equally important, it would be an indispensable instrument for revitalising future efforts regarding the governance of the outer space environment. Indeed, as the experience of the ill-fated ICoC clearly shows, proactive diplomatic engagement with like-minded partners is a necessary step to reaching consensus on norms regulating responsible space behaviour. While India had an interest in such a Code, it expressed reservations because it felt that the EU should have engaged New Delhi in preparing the Code from the start (Rajagopalan 2011). Because of this diplomatic negligence, the perception was that the EU-generated Code wanted to protect some vested interests of the West. For this reason, regular discussions at senior policy level would be extremely helpful to reduce possible miscommunication and misconceptions between the two actors. In addition, teaming up with India through bilateral level consultations and in multilateral fora such as UNCOPUOS in the future might be key to solidifying or advancing new normative solutions on the international stage.

Cooperation in space security does not necessarily have to strictly limit itself to bilateral discussions or diplomatic initiatives. Nowadays, the ability to monitor the space environment has become crucial for any space actor aiming to obtain independent information on various kinds of potential threats. Considering that a fully fledged cooperation scheme with established space powers such as the United States, Russia and China might be difficult to achieve, but that both India and Europe have strong incentives to improve their awareness of the space domain, the pursuit of cooperative strategies for improved SSA and sharing of data should be carefully considered.

In Europe, some national capabilities already exist, and the process of developing a common network is in motion both within ESA and the EU framework, with the aim of covering space weather, near-Earth objects and Space Surveillance and Tracking (SST). India, on its side, has begun to consider the acquisition of such a system, as well as the pooling of systems with Europe, whose geographical position is advantageous in terms of complementarity.

The extent and directions in which joint efforts could go depend on the respective leadership. It is clear that Europe and India will have to strike a balance between multiplying the benefits of enhanced cooperation and the level of independence/autonomy they can achieve. In any case, given the strategic sensitiveness of cooperating on SST, the creation of a space weather monitoring system as a predominantly scientific endeavour could be envisaged as a first step to enable further cooperation. Along the same lines, the EU/ESA and India could consider the joint development of a ground-based NEO monitoring system within their territories. The EU-India Agenda for Action 2020 has already listed "enhanced cooperation for joint scientific undertakings" as one of the areas of cooperation. By sharing and pooling technical, scientific and human resources to carry out joint scientific activities with respect to space weather and NEOs, the two actors would not only be able to accrue valuable scientific outcomes but also to make important steps towards gradually expanding cooperation in the geopolitically sensitive SST domain.

8.4   Conclusions: Towards a New India Policy

India matters to Europe. In fact, it has always mattered, having been, for over 2000 years, the source of invaluable ideas, knowledge and of so many of the things Europe strived for, from pepper and spices to precious textiles, silk and even high-quality steel (Hobson 2004). After all, it was Europeans' desire to find new routes to India and have direct trade links with the Golden Bird (*Sone ki Chidiya*) – as ancient India was known due to its majestic wealth – that gave birth to the Great Age of Exploration, to the arrival of Europeans in the Americas and, eventually, to the rise of the modern world.

Today, the Golden Bird is once again preparing to unfold its wings in a perfect example of "leapfrogging development". Europe, however, has largely remained aloof, probably in the face of China's more astonishing resurgence. Certainly, greater acknowledgement of India's importance is gradually re-emerging, as inter alia evidenced by the outcomes of the 13th EU-India Summit in 2016. But ministerial declarations alone will not suffice to seize India's full potential. Actionable policy measures and new cooperative efforts in key areas of mutual interest are imperative if Europe is to re-energise its relations with the subcontinent (Aliberti 2017).

Among the various playing fields that could be explored for a new level of mutually beneficial cooperation, space figures, as a very promising and effective one, possesses unique transversal qualities with the potential to accrue a myriad of political, economic, technological and societal payoffs. Indeed, the modalities through which Europe and India interact in space could likely become one of the most important arenas for potential cooperation in the future.

There are many ways India and Europe can collaborate on space efforts and build a strategic relationship that still looks in need of an urgent reboot. Evaluating possible space cooperation in four major areas of space activities has revealed compelling rationales for closer space ties between Europe and India. Scientific research in climatology, downstream space applications, robotic exploration and human space flight and joint efforts in the field of security (be it from space or in space) are all loaded with the potential to achieve a plethora of common policy objectives between Europe and India – objectives which would in turn bolster the overarching goal of deepening further political, economic, commercial, technological and societal relations and realising the full potential of the India-EU Strategic Partnership.

If there appears to be a multitude of opportunities to unfold mutually beneficial space cooperation, it is also clear that both partners need to develop a coherent strategy concerning their future space relations that can contribute to mutual political, economic and security interests. It appears in fact evident that the whole cooperation scheme – as conceptualised in the current Europe–India dialogue mechanisms – seems to ignore the wide range of opportunities that space technology can offer, not only in science and technology but in most of the issue areas the two actors have committed to working together, including research and innovation, business and sustainable development and even global security issues such as climate change, water management, migration, piracy and terrorism (Sourbès-Verger 2016). Indeed, an extensive review of the past cooperation experience and current policy documents reveals that opportunities tend to be limited to S&T issues. For that reason, the political approach to space cooperation opportunities – on both the EU and India sides – should be re-examined and adapted to take into account the transversal nature of the contributions offered by space. Arguably, inserting key space issue areas into existing Europe–India bilateral consultations and decision-making would provide the two actors with an important starting point to deepen both their space and political ties.

Clearly, some institutionally entrenched mechanisms for cooperation will also be required to move beyond policy dialogue and make things happen. An ad hoc space cooperation committee should thus be set up within the broader architecture of EU–India dialogue, with regular meetings among *all* concerned stakeholders.[24] Indeed, given that the EU is not a programme-implementing body with regard to space programmes and that space cooperation entails not only diplomatic but also technical, scientific and industrial dimensions, it is essential that all stakeholders (including ESA, industry and academia) are represented in such a committee.[25] It should go without saying that for a unified space policy towards India to be effective, better integration of the competencies and experiences of the various European stakeholders should be achieved. Specifically, it is essential that the posture and action of the major European constituencies (ESA, the EU and their Member States) are not only coordinated but also mutually enforcing. Achieving concerted policy actions among all European constituencies might prove a Sisyphean challenge, given the ever-present fragmentation of European efforts. But a unified and coherent India policy (be it in space and elsewhere) can certainly expect to see the benefits far outpace the burdens of doing so.

All in all, if Europe and India seem well-positioned for a new level of space-related cooperation, both sides will need to have the right mix of patience, persistence, resources, managerial coordination and, more importantly, *real* political will.

International cooperation is inherently difficult, and in the case of India, it can appear even more so because India, just like Europe, remains a Gordian knot. Alexandrian solutions will, however, prove unsuitable to unravel the hurdles standing in the way of more substantial cooperative undertakings. Only by fostering a

---

[24] One such committee has been for instance established within the architecture of EU-Japan partnership.

[25] Ideally, various subcommittees to be then established for each of the identified cooperation areas, as shown in other cooperation experiences (e.g. CNES' cooperation). Furthermore, in order to promote closer ties between Europe and India, the possibility of opening an ESA representation office in New Delhi should be considered. As in the case of CNES' presence in Bangalore, the payoffs to be harvested by such a decision should not be overlooked as local representation would in fact help to progressively build mutual trust, increase opportunities for Europeans to liaise with Indian stakeholders, identify potential areas of cooperation, promote ESA activities and image in India as well as signal to New Delhi that it is regarded as a valuable partner of Europe.

new level of mutual understanding will the two unions be in a position to transform shared interests into common paths. And for this to happen, strong political commitment at the highest level will be required. There is no doubt that space cooperation ultimately needs to be achieved at a technical and operational level, but it first requires statesmen to forge a grand vision and commit to it. Clearly, any European grand vision cannot afford to miss out the India Opportunity.

References

Al-Ekabi, C. (2016). *The future of European commercial spacecraft manufacturing.* Vienna: European Space policy Institute.

Aliberti, M. (2017). Revitalising the EU-India strategic partnership: the role of space. *ORF Space Alert, 5*(2).

Aliberti, M., Nader, D., & Vernile, A. (2017). Understanding India's new space potential: Implications and prospects for Europe. In *68th International Astronautical Congress (IAC).* Adelaide: International Astronautical Federation.

Annadurai, M. (2017). Future exploration missions of ISRO. In *60th Session of the United Nations Committee on Peaceful Uses of Outer Space.* Vienna: UNOOSA.

Basin, M. (2009). The EU-India partnership: Strategic alliance or political convenience. In A. Ghosh (Ed.), *India's foreign policy* (pp. 206–226). New Delhi: Dorling Kindersley.

Benaglia, S. (2016, March). How to boost EU-India relations. *CEPS Policy Briefs* (341).

Blamont, J. (2017). Cooperation in space between India and France. In R. P. Rajagopalan & N. Prasad (Eds.), *Space India 2.0 commerce, policy, security and governance perspectives* (pp. 215–233). New Delhi: Observer Research Foundation.

CEOS. (2016, April 3). *Heads of space agencies decide to join efforts in support of COP 21 decisions.* Retrieved April 8, 2017, from CEOS: http://ceos.org/document_management/Meetings/SIT/SIT-31/160408%20Declaration%20NewDelhi_V3.docx

European Commission. (2016, March). *EU India agenda for action 2020.* Retrieved from Press Release Database. http://europa.eu/rapid/press-release_IP-16-1142_en.htm

European Commission. (2017). *Climate action.* Retrieved from European Commission. https://ec.europa.eu/clima/policies/international/cooperation/india_en

European External Action Service. (2016, October 17). *India and the EU.* Retrieved from Delegation of the European Union to India and Bhutan:https://eeas.europa.eu/delegations/india/670/india-and-eu_en

European Space Agency. (2016, October 26). *Shared vision and goals for the future of European space.* Retrieved from http://www.esa.int/About_Us/Welcome_to_ESA/Shared_vision_and_goals_for_the_future_of_Europe_in_space

European Space Agency. (2017, December 15). *Athena.* Retrieved December 21, 2017, from European Space Agency: http://sci.esa.int/athena/59896-mission-summary/

GNSS.Asia. (2016). *GNSS opportunities in India.* Retrieved from GNSS.Asia. http://india.gnss.asia/gnss-opportunities-india

Hobson, J. M. (2004). *The Eastern origins of the modern world.* Cambridge: Cambridge University Press.

Jaffrelot, C. (2006). *India and the European union: The charade of a strategic partnership.* Paris: Centre de Reserches Internationales.

Jain, R. K. (2014, June). *India-EU strategic partnership: Perceptions and perspectives.* NFG Working Paper Series, 10|2014.

Khandekar, G. (2011, August). The EU and India: A loveless arranged marriage. *FRIDE Policy Brief,* 90|2011.

Khandekar, G. (2015). Towards an EU-India partnership on urbanisation. *Global Relations Forum Policy Brief, 27*(1).

Lahcen, A. (2013). *EUMETSAT-NOAA collaboration in meteorology from space. Review of a long-standing Trans-Atlantic partnership*. Vienna: European Space Policy Institute.

Leffler, C. (2017, January 17). *Speech by Christian Leffler on EU foreign policy & future of EU-India relations*. Retrieved from European External Action Service. https://eeas.europa.eu/sites/eeas/files/christian_lefflers_speech_at_think_tank_twinning_initiative.pdf

NASSCOM. (2016). *Start-up report – Momentous rise of the Indian start-up ecosystem*. Retrieved from NASSCOM. http://www.nasscom.in/knowledge-center/publications/start-report-momentous-rise-indian-start-ecosystem

Observer Research Foundation. (2016). *2nd ORF Kalpana Chawla annual space policy dialogue*. New Delhi: Observer Research Foundation.

Prasad, N. (2017a). Traditional space and new space in industry in India: Current outlook and perspectives for the future. In N. Prasad & R. P. Rajagopalan (Eds.), *Space India 2.0 commerce, policy, security, and governance perspectives*. New Delhi: Observer Research Foundation.

United Nations. (2016, June 14). *"Open Universe" proposal, an initiative under the auspices of the Committee on the Peaceful Uses of Outer Space for expanding availability of and accessibility to open source space science data*. Retrieved July 7, 2017, from UNOOSA: http://www.unoosa.org/res/oosadoc/data/documents/2016/aac_1052016crp/aac_1052016crp_6_0_html/AC105_2016_CRP06E.pdf

Radjou, N., & Prabhu, J. (2010). *Jugaad innovation*. New Delhi: Jossey-Bass.

Rajagopalan, R. P. (2011, October). *Debate on space code of conduct: An Indian perspective* (ORF Occasional Paper 26).

Rao, M. K., Murthi, S., & Raj, B. (2015). *Indian Space – Towards a "National Eco-System" for future space activities*. Jerusalem: 66th International Astronautical Federation.

Sourbès-Verger, I. (2016, December). *EU-India cooperation on space and security* (IAI Working Papers, 16|38).

Start-Up Europe. (2017). *Start-up Europe India network*. Retrieved from Start-Up Europe India Network. https://startupeuropeindia.net

Zinnov. (2015). *The Indian promise*. Retrieved October 12, 2016, from Zinnov.com: http://zinnov.com/the-indian-promise/

Annex

Fact Sheet on India 2015 and 2035

INDIA
today
(2015 data)

			Global Rank	
Population	1,311 Billion		2nd	/ 193
GDP (in USD, 2010 PPP)	$ 4,751 Trillion		3rd	/ 193
Urban Population	32,7%		156th	/ 193

Largest Cities:

1. Delhi   25,7 Million
2. Mumbai   21 Million
3. Kolkata   14,9 Million
4. Bangalore   10,1 Million
5. Chennai   9,9 Million
6. Hyderabad   8,9 Million
7. Ahmadabad   7,3 Million

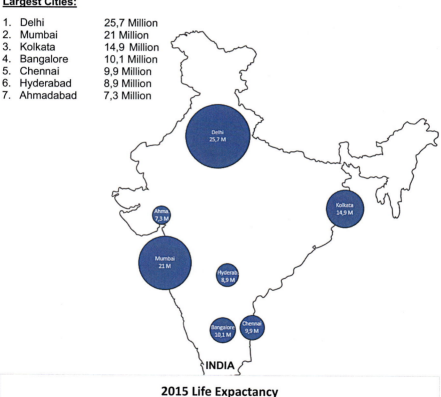

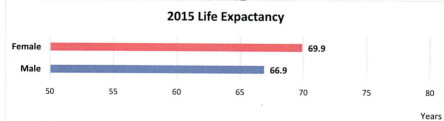

in 2035

		Global
Rank		
Population	1,585 Billion	1st/ 193
GDP (in USD, 2010 PPP)	$ 14,504 trillion	3rd /193
Urban Population	42,1%	156th/ 193

Largest Cities (2030):

8. Delhi   36,0 Million
9. Mumbai   27,8 Million
10. Kolkata   19,1 Million
11. Bangalore   14,8 Million
12. Chennai   13.9 Million
13. Hyderabad   12.8 Million
14. Ahmadabad   10,5 Million

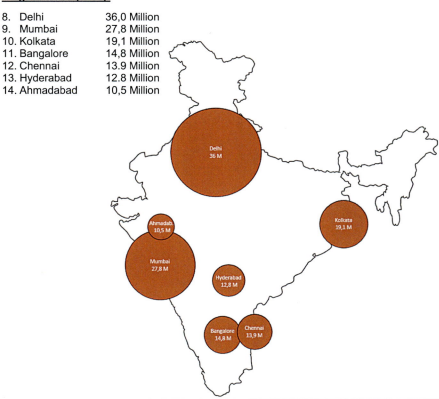

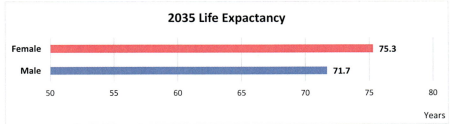

Timeline of ISRO's Achievements

Jan	1962	Indian National Committee for Space Research (INCOSPAR) formed
Nov	1963	First rocket launch from Thumba Equatorial Rocket Launching Station (TERLS)
Jan	1965	Space Science and Technology Centre (SSTC) established in Thumba
Jan	1967	Satellite Telecommunication Earth Station set up by Ahmedabad
Aug	1969	Indian Space Research Organisation (ISRO) established
June	1972	Space Commission and Department of Space setup
Apr	1975	ISRO becomes a government organisation
Apr	1975	First Indian satellite (Aryabhata) launched
July	1980	Rohini satellite successfully placed in orbit (first satellite successfully launched with Indian Satellite Launch Vehicle)
Apr	1982	Indian National Satellite System started by launching INSAT-1A
Apr	1984	Manned space mission with Soviet Union
Mar	1988	Launch of first operational Indian Remote Sensing Satellite (IRS-1A)
July	1992	Second generation of INSAT is born with launch of INSAT-2A
Sept	1997	First operational launch of PSLV (with IRS-1D on board)
Mar	2000	Third generation of INSAT is born with launch of INSAT-3B
May	2003	GSAT series (communication satellites) started with successful launch of GSAT-2
Sept	2004	First operational flight of GSLV (GSLV-F01 with EDUSAT on board)
Jan	2007	Successful recovery of SRE-1 after manoeuvring it to re-enter atmosphere
Oct	2008	Launch of first indigenous moon exploration technology Chandrayaan-1
Nov	2010	Launch of HYLAS (Highly Adaptable Satellite); advanced communication satellite built in partnership with EADS-Astrium
July	2013	First satellite of Indian Regional Navigation Satellite System (IRNSS-1A) launched
Nov	2013	Mars Orbiter Mission spacecraft launched (first nation to successfully enter Martian orbit on first attempt)
Dec	2014	Experimental flight of next generation launch vehicle GSLV Mk. III
May	2016	RLV-TD (reusable launch vehicle technology demonstrator) launch
Aug	2016	Scramjet Engine Technology Demonstrator test flight
Feb	2017	Commercial Launch of 104 satellites (world record)

List of Foreign Satellite Launched by India
(as of 1 March 2017)

#	Country	Satellite	Mass (in kg)	Launch date
1	Germany	DLR-TUBSAT	45	26.05.1999
2	Republic of Korea	KITSAT-3	110	26.05.1999
3	Belgium	PROBA	94	22.10.2001
4	Germany	BIRD	92	22.10.2001
5	Argentina	PEHUENSAT-1	6	10.01.2007
6	Indonesia	LAPAN-TUBSAT	56	10.01.2007
7	Italy	AGILE	350	23.04.2007
8	Israel	TECSAR	300	21.01.2008
9	Canada	NLS-5	16	28.04.2008
10	Canada	CAN-X2	7	28.04.2008
11	Denmark	AAUSAT-II	3	28.04.2008
12	Germany	RUBIN-8	8	28.04.2008
13	Germany	COMPASS-I	3	28.04.2008
14	Japan	CUTE-1.7	5	28.04.2008
15	Japan	SEEDS	3	28.04.2008
16	The Netherlands	DELFI-C3	6,5	28.04.2008
17–18	Germany	CUBESAT-1 and CUBESAT-2	1 each	23.09.2009
19–20	Germany	RUBIN-9.1 and RUBIN-9.2	1 each	23.09.2009
21	Switzerland	CUBESAT-4	1	23.09.2009
22	Turkey	CUBESAT-3	1	23.09.2009
23	Algeria	ALSAT-2A	116	12.07.2010
24	Canada	NLS-6.1	6,5	12.07.2010
25	Switzerland	NLS-6.2	1	12.07.2010
26	Singapore	X-SAT	106	20.04.2011
27	Luxembourg	VESSELSAT-1	28,7	12.10.2011
28	France	SPOT-6	712	09.09.2012
29	Japan	PROITERES	15	09.09.2012
30–31	Austria	NLS-8.1 and NLS-8.2	14 each	25.02.2013
31	Canada	SAPPHIRE	148	25.02.2013
33	Canada	NEOSSAT	74	25.02.2013
34	Denmark	NLS-8.3	3	25.02.2013
35	United Kingdom	STRAND-1	6,5	25.02.2013
36	Canada	NLS-7.1 (CAN-X4)	15	30.06.2014
37	Canada	NLS-7.2 (CAN-X5)	15	30.06.2014
38	France	SPOT-7	714	30.06.2014
39	Germany	AISAT	14	30.06.2014
40	Singapore	VELOX-1	7	30.06.2014
41–43	United Kingdom	DMC 3-1, 3-2 and 3-3	447 each	10.07.2015

(continued)

(continued)

#	Country	Satellite	Mass (in kg)	Launch date
44	United Kingdom	CBNT-1	91	10.07.2015
45	United Kingdom	DeOrbitSail	7	10.07.2015
46	Canada	NLS-14 (Ev9)	14	28.09.2015
47	Indonesia	LAPAN-A2	76	28.09.2015
48–51	United States	4 LEMUR satellites	28 together	28.09.2015
52	Singapore	TeLEOS-1	400	16.12.2015
53	Singapore	VELOX-CI	123	16.12.2015
54	Singapore	Kent Ridge-1	78	16.12.2015
55	Singapore	VELOX-II	13	16.12.2015
56	Singapore	Athenoxat-1	< 5	16.12.2015
57	Singapore	Galassia	3,4	16.12.2015
58	Canada	M3MSat	85	22.06.2016
59	Canada	GHGSat-D	25,5	22.06.2016
60	Germany	BIROS	130	22.06.2016
61	Indonesia	LAPAN-A3	120	22.06.2016
62–73	United States	12 Dove satellites	4,7 each	22.06.2016
74	United States	SkySat Gen2-1	110	22.06.2016
75	Algeria	ALSAT-2B	117	26.09.2016
76	Algeria	ALSAT-1B	103	26.09.2016
77	Algeria	ALSAT-1N	7	26.09.2016
78	Canada	NLS-19 (CAN-X7)	8	26.09.2016
79	United States	Pathfinder-1	44	26.09.2016
80	Israel	BGUSat	4,3	15.02.2017
81	Kazakhstan	Al-Farabi-1	1,7	15.02.2017
82	Switzerland	DIDO-2	4,2	15.02.2017
83	The Netherlands	PEASS	3	15.02.2017
84	United Arab Emirates	Nayif-1	1,1	15.02.2017
85–172	United States	88 Flock-3p satellites	4,7 each	15.02.2017
173–180	United States	8 LEMUR satellites	4,6 each	15.02.2017

Timeline India–EU Relations

1962 India and European Economic Community (EEC) establish formal diplomatic relations

1970 Operation Flood initiated to boost Indian milk production (62% financed by EEC)

1973 Commercial Cooperation Agreement signed (framework for trade cooperation)

1976 EEC begins contributing to development of drought-prone Indian areas

1980 Indian Trade Centre opened in Brussels

1981 New Commercial Cooperation Agreement signed (setting up the Joint Commission)

1983 EU Delegation established in Delhi

1985   Commercial and Economic Cooperation Agreement signed
1988   First Joint Commission meeting
1989   European Commission-India Cooperation and Exchange Programme (EICEP) launched
1992   Indian and European business community create the Joint Business Forum
1993   European Union supports District Primary Education Programme; India and the European Commission (EC) sign joint political statement
1994   New cooperation agreement on partnership and development takes relations beyond economic cooperation
1996   EC helps fund Indian health and family welfare programme
1998   Asia-Invest founded to improve mutual business communication; Asia Urbs founded to promote cooperation between Asian and European cities
2000   First-ever EU–India summit in Lisbon: joint declaration + agenda for action EU–India Civil Aviation Cooperation Agreement signed
2001   Second summit introduces declaration against terrorism, vision statement on information technology and agreement on science and technology
2002   Third summit presents new 5-year strategic agenda
2004   India and EU agree to strategic partnership at summit
2005   Road map/Joint Action Plan mutually agreed upon; India agrees to partake in the Galileo programme
2006   State Partnership Programme
2007   EU countries agree to start negotiating a Free Trade Agreement with India
2008   Joint Action Plan from 2005 updated and fortified
2011   European Investment Bank (EIB) provides loan for Indian renewable energy projects
2012   Joint Declaration on Energy agreed upon; India–EU Skills Development Project started
2014   Non-Proliferation and Disarmament dialogue held
2016   13th summit in Brussels introduces EU–India Agenda for Action 2020

EU–India Political Relations

Evolution of EU–India Relations

India's formal diplomatic relations with the precursor of the European Union – the European Economic Community (EEC) – date back to 1962; in that year, India appointed an ambassador to the EEC, marking the starting point of official relations between the two. Predominantly economic ties were initially created between the two entities, highlighted by the signing of the Commercial Cooperation Agreement in 1973. Five years later, this agreement was to be expanded and fortified. Just that happened in 1981 when a joint commission was put in place to further Indo–European economic cooperation, to which the most-favoured nation clause of

GATT applied. Reduced or eliminated trade barriers played key roles in incentivising bilateral trade and growing the commercial partnership (Delegation of the European Union to India 2013). The political interplay remained, however, very limited throughout the duration of the Cold War.

The collapse of the Soviet Union in 1991 and India's foreign policy realignment brought about new possibilities to push forward Indo–European relations beyond the scope of commerce and economy. In 1994, India and the European Community forged what is considered the basis for today's comprehensive relationship between the two actors. Named "Cooperation Agreement between the European Community and the Republic of India on partnership and development", it built on the aforementioned 1973 and 1981 agreements. As the document alluded, cooperation with affluent and societally advanced European nations helped India further develop its own economic strength, the scope and tremendous potential of which cannot be overstated. Additionally, India's newly liberated economy helped boost its global economic might which positioned it as a prolific balancing power to China in Asia and thus a valuable economic partner to Europe. Among other factors, the economic upside coupled with the fact that India is the single most populous democracy in the world makes India an indisputably important partner for Europe. Beyond mere economic cooperation, the 1994 agreement in conjunction with the 1993 joint political statement helped pave the way for a closer political alliance for years to come (Council of the European Union 2005).

The relationship between India and the EU was once again strengthened and first declared a strategic partnership at the 5th India–EU summit in 2004, at the prime of India's economic growth. As outlined in the 2005 joint action plan, India and the EU:

> Share a common commitment to democracy, pluralism, human rights and the rule of law, to an independent judiciary and media. India and the EU also have much to contribute towards fostering a rule-based international order – be it through the United Nations (UN) or through the World Trade Organization (WTO). We hold a common belief in the fundamental importance of multilateralism in accordance with the UN Charter and in the essential role of the UN for maintaining international peace and security, promoting the economic and social advancement of all peoples and meeting global threats and challenges. (Council of the European Union 2005)

After 3 years under the joint action plan of 2005, India and the EU fortified their agreement with an updated Joint Action Plan in 2008 "with the objective of promoting international peace and security and working together towards achieving economic progress, prosperity and sustainable development" (European External Action Service 2008). While negotiations regarding a Free Trade Agreement between India and the EU started in 2007, various differences have yet to be resolved before such an agreement can be put in place. As of recently, negotiations regarding such a Bilateral Trade and Investment Agreement (BTIA) have stalled and seem to be put aside for the time being. Among other hurdles, it could be very challenging and costly to implement new standards and regulations, especially so in the meat industry, for instance. However, a Free Trade Agreement has the potential to bring about significant economic, societal and environmental benefits. According to a

Trade Sustainability Impact Assessment study published in 2009, "the more ambitious the FTA, the more poverty is reduced as income effects increasingly dominate price effects" in India (Ecorys 2009) (Table 1).

Table 1 EU–India summits overview

Summit	Date	Venue	Key outcome(s)
1	28 June 2000	Lisbon, Portugal	Strong bilateral commitment to mutually beneficial partnership and regular summit meetings established
2	23 November 2001	New Delhi, India	Agreement on Cooperation in Science and Technology
			Joint Vision Statement on ICT
			Creation of an EU–India Cultural Forum in 2002
3	10 October 2002	Copenhagen, Denmark	Action Plan for the EU–India Joint Initiative for Enhancing Trade and Investment
4	29 November 2003	New Delhi, India	EU Scholarship Programme (masters and doctorate)
			Disaster Preparedness Programme
			Trade and Investment Development Programme
			Partnership for Progress in Social Development
5	08 November 2004	The Hague, Netherlands	Relations are declared a strategic partnership
			First EU–India Cultural Declaration
			India–EU Energy Panel
			Creation of EU–India Environmental Conference
			Erasmus grant for Indian post-graduate students
			Creation of joint workshops in automotive engineering, genomics/life sciences and nanotechnology
			Declaration for ESA-ISRO cooperation on lunar exploration mission Chandrayaan-1
6	07 September 2005	New Delhi, India	Adoption of Joint Action Plan
7	13 October 2006	Helsinki, Finland	Welcoming of adoption of UN Counter-Terrorism Strategy by General Assembly
			Agreement to establish regular macro- economic dialogue
8	30 November 2007	New Delhi, India	Welcoming Memorandum of Understanding on the Country Strategy Paper for India for 2007–2010
9	29 September 2008	Marseille, France	Comprehensive review and further fortification of 2005 Joint Action Plan
			EU–India Civil Aviation Agreement
10	06 November 2009	New Delhi, India	Push for all WTO members to close Doha Round in 2010
			Launch of call for proposals on solar power technology
			Launch of joint call for proposals in biotechnology

(continued)

Table 1 (continued)

Summit	Date	Venue	Key outcome(s)
11	10 December 2010	Brussels, Belgium	Joint declaration on international terrorism
			Joint declaration on culture
12	10 February 2012	New Delhi, India	Joint declaration on international terrorism
			Joint Declaration for Enhanced Cooperation in Energy
			Joint Declaration on Research and Innovation Cooperation
			Memorandum of Understanding on Statistical Cooperation
13	30 March 2016	Brussels, Belgium	EU–India Agenda for Action 2020
			Joint Declaration on the India–EU Water Partnership
			Joint Declaration on a Clean Energy and Climate Partnership
			Common Agenda on Migration and Mobility (CAMM)
			Renewal of 2010 Joint Declaration on International Terrorism
			Science and Technology Agreement 2020 extension
			Adoption of 2030 Agenda for Sustainable Development and Addis Ababa Action Agenda

In the spring of 2016, Indian Prime Minister Modi met with European Council President Tusk and European Commission President Juncker in Brussels for the 13th EU–India Summit. The so-called EU–India Agenda for Action 2020 – a follow-up to the 2005 and 2008 action plans – was the main focal point of the summit. Also heavily discussed was a topic for which the EU–India summit of 2016 brought about a joint declaration: the global threat of terrorism. The issue was particularly noteworthy just days after suicide bombings shook the city of Brussels, killing more than 30 and injuring hundreds of civilians. A similarly gruesome terror attack left 130 dead and hundreds more injured in Paris in 2015. The list of terror attacks around Europe and the world in recent history continues rather extensively, thus pushing the fight against terrorism to the top of many world leaders' lists of priorities in our day and age (European External Action Service 2016).

The Agenda for Action 2020 entails joint objectives across the board:

- *Foreign Policy*: Mutual interests are sought to be pursued through regular contact between the European External Action Service and the Indian Ministry of External Affairs, while the EU and India also consider the option of creating triangular relations with third parties and promoting an international dialogue that addresses persisting human rights issues and the mitigation of natural disaster risk.
- *Security*: As mentioned in the previous paragraph, security was a key topic with much common ground between Europe and India. The aforementioned fight

against terrorism and radicalization but also nuclear disarmament and non-proliferation as well as counter-piracy and cybersecurity were vital points of discussion. As accuracy and depth of available intelligence is arguably more important than ever in today's digital and data-centric world, possibilities for information sharing between Indian agencies and EUROPOL for security purposes were addressed.

- *Economy, Business and Trade*: Existing cooperation commitments were reaffirmed, and different ways of deepening these ties were discussed. Reoccurring emphasis was placed on strengthening bilateral trade and investment activity as well as working together in support of common ideals regarding intellectual property regulations and other issues on a multilateral, international, level.
- *Climate Change, Energy and Environment*: Discussions indicated an aim for a closer future partnership in combatting climate change in line with the framework of the Paris Climate Agreement and the Montreal Protocol of 2015 as well as exploring various options for partnering in pursuit of cleaner, more sustainable energy in support of the United Nation's "Sustainable Energy for All" initiative. Envisioning a prolific Indo–European Water Partnership, improved research efficiency and bilateral work to develop ties related to environmental issues seem poised to take a big step forward until 2020.
- *Urban and Sustainable Development*: In support of the UN 2030 Agenda for Sustainable Development, discussions were held regarding a partnership that would help achieve the agenda's social, environmental and economic objectives. Developing the massive urban metropoles of India is not new to the EU, which has supported this through the "100 smart cities" programme, for instance. In particular, the stronger involvement of Indian as well as European cities, regions and states in partnership programmes is sought in conjunction with regular dialogue.
- *Research and Innovation*: Research and innovation cooperation, in the form of exchange of researchers, for instance, is to be fostered through joint work within the framework of the 10th India–EU Science and Technology Steering Committee Meeting of 2015 as well as programmes such as EU Horizon 2020, the agreement between EURATOM and India and the International Thermonuclear Experimental Reactor (ITER).
- *Information and Communications Technology (ICT)*: Primarily highlighted was the aim to establish collaboration between Europe's "Digital Single Market" and India's "Digital India", especially so with regard to legislative and economic matters. Within the India–EU ICT joint working group, this and other objectives such as the drafting of a joint declaration for cooperation on 5G communication networks are to be achieved. Also on the extensive list of ICT cooperation was a push to debate simpler mechanisms for mutual financing in related areas, knowledge exchange and a general prompt to further talks that will help facilitate IT industry partnership.
- *Space and Transport*: In space, the Agenda for Action 2020 highlights cooperation on earth observation and navigation satellites to facilitate interaction between Galileo and IRNSS and partnerships on payloads launched for scientific pur-

poses. Concerning transport, it is a key objective of India and Europe to put in place the 2008 signed EU–India horizontal agreement, easing restrictions for civil air travel.

- *Migration and Mobility – Skills, Employment and Social Policy*: India and the EU expressed the desire to put a Common Agenda on Migration and Mobility (CAMM) in place that would help constructively address issues related to migration and human trafficking. In the cooperation for skills development, the stronger involvement of EU educational facilities, businesses and member states is seen as an important step forward. The organisation of high-profile skill development events and the promotion of employment-related mutual objectives within the G20 were mentioned.
- *Education and Culture*: An all-around strengthening of communication and interaction between the EU and India in regard to education is desired by the two parties, especially so through the platforms of Europe's Erasmus+ and India's GIAN. An exchange between the rich cultures of India and Europe is sought to be facilitated and encouraged through partnerships in several artistic fields including cultural heritage, for instance, among other creative areas.
- *Authorities, Civil Society and the EU–India Strategic Partnership*: To allow for fluent political relations, regular meetings of parliamentary representatives from the EU as well as India are to be instated, along with exchange and interaction between various decentralized organisations, institutions and authorities. Pertaining to the strategic partnership forged between India and the EU in the mid-2000s, security and foreign policy discussions are to be streamlined into Foreign Policy and Security Consultations (FPSC) (European Commission 2016a, b).

The agenda's key areas as outlined above are also reflected in a similar structure in the 2016 summit's joint statement. There, the wording calls for a stronger EU–India partnership, an advancement of foreign policy, human rights and security cooperation, growth and jobs through the fostering of trade and economic cooperation, global prosperity for coming generations and finally enhancing citizens' involvement in the strategic partnership (European Commission 2016a, b). More broadly, the 13th EU–India Summit leaders were intended to confer a new momentum to the relationship and achieve the full political potential of EU–India Strategic Partnership (Fig. 1).

Political and Security Dialogue

Digging into the sectorial dimension of EU–India relations, security partnership has proven to be especially challenging to implement due to the diversified geopolitical standing and alliances of the various powers. Taking the role of Pakistan as an example, military support provided to India's neighbour and foe by parts of the western world somewhat alienated India to those advanced nations in Europe and North America. What united India and these nations however was the emergence of

Fig. 1 The 13th EU–India summit (Credit: European External Action Service)

terrorism in the region and around the globe, carried out by radical groups who seemed to have had links to Pakistan. With global tensions and security challenges having shifted and taken on new forms over these past decades, India and the European Union have most recently shown unwavering bilateral support to counter terrorism (Sachdeva 2015).

The Development of Security Cooperation

What started as an economic alliance in the 1960s was upgraded to include broader political dialogue in the 1990s before the launch of annual summit meetings in 2000, where the necessity of joint work to ensure security and peace was addressed at the very first of 13 summits to date. As the relationship grew stronger and broader in scope, the 2004–2005 strategic partnership and Joint Action Plan took the political dialogue to unprecedented heights, with the clear aim of "meeting common threats and global challenges, starting with terrorism" (Council of the European Union 2005). In 2013 the two strategic partners first engaged in a non-proliferation and disarmament talks (Gateway House: Indian Council on Global Relations; Istituto Affari Internazionali 2017).

In the absence of consolidated European armed forces, Indian security cooperation has been established with individual EU member states on bilateral bases. In security and defence – as in space – France is considered as one of India's strongest European partners. Regular military exercises on land, on water and in the air are the result of nearly two decades of official security partnership. Following India's purchase of 36 French fighter jets, Indian Prime Minister Modi's visit to France in

2015 brought 20 new bilateral agreements between the two nations, which are set to strengthen and extend the security partnership established in 1998.

The fact that India turns to Europe for weapon systems to this day perfectly exemplifies the strong bond and steadfast trust between India and Europe and the optimistic outlook for such a relationship in the future (see Fig. 2 and Table 2).

It must be kept in mind that the acquisition of three dozen fighter jets is not a mere one-time transaction; it creates Indian reliance on the European industry for replacement parts, servicing and similar by-products and services of the initial arms purchase. In this sense, the arms trade really serves the bigger purpose of strengthening the entire India–EU relationship from several, mutually beneficial angles (Sachdeva 2015).

Countering Terrorism

As briefly mentioned earlier in this subchapter, countering terrorism is an objective both India and the EU take very seriously and one that has been pushed to the forefront of their joint mission in the last decade or so. Through the Joint Action Plan of 2005, several instruments have been put in place to eradicate the threat of terror attacks such as joint working groups not only in counter-terrorism but also in cybersecurity and counter-piracy. Cybersecurity in particular has strong ties to terrorism and is often debated in this context, as the Internet and secure messaging apps provide terror organisations a prolific platform to spread their hateful ideologies to people around the world. Keeping sensitive information out of the hands of hackers or people affiliated with terrorist groups who intend to use it maliciously is another key objective of course (Sachdeva 2015).

Later, in 2010, the EU and India issued a Joint Declaration on International Terrorism, once again clearly asserting their condemnation of terrorism and calling for united action to fight terrorism in all its shapes and forms. What is more, the Joint Declaration specifically called out any and all sponsors and direct or indirect

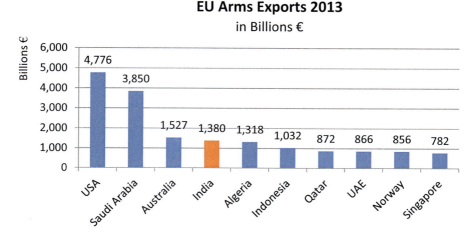

Fig. 2 EU arms exports (*Source*: Sabbati and Cirlig 2015)

Table 2 The European defence industry's links to India

Nation	Major enterprises	Cooperation with India	Bilateral agreements with India
Bulgaria	>Arsenal AD >Kintex >TEREM >VMZ Sopot	>Arsenal AD supplied 67.000 AK-47 rifles to Indian paramilitary forces from 2010–2012	>1993 Memorandum of Understanding on Defence Cooperation
	>Samel 90 >Apollo GmbH >THOR Global Defense Group	>KAS Engineering Consortium (Bulgarian state-owned agency) allegedly supplied weapons dropped over Purulia (West Bengal) in 1995	
Czech Republic	>TATRA Trucks >OMNIPOL a.s. >Gearspect Group a.s.	>TATRA Trucks have a partnership with Bharat Earth Movers Limited (India); delivered 100 all-wheel drive vehicles	>2003 Agreement on Defence Cooperation
		>OMNIPOL partners with Indian entities: Ordnance Factories Board (OFB), Heavy Vehicle Factory Avadi and Heavy Engineering Corporation Ranchi	
France	>Dassault Aviation >MBDA >Thales > DCNS	>Multifaceted involvement in India	>2006 Agreement on Defence Cooperation
		>Several joint ventures with private defence companies	
Germany	>Krauss-Maffei Wegmann >Diehl Remscheid GmbH	>Joint Venture between Ashok Leyland (India) and Krauss-Maffei Wegmann >Tracks and accessories for Arjun tank supplied by Diehl Remscheid >Rheinmetall blacklisted by India on corruption charges >Indian Navy operates fleet of Howaldtswerke-Deutsche Werft (HDW) submarines	>2006 Bilateral Defence Cooperation >2001 Strategic Partnership Agreement
	>Rheinmetall AG >ThyssenKrupp Marine Systems		
Poland	>Polish Bumar (Polish Defence Holding)	>Contract between Polish Bumar and Bharat Earth Movers Limited (India)	>2003 Memorandum of Understanding on Defence Cooperation (+2011 Addendum) >1996 Bilateral Agreement for the Promotion and Protection of Investments
Spain	>Indra Sistemas >Navantia >Instalaza SA	>Navantia is part of contract originally awarded to DCNS (France) for Scorpene submarines	>2012 Memorandum of Understanding on Defence Cooperation >1972 Agreement on Trade and Cooperation

(continued)

Table 2 (continued)

Nation	Major enterprises	Cooperation with India	Bilateral agreements with India
United Kingdom	>BAE Systems >Cobham	>Indian Navy (BAE's Sea Harrier aircraft) >Licenced production of BAE's Hawk Trainer Jet in India >Assembly, integration and testing facility in India for M777 ultra lightweight Howitzer >TASL (India) manufactures for Cobham's 5th generation air-to-air refuelling equipment	>2004 Strategic Partnership Agreement

Source: (Gateway House: Indian Council on Global Relations; Istituto Affari Internazionali 2017)

supporters of terrorism and singled out Pakistan, urging India's neighbour to actively stand against terrorism carried out by their countrymen by bringing those to justice who were in any way involved in the Mumbai attacks of 2008 that left more than 160 people dead. Attacks on European soil in Madrid and London in 2004–2005 made for a situation where both India and the EU were now facing imminent threats to their national security, which gave rise to the strong antiterror agenda.

Obvious difficulties that remain focal points for intelligence agencies around the globe to this day include the reliable identification of potential perpetrators and monitoring of their communication and networking activities, especially their online presence. This is where close cooperation between internal security agencies such as Europol and Indian equivalents, namely, the National Investigation Agency (NIA) or the Central Bureau of Investigation (CBI), becomes absolutely paramount (Saran et al. 2016). It is important however to also be aware of the fact that only roughly 5% of counter-terror activity is actually dealt with by the European Union; the remaining 95% is the responsibility of each individual member state and their judicial and executive branches (EU Institute for Security Studies 2012).

Remarks

Even as India and Europe seem to find common ground on security issues, threat assessments on China as well as Pakistan and the conflicts in the Kashmir region, to name some examples, have caused variance between Europe's views and those of India (Gateway House: Indian Council on Global Relations; Istituto Affari Internazionali 2017). After failing to identify the security relationship with India as one that could be further fortified in the EU's Foreign and Security Policy report released mid-2016, it seems after all that as things currently stand, India's economic might remain Europe's primarily interest in the strategic partnership. Indeed, India is mentioned in context of the EU's pursuit of a Free Trade Agreement with their strategic partner; however India is left out when the document elaborates on the European Union's security strategy, asserting: "We will expand our partnerships, including on security, with Japan, the Republic of Korea, Indonesia and others" (European Union Global Strategy 2016), in context of a peaceful, stable and secure

Asia. This certainly does not however mean that this cannot or will not change as time passes and geopolitical shifts occur.

Bi- or multilateral partnerships concerning security and defence are always directly tied to shared values, concerns, threats, interests and enemies. They are always dependent on foreign policy and the geopolitical landscape at any given time as seemingly unrelated activity in the global political and diplomatic landscape can have a profound and direct influence on India, the EU and their security partnership. What used to be an ideological conflict between capitalism and communism displayed to the world during the Cold War as well as regional confrontations in South Asia, today, is the threat of terrorism and ongoing war in the Middle East. The future may present vastly different circumstances or enemies that will either strengthen or divide today's global partnerships. What is clear is that as of today, India and the EU are on the same page regarding our age's looming threats, thus providing the necessary driving force and stability that create the key building blocks of what could be a prolific transcontinental security partnership.

Trade and Economics

Despite the political stagnation during the past 10 years, economic interaction has continued to flourish. Data from 2015 indicates that the European Union as a whole is India's number one trading partner, accounting for 13% of India's trade while India was the EU's ninth-ranked trading partner at 2.2% of the European Union's total trade that year. Furthermore, the value of the European Union's exports to India grew by almost €17 billion (79%) in the decade leading up to 2015, while EU imports from India grew by just over €20 billion (106%) from 2002 to 2015. The service sector trade between India and the EU grew threefold in that same 14-year period, reaching €14 billion in 2015. Foreign direct investment by the EU in India was worth €38,5 billion in 2014, up 11% from 2013 (European Commission n.d.-a, b).

Figures 3 and 4 give overviews of goods and services traded between the European Union and India in the past years:

The goods trade balance was in Europe's favour until 2013, when India came out on the positive side with their exports to the EU exceeding those from the EU by €884 million. This number rose by 72% to €1,5 billion in 2014. In 2015, the European Union recorded a €1,465 billion goods trade deficit with India. The exact opposite can be observed with services traded between India and the EU, where European imports exceeded exports by €1,2 billion in 2012 and €900 million in the subsequent year; however, the 2014 increase in service exports and simultaneous decrease in imports from the European Union's perspective gave Europe an €800 million service trade surplus that year. The service trade balance remained on the positive side for Europe in 2015 with €700 million in surplus.

Putting goods trade numbers into perspective, Fig. 5 shows the trade balance for all products between the 28 European Union member states and select global economies from January 2016 to January 2017 (Eurostat 2017).

Fig. 3 EU goods trade with India (*Source*: European Commission 2017a, b, c, d)

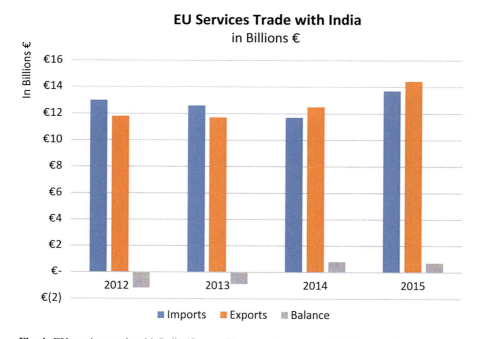

Fig. 4 EU services trade with India (*Source*: European Commission 2017a, b, c, d)

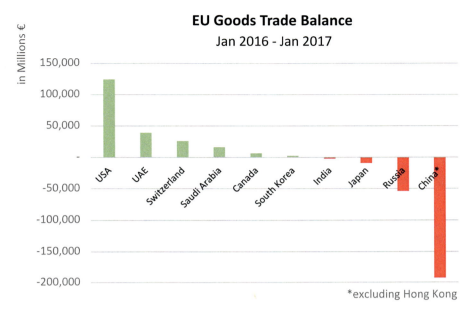

Fig. 5 EU goods trade balance (*Source*: Eurostat*)*

Over these recent 13 months leading into 2017, India has accumulated a fairly small €2,2 billion goods trade surplus with the EU in comparison to China's surplus of an astonishing €193 billion, Russia's €54 billion and Japan's €9,2 billion. In contrast, the EU has had very profitable trade relations with the United States, the United Arab Emirates and Switzerland, recording positive goods trade balances of €124,3 billion, €39,1 billion and €25,7 billion for each of the respective aforementioned countries in the given timespan.

In further analysing goods trade between the EU and India, Figs. 6 and 7 break down the flow of goods by Standard International Trade Classification (SITC) to help paint a clearer picture of exactly what the trade partners were exchanging in 2016 (European Commission 2017a, b, c, d):

As illustrated, half of all goods imported to the EU from India are either miscellaneous manufactured articles or manufactured goods classified chiefly by material, both sharing almost an equal 25% split. The third and fourth most imported into the European Union is machinery and transport equipment closely followed by chemicals and related products. Other sub-10% categories make up the remaining 15% of imports from India.

As far as exports from the EU to India are concerned, nearly 40% is comprised of machinery and transport equipment. Adding the roughly 29% of manufactured goods which are classified chiefly by material to that enormous share, the top two categories cover around two thirds of all goods exported to India from the European Union. The remaining third consist of chemicals, miscellaneous manufactured articles and other smaller-scale exports.

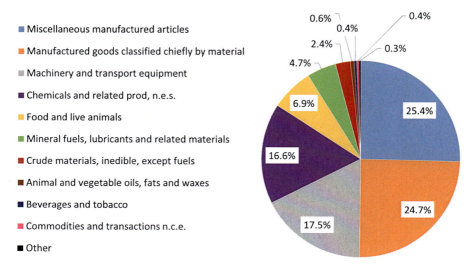

Fig. 6 EU exports to India (*Source*: European Commission 2017a, b, c, d)

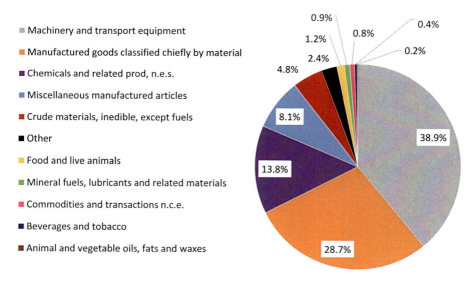

Fig. 7 EU exports to India (*Source*: European Commission)

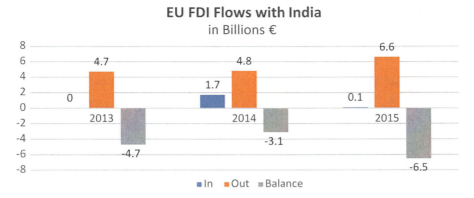

Fig. 8 EU–India FDI flows (*Source*: European Commission)

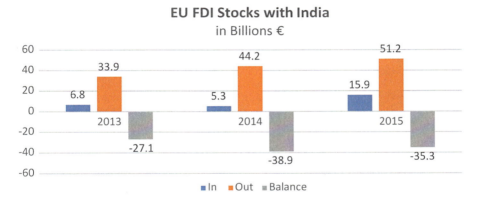

Fig. 9 EU–India FDI stocks (*Source*: European Commission 2017a, b, c, d)

Trade aside, India and the EU cumulatively conducted €6,7 billion in foreign direct investment (FDI) flows and $ 67,1 billion in foreign direct investment stocks in 2015. The inflows, outflows and balance values for flows and stocks between 2013 and 2015 are detailed in Figs. 8 and 9 (European Commission 2017a, b, c, d).

Protectionist Measures
A vital factor regarding economic trends and developments over the next few years and possibly decades will be the prevalence of protectionist policies around the world. Free trade – at least the way it had been conducted in the past – has recently come under scrutiny in several nationalistic movements, politically rather conservative, often referred to as "populist movements", including that of US President Donald Trump, Britain's push to exit the European Union and several other European freedom and people's parties. An increased focus on issues such as national security and stricter immigration policy along with the protection of domestic industries and

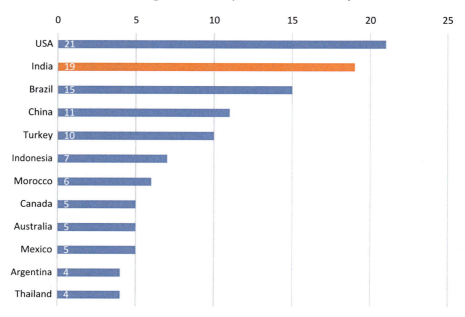

Fig. 10 Trade defence measures against EU exports (*Source*: European Commission n.d.-a, b)

workers through economic reform are falling on receptive ears in several countries where people feel that globalisation and free trade have harmed their economies. Keeping in mind of course that not all protectionist policies exclusively stand in connection with the geopolitical and economic trends mentioned above, the below chart shows the nations aiming the highest numbers of trade defence measures at EU exports (Fig. 10):

As illustrated, India has a total of 19 trade defence measures in force against EU exports, second only to the United States' 21 measures. Of those 19 cases, 14 (or 74%) are considered anti-dumping measures, which are meant to prevent the sale of goods under their cost of production or domestic price in foreign markets, i.e. India. The remaining five measures (or 26%) are safeguard actions which nations enforce with the aim of protecting their domestic industries from imports of any specific product or product group which they consider to present a threat to the flourishing of that nation's own domestic industry.

Trade defence measures going the other way – that is, implemented by the EU against Indian exports – add up to a total of nine measures definitively in force as of March 2017, as illustrated in Fig. 11.

Uncontested, as displayed, stand the 61 measures in force against goods coming into the EU from China. Of the nine measures enforced on second-ranked India, five (or 56%) are antisubsidy measures meant to halt the import of goods of which the

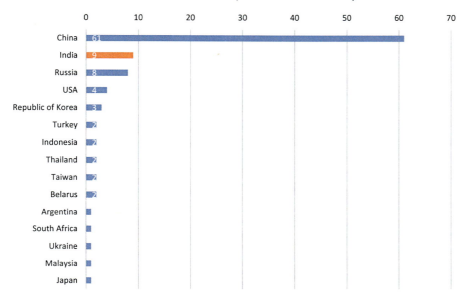

Fig. 11 EU trade defence measures against foreign imports (*Source*: European Commission 2017a, b, c, d)

price is artificially low through subsidies, which simultaneously increases that goods' competitiveness in Europe, while the other four (or 44%) are anti-dumping measures.

Free Trade Agreement

Trade defence measures going both ways present an apparent impediment not just to the trade of specific products or product groups but also to the overall bi- and multilateral trade climate. The Indo–European economic alliance remains significant, but an upgrade to this relationship has long been sought out. A Free Trade Agreement (FTA), also referred to as a Bilateral Trade and Investment Agreement (BTIA), has been discussed since 2007, so far without much progress or success. A 2007 economic impact study showed that India would greatly benefit from such an agreement as the country's economy would see a €17–18 billion gain. Most significantly, India would have been expected to increase textile, leather and wearing apparel exports by $3,6 billion to the EU alone until 2020. Additionally, European investment and increased bilateral economic activity would help India make great strides in productivity – a benefit not quantitatively analysed or expressed in this European Commission-financed study yet a factor vital to India's developing economy (Decreux and Mitaritonna 2007).

The European Union on the other hand had not been projected to gain nearly as much as India through the implementation of a bilateral FTA. An extensive economic analysis conducted in support of then ongoing negotiations in 2009 found that the EU's long-term gains would be close to €1,6 billion; just a fraction of India's aforementioned €17–18 billion. In the short term, both would gain roughly €4,5–5 billion as a result of less restrictive regulations and trade terms. Moreover, increased foreign direct investment flows and the downsizing of non-tariff barriers were estimated to generate a cumulative €18 billion for both India and the EU. From a social perspective, a BTIA could indirectly help counter poverty in India through job creation and higher real wages, which in turn would positively impact education and health in the country (Ecorys 2009).

While the absence of significant demonstrable long-term economic benefit for Europe in a trade deal of this nature can be considered an important factor in recently stagnating negotiations, many other disparities between India and the EU remain to stand in the way of a Free Trade Agreement. India, for instance, has a 14.5% average tariff on goods entering the country as opposed to the more than 3,5 times lower EU average of 4,1%. Furthermore, small- and medium-sized Indian enterprises face the non-tariff barrier of high EU quality standards, which are difficult and costly to meet for these ventures. Other key impediments to the implementation of a FTA between India and the EU include differences in data policy, India's hesitance to sign the WTO's Information Technology Agreement 2 (ITA II), the fact that there is no single market for services within the EU, Indian FDI restrictions, and as mentioned and illustrated earlier of course the steadily issued trade defence measures among several other barriers (Mukherjee et al. 2013).

Science, Technology and Innovation

Science and technology is another issue area in which Europe and India have nurtured positive – yet still marginal – results. Indo-European science, technology and innovation cooperation aims to achieve mutual benefit by increasing interaction in government, science agency, commercial and academic partnerships. This includes cooperation in the form of joint projects, workshops, seminars, meetings, programmes and research centres as well as the exchange of scientists and information. Apart from bilateral India–EU cooperation, both India and Europe also join forces in science and technology through a variety of multilateral framework programmes such as UNESCO, UNFCC, UNDP, ASEAN, BIMST-EC, the Asia-Pacific Economic Forum, the Indian Ocean Rim-EC, Asia-Pacific Climate Change Council and the Third World Academy of Sciences. Table 3 highlights key programmes and projects between India and the EU in science and technology:

Indian participation in EU Framework Programmes grew substantially between 1994 and 2013 with a total project count of 33 in FP4, 32 in FP5, 60 in FP6 and 175 in FP7, for a total of 300 projects in nearly two decades. The health and medicine field saw the highest amount of cooperation with roughly 22% of projects having taken place in this sector, followed closely by the Marie Curie action programme

Table 3 EU–India cooperation in STI

Framework/project	Remarks
India–EU bilateral cooperation	
EU International Thermonuclear Experimental Reactor (ITER)	India is full member in ITER (designed to prove scientific/ technological feasibility of a full-scale fusion power reactor)
EU–India Science and Technology Cooperation Agreement	Launched in 2001, renewed in 2009
Joint Action Plan, 2005–2008 India–EU summit	Relations upgraded to "strategic partnership": co-sponsoring of joint projects, joint workshops, increased researcher exchange/ mobility
India Pilot Initiatives, Strategic Forum for International Cooperation (SFIC)	Focused on the areas of water and bioenergy resources
Brussels Communiqué, Joint Declaration at the 2012 India–EU summit	Areas of cooperation: climate change mitigation, clean energy, energy efficiency and renewable energy, computational materials, food and nutrition research and water technologies
Multilateral cooperation	
6th EU Framework Programme (FP6)	Indian contribution: €11M; 100 Indian science and technology institutions, 80 projects; ~100 Indian applicants signed short-listed proposal grant agreements
7th EU Framework Programme (FP7)	India contribution: €41M; 300 Indian science and tech. institutions, 175 projects; India = EU's 4th largest international partner; 135 Indian applicants signed short-listed proposal grant agreements
Marie Curie fellowships for mobility and training of researchers	Staff exchanges and networking activities between European research organisations and Indian research organisations; 2007–13: EU contributions: €4.5M; ~900 Indian researches participated
Erasmus Mundus programme	Includes EU Promotion of Health through Research, Applied Technology, Education and Science (EUPHRATES) programme; 3,400 Indian graduate and Ph.D. Erasmus participants (2004–2013)
New INDIGO	Supported Indo-European multilateral research and networking projects
INNO INDIGO	Successor of New INDIGO; pooling of funds for joint projects between EU member states and Indian science/tech institutions
Horizon 2020	India funds own participation costs, has the option to seek third-party funding

which accounted for about 21% of FP7 cooperation projects (Krishna and Mishra 2016).

Apart from the many smaller cooperation projects in science and technology, "Big Science" projects on which India collaborates with the EU include the European Union's International Thermonuclear Experimental Reactor (ITER) nuclear fusion energy project, the European Galileo satellite-based navigation system, the Facility for Antiproton and Ion Research (FAIR) project and the European Organization for Nuclear Research (CERN).

Outside the Framework of the European Union, India has a plethora of bilateral cooperation agreements with the United States and Russia, for instance, also however with European nations such as Germany, France, the United Kingdom and Belgium. Three binational science and technology centres with France, Germany and the United States are being co-funded by India. India collaborated most with the United States between 2011 and 2015, recording 104 projects compared to 86 with Japan, 72 with France, 70 with Germany and 65 with Canada. Fifty-nine of the Canadian and US cooperation projects were in a multilateral setting, while the same holds true only for two projects with France and Japan and three with Germany (Krishna and Mishra 2016).

List of Acronyms

A
AARS	Asian Association for Remote Sensing
ADCOS	Advisory Committee on Space Research
ADS	Airbus Defence and Space
AICTE	All India Council for Technical Education
AIM	Atal Innovation Mission
AIT	Assembly, Integration and Testing
ALTM	Airborne Laser Terrain Mapper
AMRUT	Atal Mission for Rejuvenation and Urban Transformation
API	Application Programming Interface
APPLE	Ariane Passenger Payload Experiment
APRSAF	Asia-Pacific Regional Space Agencies Forum
APSCO	Asia-Pacific Space Cooperation Organisation
ARISE	Agricultural Resources Inventory and Survey Experiment
ARMUT	Atal Mission for Rejuvenation and Urban Transformation
ASAR	Airborne SAR
ASAT	Anti-Satellite Weapon
ASEAN	Association of South East Asian Nations
ASI	Agenzia Spaziale Italiana (Italian Space Agency)
ASLV	Augmented Satellite Launch Vehicle
ATS-2	American Technology Satellite-2
ATV	Automated Transfer Vehicle

B
B2B	Business-to-Business
B2C	Business-to-Customer
BEL	Bharat Electronics Limited
BJP	Bharatiya Janata Party
BMD	Ballistic Missile Defence
BRICS	Brazil–Russia–India–China–South Africa
BTIA	Bilateral Trade and Investment Agreement

© Springer International Publishing AG 2018

M. Aliberti, *India in Space: Between Utility and Geopolitics*, Studies in Space Policy 14, https://doi.org/10.1007/978-3-319-71652-7

C

C4ISR	Command, Control, Communications, Computing, Intelligence, Surveillance and Reconnaissance
CAPE	Crop Acreage and Production Estimation
CARE	Crew Module Atmospheric Re-entry Experiment
CD	Conference on Disarmament
CEC	Consortium for Educational Communication
CEOS	Committee on Earth Observation Satellites
CGMS	Coordination Group for Meteorological Satellites
CIET	Central Institute of Educational Technology
CII	Confederation of Indian Industries
CMA	China Meteorological Administration
CME	Continuing Medical Education
CNES	Centre National D'Etudes Spatiales (French National Centre for Space Studies)
COMINT	Communication Intelligence
COP21	21st Conference of the Parties
COSPAR	Committee on Space Research
COSPAS-SARSAT	International Search and Rescue Satellite-Aided Tracking System
CSSTE-AP	Centre for Space Science and Technology Education in Asia and the Pacific

D

DAE	Department of Atomic Energy
DECU	Development and Educational Communication Unit
DISHA	Distance Healthcare Advancement Project
DLR	Deutsches Zentrum für Luft- und Raumfahrt (German Aerospace Centre)
DMS	Disaster Management Support
DMS	Disaster Management System
DOS	Department of Space
DOS-ISRO HQ	Department of Space – ISRO Headquarters
DRDO	Defence Research Development Organisation
DST	Department of Science and Technology
DTH	Direct to Home

E

EC	European Commission
ECIP	European Community Investment Partners
ECMWF	European Centre for Medium Range Weather Forecasts
EDTIB	European Defence Technological and Industrial Base Strategy
EEC	European Economic Community
EGNOS	European Geostationary Navigation Overlay Service

EIB	European Investment Bank
EICEP	Euro-India Cooperation and Exchange Programme
ELINT	Electronic Intelligence
EMMRC	Educational Multi-Media Research Centres
EO	Earth Observation
ESA	European Space Agency
ESCES	Experimental Satellite Communication Earth Station
ESPI	European Space Policy Institute
EU	European Union
EUGS	European Union Global Strategy
EUMETSAT	European Organisation for the Exploitation of Meteorological Satellites
F	
FAO	Food and Agriculture Organisation
FDI	Foreign Direct Investment
FIR	Flight Information Region
FPSC	Foreign Policy and Security Consultations
FTA	Free Trade Agreement
G	
G2G	Government-to-Government
GAGAN	GPS Aided Geo Augmented Navigation
GCOS	Global Climate Observing System
GEO	Geostationary Earth Orbit
GIS	Geographical Information System
GNSS	Global Navigation Satellite System
GSA	Global Navigation Satellite Systems Agency (EU)
GSAT	Geosynchronous Satellite System
GSLV	Geosynchronous Satellite Launch Vehicle
GSO	Geosynchronous Orbit
GTO	Geostationary Transfer Orbit
H	
HAL	Hindustan Aeronautics Limited
HMS	Hydro-Meteorological Service
HoW	Hospital on Wheels
HSF	Human Spaceflight
HTS	High Throughput Satellite
I	
IAA	International Academy of Astronautics
IAC	International Astronautical Congress
IADC	Inter-Agency Debris Coordination Committee
IAF	International Astronautical Federation

IARI	Indian Agriculture Research Institute
ICBM	Inter-Continental Ballistic Missile
ICC	INSAT Coordination Committee
ICoC	International Code of Conduct
ICT	Information and Communications Technology
IEA	International Energy Agency
IFS	Indian Foreign Service
IGOS	International Global Observing Strategy
IIRS	Indian Institute of Remote Sensing
IISc	Indian Institute of Science
IISL	International Institute of Space Law
IIST	Indian Institute of Space Science and Technology
IISU	ISRO Inertial Systems Unit
IITs	Indian Institute of Technologies
IMF	International Monetary Fund
IMINT	Image Intelligence
IMSD	Integrated Mission for Sustainable Development
INCOSPAR	Indian National Committee for Space Research
INLUS	Indian Land Uplink Stations
INMCC	Indian Master Control Centres
INRES	Indian Reference Stations
INSAT	Indian National Satellite System
IODC	Indian Ocean Data Coverage
IOR	Indian Ocean Region
IoT	Internet of Things
IPRC	ISRO Propulsion Complex
IRNSS	Indian Regional Navigation Satellite System
IRS	Indian Remote Sensing
ISAC	ISRO Satellite Centre
ISC	Integrated Space Cell
ISECG	International Space Exploration Coordination Group
ISITE	ISRO Satellite Integration Test Establishment
ISRO	Indian Space Research Organisation
ISRO-GBP	ISRO Geosphere Biosphere Programme
ISTRAC	ISRO Telemetry, Tracking & Command Network
ISU	International Space University
IT	Information Technology
ITAR	International Traffic in Arms Regulations
ITER	International Thermonuclear Experimental Reactor
ITU	International Telecommunications Union
ITU	International Telecommunication Union
J	
JAM	Jan Dhan Yojana, Aadhaar and Mobile number
JAP	Joint Action Plan
JAXA	Japanese Aerospace Exploration Agency

| JRP | Joint Research Programme |
| JV | Joint Venture |

K

| KCP | Kheda Communications Project |

L

LEO	Low Earth Orbit
LEOS	Laboratory for Electro-Optics Systems
LIGO	Laser Interferometer Gravitational-Wave Observatory
LPSC	Liquid Propulsion Space Centre

M

M&A	Merger and Acquisition
MAVEN	Mars Atmosphere and Volatile Evolution Spacecraft
MCF	Master Control Facility
MEA	Ministry of External Affairs
MEO	Medium Earth Orbit
MIDHANI	Mishra Dhatu Nigam Limited
MNCFC	Mahalanobis National Crop Forecasting Centre
MOM	Mars Orbiter Mission
MOX	Mission Operations Complex
MoU	Memorandum of Understanding
MSAS	MTSAT Satellite Augmentation System (Japan)
MTCR	Missile Technology Control Regime
MTSAT	Multifunctional Transport Satellites (Japan)

N

NARL	National Atmospheric Research Laboratory
NASA	National Aeronautics and Space Administration
NATO	North Atlantic Treaty Organization
NavIC	Navigation Indian Constellation
NCERT	National Council of Educational Research and Training
NDEM	National Database for Emergency Management
NESAC	North Eastern Space Applications Centre
NIAS	National Institute of Advanced Studies
NICES	National Information System for Climate and Environment Studies
NISAR	NASA-ISRO Synthetic Aperture Radar
NITI	National Institution for Transforming India
NPTEL	National Project for Technology Enhanced Learning
NRMS	Natural Resource Management System
NRSA	National Remote Sensing Agency
NRSC	National Remote Sensing Centre
NSSP	Next Steps in Strategic Partnership

O
OECD   Organisation for Economic Cooperation and Development
ORF   Observer Research Foundation

P
P&T   Post and Telegraphs Department
PC-NNRMS   Panning Committee on National Natural Resource
                        Management System
PLANEX   Planetary Science and Exploration
PM   Prime Minister
PMB   Project Management Board
PMC   Project Management Council
PNL   Physical Research Laboratory
PNT   Positioning, Navigation and Timing
PPP   Purchasing Power Parity
PRL   Physical Research Laboratory
PSLV   Polar Satellite Launch Vehicle
PSU   Public Sector Units

R
RESPOND   Research Sponsored
RRSCs   Regional Remote Sensing Centres
RS   Restricted Service
RSDP   Remote Sensing Data Policy

S
SAARC   South Asian Association for Regional Cooperation
SAC   Space Application Centre
SAR   Synthetic Aperture Radar
SARAL   Satellite for Argos and Altika
Satcom   Satellite Communications
SCL   Semi-Conductor Laboratory (Semi-Conductor Complex
                        Limited)
SCO   Shanghai Cooperation Organisation
SDSC/SHAR   Satish Dhawan Space Centre/Sriharikota High Altitude
                        Range
SEP   Société Européenne de Propulsion
SFCG   Space Frequency Coordination Group
SIATI   Society of Indian Aerospace Technologies and Industries
SIET   State Institutes of Educational Technology
SIGINT   Signals Intelligence
SITC   Standard International Trade Classification
SITE   Satellite Instructional Television Experiment
SLV   Satellite Launch Vehicle
SMEs   Small and Medium Enterprises
SPEs   Special Purpose Entities

SPPU	Savitribai Phule Pune University
SPS	Standard Positioning Service
SRE-1	Space Capsule Recovery Experiment-1
SROSS	Stretched Rohini Satellite Series
SSA	Space Situational Awareness
SST	Space Surveillance and Tracking
STC	Space Technology Cells
STEP	Satellite Telecommunication Experiments Project
SVAB	Second Vehicle Assembly Building
T	
TAS	Thales Alenia Space
TCBMs	Transparency and Confidence-Building Measures
TE	Tele-Education
TERLS	Thumba Equatorial Rocket Launching Station
TM	Tele-Medicine
TPCR	Technology Perspective and Capability Roadmap
TPE	Transponder Equivalent
TT&C	Telemetry, Tracking and Command
U	
UAV	Unmanned Aerial Vehicle
U.S.	United States
UGC	University Grants Commission
UN	United Nations
UN-COPUOS	United Nations Committee on the Peaceful Uses of Outer Space
UN-SPIDER	United Nations Platform for Disaster Management and Emergency Response
UNDP	United Nations Development Programme
UNESCAP	United Nations Economic and Social Commission for Asia and the Pacific
USD	United States Dollar
USSR	Union of Soviet Socialist Republics
V	
VAB	Vehicle Assembly Building
VHRR	Very High-Resolution Radiometer
VRC	Village Resource Centre
VSAT	Very Small Aperture Terminal
VSSC	Vikram Sarabhai Space Centre
W	
WEF	World Economic Forum
WMO	World Meteorological Organization
WTO	World Trade Organisation

Bibliography

Abhijeet, K. (2016). Privatisation of the PSLV: What the law of outer space demands? In *67th International Astronautical Congress* (IAC-16,E7,5,8,x32950). Guadalajara: IAF.

Agenzia Spaziale Italiana. (2014). *Piano Triennale delle Attività 2015–2017*. Rome: Agenzia Spaziale Italiana.

Agraval, N. (2016, October 4). *Inequality in India: What's the real story?* Retrieved November 2, 2016, from World Economic Forum: https://www.weforum.org/agenda/2016/10/inequality-in-india-oxfam-explainer/

Agrawal, P., Sreekantan, B., & Bhandari, N. (2007, December). Space astronomy and interplanetary exploration. *Current Science, 93*(12), 1767–1778.

Airbus. (2017). *Airbus in India*. Retrieved February 2, 2017, from Airbus: http://www.airbusgroup.com/int/en/group-vision/global-presence/india.html

Alawadhi, N. (2016, June 6). *Government must use satellite technology for Digital India initiative: TV Ramachandran*. Retrieved November 15, 2016, from The Economic Times: http://economictimes.indiatimes.com/tech/ites/government-must-use-satellite-technology-for-digital-india-initiative-tv-ramachandran/articleshow/52615913.cms

Al-Ekabi, C. (2015). *Space policies, issues and trends in 2015–16*. Vienna: European Space Policy Institute.

Al-Ekabi, C. (2016). *The future of European commercial spacecraft manufacturing*. Vienna: European Space Policy Institute.

Alexandratos, N., & Bruinsma, J. (2012). *World agriculture. Towards 2030/2050: The 2012 revision*. Rome: FAO.

Aliberti, M. (2013, April). Regionalisation of space activities in Asia? *ESPI Perspectives 66*. Vienna: European Space Policy Institute.

Aliberti, M. (2015). *When China goes to the Moon*. Vienna: Springer.

Aliberti, M. (2017). Revitalising the EU-India strategic partnership: The role of space. *ORF Space Alert, 5*(2).

Aliberti, M., Ferretti, S., Hulsroj, P., & Lahcen, A. (2016). *Europe in the future and the contributions of space*. Vienna: European Space Policy Institute.

Aliberti, M., Nader, D., & Vernile, A. (2017). Understanding India's new space potential: Implications and prospects for Europe. In *68th international Astronautical Congress (IAC)*. Adelaide: International Astronautical Federation.

Andrés, L., Biller, D., & Dappe, M. (2013, December). *Reducing poverty by closing South Asia's infrastructure gap*. Retrieved October 30, 2016, from The World Bank: http://documents.worldbank.org/curated/en/881321468170977482/pdf/864320WP0Reduc0Box385179B000PUBLIC0.pdf

Annadurai, M. (2017a). Future exploration missions of ISRO. In *60th session of the United Nations Committee on peaceful uses of outer space*. Vienna: UNOOSA

Annadurai, M. (2017b). Recent Indian space missions. In *54th Session of the Scientific and Technical Subcommittee of United Nations Committee on peaceful uses of outer space*. Vienna: UNOOSA.

Annadurai, M. (2017c). South Asian Satellite – A new approach to regional co-operation. In *60th Session of the United Nations Committee on the peaceful uses of outer space*. Vienna: UNOOSA.

Antrix Corporation. (2015). *Satellite platform*. Retrieved August 22, 2016, from Antrix: http://www.antrix.gov.in/business/satellite-platforms

Antrix Corporation. (2016). *About us*. Retrieved September 10, 2016, from Antrix Corporation Limited: http://www.antrix.gov.in/about-us

Antrix Corporation Limited. (2015). *Antrix annual report 2014–15*. Bangalore: Antrix.

Antrix Corporation Limited. (2016a). *Services*. Retrieved December 3, 2016, from Antrix Corporation: http://www.antrix.gov.in/business/services

Antrix Corporation Limited. (2016b). *Remote sensing services*. Retrieved November 30, 2016, from Antrix Corporation Limited: http://www.antrix.gov.in/business/remote-sensing-services

Antrix Corporation Limited. (2016c). *Satellite platforms*. Retrieved December 9, 2016, from Antrix Corporation Limited: http://www.antrix.gov.in/business/satellite-platforms

Aravamudan, R. (2015). Evolution of ISRO: A personal account. In P. M. Rao (Ed.), *From fishing hamlet to red planet: India's space journey*. New Delhi: Harper Collins.

Arnold, D. (2000). *Science, technology and medicine in colonial India*. Cambridge: Cambridge University Press.

Ashok, G. V., & D'souza, R. (2017). SATCOM policy: Bridging the present and the future. In R. P. Rajagopalan & N. Prasad (Eds.), *Space India 2.0 commerce, policy, security and governance perspectives* (pp. 119–139). New Delhi: Observer Research Foundation.

Atomic Energy Commission. (1970). *A profile for the decade 1970–1980 for Atomic Energy and Space Research*. New Delhi: Government of India.

Bagchi, I. (2011, March 6). India working on tech to defend satellites. *Times of India*.

Bajaj, V. (2010, October 15). In Mumbai, adviser to Obama extols India's economic model. Retrieved May 10, 2017, from *The New York Times*: http://www.nytimes.com/2010/10/16/business/global/16summers.html

Basin, M. (2009). The EU-India partnership: Strategic alliance or political convenience. In A. Ghosh (Ed.), *India's foreign policy* (pp. 206–226). New Delhi: Dorling Kindersley.

Basu, P. (2016, December 1). *New horizon for satellite Earth observation*. Retrieved December 2, 2016, from Observer Research Foundation: http://www.orfonline.org/expert-speaks/new-horizons-satellite-earth-observation/

Benaglia, S. (2016, March). *How to boost EU-India relations, CEPS Policy Briefs(341)*. Brussels: Centre for European Policy Studies.

Bhatia, B. S. (2015). Satcom for development education. The indian experience. In P. M. Rao (Ed.), *From fishing hamlet to red planet: India's space journey*. New Delhi: Harper Collins.

Bhatia, B. (2016). Space applications for development – The indian approach. In *ESPI 10th Autumn conference*. Vienna: European Space Policy Institute.

Bhattacharjee, S. (2017, May 20). TRAI bats for private players in satellites. *Business Standard*.

Bigot, J.-C. (2017). 40 years of cooperation between ESA and ISRO. In *India in space: The forward look to international cooperation*. Vienna: European Space Policy Institute.

Blamont, J. (2015). Starting the Indian Space Programme. In P. M. Rao (Ed.), *From fishing hamlet to red planet: India's space journey*. New Delhi: Harper Collins.

Blamont, J. (2017). Cooperation in space between India and France. In R. P. Rajagopalan & N. Prasad (Eds.), *Space India 2.0 commerce, policy, security and governance perspectives* (pp. 215–233). New Delhi: Observer Research Foundation.

Bommakanti, K. (2009). Satellite integration and multiple independently retargetable reentry vehicles technology: Indian–United States Civilian Space Cooperation. *Astropolitics: The International Journal of Space Politics & Policy, 7*(1), 7–31.

Bommakanti, K. (2017). The significance of an Indian direct ascent kinetic capability. *ORF Space Alert, 2*(2).

Bonnefoy, A. (2014). *Humanitarian telemedicine.* Vienna: European Space Policy Institute.

Bouhey-Klapisz, C. (2017). Space cooperation between India and France. In *India in space: The forward look to international cooperation.* Vienna: European Space Policy Institute.

BP. (2017). *BP energy outlook. Country and regional Insights – India.* New Delhi: BP P.iC.

Brown, P. J. (2010a, January 22). *India's targets China's satellites.* Retrieved November 18, 2016, from Asia Times: Brown, Peter J "India's Targets China's Satellites. Asia Times" http://www.atimes.com/atimes/South_Asia/LA22Df01.html

Brown, P. J. (2010b, May 1). *India's space program takes a hit.* Retrieved November 10, 2016, from Asia Times Online: http://www.atimes.com/atimes/South_Asia/LE01Df01.html

Busvine, D., & Pinchuk, D. (2016, October 15). *India, Russia agree to missile sales, joint venture for helicopters.* Retrieved May 20, 2017, from Reuters: https://www.reuters.com/article/us-india-russia-helicopters/india-russia-agree-to-missile-sales-jointventure-for-helicopters-idUSKBN12F058

Bute, S. (2013, March 7). *Innovation: The new mantra for science and technology policy in India, Pakistan and China.* Retrieved October 10, 2016, from Institute for Defence Studies and Analyses: http://www.idsa.in/backgrounder/ScienceandTechnologyPoliciesinIndiaPakistanandChina

Centre National D'Etudes Spatiales. (2015, April 10). *CNES and ISRO sign new cooperation agreement at the Elysée palace.* Retrieved October 20, 2016, from CNES: https://presse.cnes.fr/en/cp-9834

Centre National D'Etudes Spatiales. (2016, May 18). *New Delhi declaration comes into effect. World's space agencies working to tackles climate change.* Retrieved September 11, 2016, from CNES: https://presse.cnes.fr/en/new-delhi-declaration-comes-effect-worlds-space-agencies-working-tackle-climate-change

Centre National D'Etudes Spatiales. (2017, January 9). *France-India space cooperation. Agreement signed on future launchers and lunar exploration.* Retrieved February 18, 2017, from CNES: https://presse.cnes.fr/sites/default/files/drupal/201701/default/cp003-2017_-_cooperation_spatiale_entre_la_france_et_linde_va.pdf

Chandrashekar, S. (2015). *Space, war and security. A strategy for India.* Bangalore: National Institute of Advanced Studies.

Chandrashekar, S. (2016). Space, war, and deterrence: A strategy for India. *Astropolitics: The International Journal of Space Politics and Policy, 14*(2–3), 153–157.

Chatterjee Miller, M. (2013, May/June). India's feeble foreign policy. *Foreign Affairs, 92*(3), 14–19.

Cohen, S. (2001). *India: Emerging power.* Washington, DC: Brookings Institution Press.

Committee on Earth Observation Satellites. (2016, April 3). *Heads of space agencies decide to join efforts in support of COP 21 decisions.* Retrieved May 10, 2017, from CEOS: http://ceos.org/document_management/Meetings/SIT/SIT-31/160408%20Declaration%20NewDelhi_V3.docx

Comptroller and Auditor General of India. (2012). *Report of the Comptroller and Auditor General of India on hybrid satellite digital multimedia broadcasting service agreement with Devas.* New Delhi: Department of Space.

Consulate General of France in Bangalore. (2017, April 10). *Signing of ISRO-CNES and CNES-TeamIndus agreements.* Retrieved April 11, 2017, from Consulate General of France in Bangalore: https://in.ambafrance.org/Signing-of-ISRO-CNES-and-CNES-TeamIndus-agreements-14402

Correll, R. R. (2006). U.S.-India space partnership: The jewel in the crown. *Astropolitics: The International Journal of Space Politics & Policy, 4*(2), 159–177.

Corrigan , G., & Di Battista, A. (2015, November 5). *19 Charts that explain India's economic challenge.* Retrieved October 2, 2016, from World Economic Forum: https://www.weforum.org/agenda/2015/11/19-charts-that-explain-indias-economic-challenge/

Council of the European Union. (2005, September 7). *The India-EU strategic partnership joint action plan.* Retrieved 2017, from Council of the European Union: http://www.consilium.europa.eu/ueDocs/cms_Data/docs/pressData/en/er/86130.pdf

Council of the European Union. (2016, March 30). *EU-India agenda for action 2020.* Retrieved April 10, 2016, from Council of the European Union: http://www.consilium.europa.eu/en/meetings/international-summit/2016/03/30/

Credit Suisse. (2015). *Global wealth databook 2015.* Zurich.

Dasgupta, A. (2017). Unlocking the potential of geospatial data. In R. Rajagopalan & N. Prasad (Eds.), *Space India 2.0 commerce, policy, security and governance perspectives.* New Delhi: Observer Research Foundation.

De Cosmo, V. (2006). *La partecipazione ASI alla missione ISRO OCEANSAT-2.* Roma: Agenzia Spaziale Italiana.

de Selding, P. (2011, October 12). ISRO launches joint French-Indian satellite to study tropical monsoons. *Space News, 2016* (September), 10.

de Selding, P. (2016, June 8). Satellite operators give negative reviews of Indian regulator's satellite-TV proposal. *Space News.*

de Selding, P. (2017, January 10). *France and India to study future rocket designs, including reusable vehicles.* Retrieved February 11, 2017, from Space Intel Report: https://www.spaceintel-report.com/france-india-to-study-future-rocket-designs/

Decreux, Y., & Mitaritonna, C. (2007). *Economic impact of a potential Free Trade Agreement (FTA) Between the European Union and India.* CEPII-CIREM.

Defence Aerospace. (2002, January 9). *India joins international charter on space and disaster.* Retrieved November 10, 2016, from Defence Aerospace: http://www.defense-aerospace.com/articles-view/release/3/8175/india-signs-two-agreements-on-space-(jan.-10).html

Degelsegger, A. (2016). *Science, technology and innovation policy in India. Some recent changes.* INDIGO.

Delegation of the European Union to India. (2013). *Fifty years of partnership.* Retrieved from European External Action Service: http://eeas.europa.eu/archives/delegations/india/documents/publications/the_eu_and_india_fifty_years_of_partnership.pdf

Department of Space – Government of India. (2016). *Outcome budget 2016–17.* New Delhi: Department of Space.

Department of Telecommunications – Government of India. (2016, February 2). *Restrictions on the use of satellite phone.* Retrieved November 10, 2016, from Department of Telecommunications: http://www.dot.gov.in/restrictions-use-satellite-phone

Der Spiegel. (2015, June 19). *We could build all kinds of things with Moon concrete.* Retrieved November 20, 2016, from Spiegel Online: We could build all kinds of things with Moon concrete

Deshpande, R. (2006, October 16). *India may quit EU-led GPS project.* Retrieved November 20, 2006, from Times of India: http://timesofindia.indiatimes.com/india/India-may-quit-EU-led-GPS-project/articleshow/2172710.cms

Diplomacy & Foreign Affairs. (2013, November 12). *The EADS group in India.* Retrieved December 7, 2016, from Diplomacy & Foreign Affairs: http://diplomacyandforeignaffairs.com/the-eads-group-in-india/

DLR. (2016, January 25). *A model for energy providers in India – the DLR test and qualification laboratory for solar power plants.* Retrieved February 14, 2017, from DLR: http://www.dlr.de/dlr/en/desktopdefault.aspx/tabid-10081/151_read-16498/#/gallery/21772

Dutta, V. (2012, December 12). Apollo enters Myanmar with telemedicine service. *The Economic Times.*

Dutta, S., Lanvin, B., & Wunsch-Vincent, S. (2016). *The global innovation index report 2016.* Retrieved September 30, 2016, from Global Innovation Index: https://www.globalinnovation-index.org/gii-2016-report

Ecorys. (2009). *Trade sustainability impact assessment for the FTA between the EU and the Republic of India.* Rotterdam.

Embassy of India. (2016, November 30). *India Germany foreign office consultations.* Retrieved February 1, 2017, from Embassy of India: https://www.indianembassy.de/pages.php?id=241

EU Institute for Security Studies. (2012). *The EU-India partnership: Time to go strategic?* Paris: EU Institute for Security Studies.

EUMETSAT. (2014, February). *International cooperation: Benefits for global users.* Retrieved March 30, 2017, from EUMETSAT: http://www.eumetsat.int/website/home/InSight/NewsUpdates/DAT_2169186.html?lang=EN

EUMETSAT. (2016, April 1). *EUMETSAT discusses Asia-Pacific cooperation.* Retrieved April 3, 2017, from EUMETSAT: http://www.eumetsat.int/website/home/News/DAT_2994861.html?lang=EN

EUMETSAT. (2017a). *EUMETCAST.* Retrieved March 20, 2017, from EUMETSAT: http://www.eumetsat.int/website/home/Data/DataDelivery/EUMETCast/index.html

EUMETSAT. (2017b). *Indian Ocean data coverage.* Retrieved March 30, 2017, from EUMETSAT: http://www.eumetsat.int/website/home/Data/MeteosatServices/IndianOceanDataCoverage/index.html

EUMETSAT. (2017c, February 1). *Meteosat-8 Now Prime IODC satellite.* Retrieved March 31, 2017, from EUMETSAT: http://www.eumetsat.int/website/home/News/DAT_3363066.html?lang=EN

European Commission. (2005, September 7). *The GALILEO family is further expanding: EU and India seal their agreement.* Retrieved January 12, 2017, from European Commission.

European Commission. (2016a, March). *EU India agenda for action 2020.* Retrieved from Press Release Database: http://europa.eu/rapid/press-release_IP-16-1142_en.htm

European Commission. (2016b, March). *Joint statement following the EU-India summit, Brussels, 30 March 2016.* Retrieved from Press Release Database: http://europa.eu/rapid/press-release_IP-16-1142_en.htm

European Commission. (2017a). *Climate action.* Retrieved from European Commission: https://ec.europa.eu/clima/policies/international/cooperation/india_en

European Commission. (2017b, February 16). *European Union, trade in goods with India.* Retrieved from Trade: http://trade.ec.europa.eu/doclib/docs/2006/september/tradoc_113390.pdf

European Commission. (2017c, February 17). *India.* Retrieved from Trade: http://trade.ec.europa.eu/doclib/docs/2006/september/tradoc_111515.pdf

European Commission. (2017d, March). *Trade.* Retrieved from Trade Defence – Investigations: http://trade.ec.europa.eu/tdi/completed.cfm

European Commission. (n.d.-a). *Actions against exports from the EU – Cases.* Retrieved from Trade: http://trade.ec.europa.eu/actions-against-eu-exporters/cases/

European Commission. (n.d.-b). *India.* Retrieved 2017, from Trade: http://ec.europa.eu/trade/policy/countries-and-regions/countries/india/

European External Action Service. (2008, September 29). *The EU-India Joint Action Plan (JAP).* Retrieved March 7, 2017, from European External Action Service: http://eeas.europa.eu/archives/docs/india/sum09_08/joint_action_plan_2008_en.pdf

European External Action Service. (2016, October 17). *India and the EU.* Retrieved from Delegation of the European Union to India and Bhutan: https://eeas.europa.eu/delegations/india/670/india-and-eu_en

European Global Navigation Satellite Systems Agency. (2014, February 19). *Boost for EU-India Cooperation.* Retrieved January 5, 2017, from European Global Navigation Satellite Systems Agency: https://www.gsa.europa.eu/news/boost-eu-india-cooperation

European Global Navigation Satellite Systems Agency. (2015a, June 26). *GNSS in Asia–Support on International Activities.* Retrieved February 1, 2017, from European Global Navigation Satellite Systems Agency.: https://www.gsa.europa.eu/gnss-asia-support-international-activities

European Global Navigation Satellite Systems Agency. (2015b, June 26). *GNSS in Asia – Support on international activities.* Retrieved February 2, 2017, from European Global Navigation Satellite Systems Agency.: https://www.gsa.europa.eu/gnss-asia-support-international-activities

European Space Agency. (2005). *Opportunity for collaboration with India on Chandrayaan-1 Mission to the Moon.* Retrieved November 10, 2016, from ESA: http://sci.esa.int/science-e/www/object/doc.cfm?fobjectid=37906

European Space Agency. (2006). *Summary. Europe's first lunar adventure.* Retrieved February 16, 2017, from European Space Agency: http://sci.esa.int/smart-1/31407-summary/

European Space Agency. (2007, September 25). *ESA and India tighten relations at.* Retrieved January 8, 2017, from ESA: http://www.esa.int/About_Us/Welcome_to_ESA/ ESA_and_India_tighten_relations_at_IAC_2007/(print)

European Space Agency. (2016, October 26). *Shared vision and goals for the future of European space.* Retrieved from http://www.esa.int/About_Us/Welcome_to_ESA/ Shared_vision_and_goals_for_the_future_of_Europe_in_space

European Space Agency. (2017, December 15). *Athena.* Retrieved December 21, 2017, from European Space Agency: http://sci.esa.int/athena/59896-mission-summary/

European External Action Service. (2008, September 29). *The EU-India Joint Action Plan (JAP).* Retrieved March 7, 2017, from European External Action Service: http://eeas.europa.eu/ archives/docs/india/sum09_08/joint_action_plan_2008_en.pdf

European Union Global Strategy. (2016). *Shared vision, common action: A stronger Europe; A global strategy for the European Union's Foreign and Security Policy.* European Union Global Strategy.

Eurostat. (2017, March 17). *Euro area international trade in goods deficit of €0.6 bn.* Retrieved from http://trade.ec.europa.eu/doclib/docs/2013/december/tradoc_151969.pdf

EUTELSAT. (2017). *Eutelsat Asia, expanding reach across the Asia Pacific.* Retrieved March 1, 2017, from EUTELSAT Asia: http://www.eutelsatasia.com/home/eutelsat-asia.html

FAO, UNICEF, SaciWATERs. (2013). *Water in India: Situation and prospects.* New Delhi: FAO, UNICEF.

Foust, J. (2008, February 11). *India and the US: Partners or rivals in space?* Retrieved September 11, 2016, from The Space Review: http://www.thespacereview.com/article/1056/1

Frugal Innovation Hub. (2016). *About us.* Retrieved October 20, 2016, from Frugal Innovation Hub: http://frugalinnovationhub.com/en/signup.html?id=1&file=first_chapter

Ganguly, S. (2016, September 19). India After Nonalignment. *Foreign Affairs.*

Gateway House: Indian Council on Global Relations; Istituto Affari Internazionali. (2017). *Moving forward the EU-India security dialogue: Traditional and emerging issues.* Mumbai/Rome: Gateway House: Indian Council on Global Relations/Istituto Affari Internazionali.

Ghosh, A. (2009). *India's foreign policy.* New Delhi: Dorling Kindersley.

Global Times. (2014, February 18). India uncertain as Abe looks for anti-China alliance. *Global Times.*

GNSS Asia. (2016a). *Country profile – India.* Retrieved January 3, 2017, from GNSS Asia: http:// www.gnss.asia/sites/default/files/Countryprofiles_INDIA_2016_v4.pdf

GNSS Asia. (2016b). *GNSS opportunities in India.* Retrieved from GNSS.Asia: http://india.gnss. asia/gnss-opportunities-india

GNSS Asia. (2017). *Opportunity & partnering.* Retrieved February 10, 2017, from GNSS Asia: http://www.gnss.asia/node/27

Goel, P. (2015). Operational satellites of ISRO. In P. M. Rao (Ed.), *From fishing hamlet to red planet: India's space journey.* New Delhi: Harper Collins.

Gopalaswamy, B. (2013). India's perspective. In A. M. Tellis (Ed.), *Crux of Asia. China, India and the emerging global order.* Washington, DC: Carnegie Endowment for International Peace.

Government of India. (1997). *Policy framework for satellite communications in India.* New Delhi: Department of Space.

Government of India. (2000). *Norms, guidelines and procedures for implementation of the policy frame-work for satellite communications in India.* New Delhi: Government of India.

Government of India. (2012). *National data sharing and accessibility policy.* New Delhi: Government of India.

Government of India. (2015, September 7). *PM Modi to address national meet on Promoting Space Technology on 7 September 2015.* Retrieved September 10, 2016, from Narendra Modi: http://www.narendramodi.in/pm-to-address-the-national-meet-on-promoting-space-techno- logy-based-tools-and-applications-in-governance-and-development-at-vigyan-bhavan-on-7th- september-2015-291009

Government of India. (2016a). *Economic survey 2015–2016* (Vol. 1). New Delhi: OUP India.
Government of India. (2016b). *SPACE – Make In India*. Retrieved November 13, 2016, from Make in India: http://www.makeinindia.com/sector/space
GPS Daily. (2016, May 4). *Operation of Indian GPS will take some more time: ISRO*. Retrieved September 21, 2016, from GPS Daily: http://www.gpsdaily.com/reports/Operation_of_Indian_GPS_will_take_some_more_time_ISRO_999.html
Guha, R. (2010). *Makers of modern India*. New Delhi: Penguin Books.
Guha, R. (2014, February 24). *Why India is the most interesting country in the world?* Retrieved from Youtube: https://www.youtube.com/watch?v=6PteQsyUAbg
Gupta, S. (2015). Beginnings of launch vehicle technology in ISRO. In P. M. Rao (Ed.), *From fishing hamlet to red planet: India's space journey*. New Delhi: Harper Collins.
Harsh, V. (2016, June 14). *India's anti-satellite weapons*. Retrieved June 15, 2016, from The Diplomat: http://thediplomat.com/2016/06/indias-anti-satellite-weapons/
Harvey, B., Smid, H. H., & Pirard, T. (2010). *Emerging space powers*. Chichester: Springer – Praxis.
Haub, C. (2012, March). *2011 census shows how 1.3 billion people in India live*. Retrieved from Population Reference Bureau: http://www.prb.org/Publications/Articles/2012/india-2011-census.aspx
Headquarters Integrated Defence Staff. (2013). *Technology perspective and capability roadmap*. New Delhi: Ministry of Defence.
Headquarters Integrated Defence Staff – Ministry of Defence. (2013). *The technology perspective and capability roadmap*. Retrieved September 10, 2016, from Ministry of Defence: http://mod.gov.in/writereaddata/TPCR13.pdf
Hobson, J. M. (2004). *The Eastern origins of the modern world*. Cambridge: Cambridge University Press.
Horikawa, Y. (2017). Space security, sustainability, and global governance: India-Japan collaboration in outer space. In R. P. Rajagopalan & N. Prasad (Eds.), *Space India 2.0. commerce, policy, security and governance perspectives*. New Delhi: Observer Research Foundation.
India Space Research Organisation. (2016). *Tele-education*. Retrieved October 3, 2016, from ISRO: http://www.isro.gov.in/applications/tele-education
India Stack. (2016). *About*. Retrieved December 2, 2016, from India Stack: http://indiastack.org/about/
India Times. (2015). *15 year development agenda to replace five year plans*. Retrieved from India Times: http://economictimes.indiatimes.com/news/economy/policy/15-year-development-agenda-to-replace-five-year-plans-to-include-internal-security-defence/articleshow/52247186.cms
Indian Armed Forces. (2016). *Integrated space cell*. Retrieved from Indian Armed Forces: http://indianarmedforces.weebly.com/integrated-space-cell.html
Indian Space Research Organisation. (2002, January 9). *ISRO Signs Cooperative Agreement with European Space Agency*. Retrieved March 1, 2017, from Vikram Sarabhai Space Centre: http://www.vssc.gov.in/VSSC_V4/index.php/2002/79-press-release-articles/789-isro-cooperative-agreement-esa
Indian Space Research Organisation. (2006a). *Eleventh five year proposal*. Bangalore: ISRO.
Indian Space Research Organisation. (2006b, February 20). *EADS Astrium-ISRO alliance sealed. First contract with Eutelsat for W2M satellite*. Retrieved December 12, 2016, from ISRO: http://www.isro.gov.in/update/20-feb-2006/eads-astrium-isro-alliance-sealed-first-contract-with-eutelsat-w2m-satellite
Indian Space Research Organisation. (2008, September 30). *India signs agreements with France on cooperation in space*. Retrieved October 1, 2016, from ISRO.
Indian Space Research Organisation. (2012, February). *Global applications of Oceansat-2*. Retrieved March 31, 2017, from United Nations Office for Outer Space Affairs: http://www.unoosa.org/pdf/pres/stsc2012/tech-49E.pdf
Indian Space Research Organisation. (2015a). *Genesis*. Retrieved July 3, 2016, from ISRO: http://www.isro.gov.in/about-isro/genesis
Indian Space Research Organisation. (2015b). *Rohini Satellite RS-1*. Retrieved July 18, 2016, from ISRO: http://www.isro.gov.in/Spacecraft/rs-1-1

Indian Space Research Organisation. (2015c). *Rohini Satellite RS-D2*. Retrieved July 18, 2016, from ISRO: http://www.isro.gov.in/Spacecraft/rohini-satellite-rs-d2

Indian Space Research Organisation. (2015d). *Space applications centre celebrates ruby year of SITE*. Retrieved July 18, 2016, from ISRO: http://www.isro.gov.in/space-applications-centre-celebrates-ruby-year-of-site

Indian Space Research Organisation. (2016a). *Climate & environment*. Retrieved October 26, 2016, from ISRO: http://www.isro.gov.in/applications/climate-environment-0

Indian Space Research Organisation. (2016b). *Small satellites*. Retrieved September 7, 2016, from ISRO: http://www.isro.gov.in/spacecraft/small-satellites

Indian Space Research Organisation. (2016c). *APPLE*. Retrieved July 20, 2016, from IRSO: http://www.isro.gov.in/Spacecraft/apple

Indian Space Research Organisation. (2016d). *ASLV*. Retrieved July 20, 2016, from ISRO.: http://www.isro.gov.in/launchers/aslv

Indian Space Research Organisation. (2016e, September 4). *Bengaluru Space Expo (BSX)-2016*. Retrieved October 5, 2016, from ISRO: http://www.isro.gov.in/bengaluru-space-expo-bsx-2016

Indian Space Research Organisation. (2016f). *Disaster management support programme*. Retrieved October 20, 2016, from ISRO: http://www.isro.gov.in/applications/disaster-management-support-programme

Indian Space Research Organisation. (2016g). *International cooperation*. Retrieved September 20, 2016, from Indian Space Research Organisation: http://www.isro.gov.in/international-cooperation

Indian Space Research Organisation. (2016h). *ISRO annual report 2015–2016*. Bangalore: ISRO.

Indian Space Research Organisation. (2016i). *ISRO technology transfer*. Retrieved September 18, 2016, from ISRO: http://www.isro.gov.in/isro-technology-transfer

Indian Space Research Organisation. (2016j). *Natural resource census*. Retrieved October 27, 2016, from ISRO: http://www.isro.gov.in/applications/natural-resources-census

Indian Space Research Organisation. (2016k). *Sponsored research – ISRO*. Retrieved 2016, from ISRO: http://www.isro.gov.in/sponsored-research-respond

Indian Space Research Organisation. (2016l). *Tele-medicine*. Retrieved October 11, 2016, from ISRO: http://www.isro.gov.in/applications/tele-medicine

Indian Space Research Organisation. (2016m). *Towards ensuring water security*. Retrieved October 1, 2016, from Indian Space Research Organisation: http://www.isro.gov.in/applications/towards-ensuring-water-security

Indian Space Research Organisation. (2016n). *Vikram Sarabhai (1963–1971)*. Retrieved June 10, 2016, from ISRO: http://www.isro.gov.in/about-isro/dr-vikram-ambalal-sarabhai-1963-1971

Indian Space Research Organisation. (2017a, February 15). *PSLV-C37 successfully launches 104 satellites in a single flight*. Retrieved from ISRO: http://www.isro.gov.in/update/15-feb-2017/pslv-c37-successfully-launches-104-satellites-single-flight

Indian Space Research Organisation. (2017b). *Government of India – Department of Space – Indian Space Research Organisation*. Retrieved from List of Communication Satellites: http://www.isro.gov.in/spacecraft/list-of-communication-satellites

Indian Space Research Organisation – Deloitte. (2010). *Overview of the Indian space sector*. Bangalore: Deloitte.

IndoGenius. (2016a, August). Creative solutions to grand challenges. *The importance of India*. IndoGenius.

IndoGenius. (2016b, August). Five types of India innovation. *The importance of India*. IndoGenius.

IndoGenius. (2016c, August). India's stack and JAM trinity. *The importance of India*. IndoGenius.

IndoGenius. (2016d, August). India's Election Calendar. *The importance of India*. IndoGenius.

IndoGenius. (2016e, August). Rural Vs. Urban India. *The importance of India*. IndoGenius.

IndoGenius. (2016f, August). The most interesting country on Earth. *The importance of India*. IndoGenius.

IndoGenius. (2016g, August). The youngest population on Earth. *The importance of India*. IndoGenius.

IndoGenius. (2016h, August). *India engages. The importance of India*. IndoGenius.

Indo-German Science and Technology Centre. (2017). *About IGSTC*. Retrieved February 3, 2017, from IGSTC: http://www.igstc.org/about_us.html

International Business Publications. (2011). *India space programs and exploration handbook*. Washington, DC: International Business Publications.

International Energy Agency. (2015). *India energy outlook*. Paris: OECD-IEA.

International Monetary Fund. (2017, April). *World economic outlook*. Retrieved May 10, 2017, from International Monetary Fund: https://www.imf.org/external/pubs/ft/weo/2017/01/weo-data/index.aspx

ITU/UNESCO Broadband Commission. (2016). *The state of broadband 2016: Broadband catalyzing sustainable development*. Geneva: ITU/UNESCO Broadband Commission for Sustainable Development.

Jaffrelot, C. (2006). *India and the European Union: The charade of a strategic partnership*. Paris: Centre de Recherches Internationales.

Jain, R. K. (2014, June). *India-EU strategic partnership: Perceptions and perspectives* (NFG Working Paper Series, 10|2014).

Japan Aerospace Exploration Agency. (2005, March 31). *JAXA Vision – JAXA 2025*. Retrieved March 20, 2016, from JAXA: http://www.docstoc.com/docs/87745747/JAXA-Vision

Jayaraj, C. (2004). *India's Space Policy and Institutions. United Nations/Republic of Korea Workshop on Space Law. United Nations Treaties on Outer Space: Actions at the National Level*. New York: United Nations.

Jayaraman, K. (2006a, November 14). *ISRO seeks government approval for manned space-flight program*. Retrieved December 13, 2016, from Space News: http://spacenews.com/isro-seeks-government-approval-manned-spaceflight-program/

Jayaraman, K. (2006b, November 13). *ISRO seeks government approval for manned spaceflight program*. Retrieved September 21, 2016, from Space News: http://www.spacenews.com/article/isro-seeks-government-approval-manned-spaceflight-program/

Joint Global Change Research Institute; Battelle. (2009). *India: The impact of climate change to 2030 a commissioned research report*. Washington, DC: National Intelligence Council.

Kalam, A. A. (2015). India's first launch vehicle. In P. M. Rao (Ed.), *From fishing hamlet to red planet: India's space journey*. New Delhi: Harper Collins.

Kale, P. (2015). Origins of INSAT-1. In P. M. Rao (Ed.), *From fishing hamlet to red planet: India's space journey*. New Delhi: Harper Collins.

Kaplan, R. (2013). *The revenge of geography. What the map tell us about future conflicts and the battle against faith*. New York: Random House Trade Paperbacks.

Karan, P. P. (1953, December). India's role in geopolitics. *India Quarterly, 9*, 160–169.

Kasturirangan, K. (2006). *India's space enterprise: A case study in strategic thinking and planning*. Canberra: Australia South Asia Research Centre.

Kasturirangan, K. (2011). *Relating science to society*. Mysore: University of Mysore.

Kaul, R. (2017). A review of India's geospatial policy. In R. P. Rajagopalan & N. Prasad (Eds.), *Space India 2.0 commerce, policy, security and governance perspectives* (pp. 141–150). New Delhi: Observer Research Foundation.

Khandekar, G. (2011, August). *The EU and India: A loveless arranged marriage* (FRIDE Policy Brief, 90|2011).

Khandekar, G. (2015). Towards an EU-India partnership on urbanisation. *Global Relations Forum Policy Brief, 27*(1).

Kissinger, H. (2014). *World order*. New York: Penguin Books.

Kochhar, R. (2015a, July 8). *A global middle class is more promise than reality*. Retrieved October 21, 2016, from Pew Research Center: http://www.pewglobal.org/2015/07/08/despite-poverty-plunge-middle-class-status-remains-out-of-reach-for-many/#poverty-retreats-in-india-but-the-middle-class-barely-expands

Kochhar, R. (2015b, July 15). *China's middle class surges, while India's lags behind.* Retrieved from Pew Research Center: http://www.pewresearch.org/fact-tank/2015/07/15/china-india-middle-class/

Korovkin, V. (2017). Evolution of India-Russia partnership. In R. P. Rajagopalan, & N. Prasad (Eds.), *Space India 2.0; commerce, policy, security and governance perspectives* (pp. 246–262). New Delhi: Observer Research Foundation.

KPMG. (2010). *Water sector in India: Overview and focus areas for the future.* New Delhi: KPMG.

Krishna, V. V. (2015). *Science, technology and innovation policy in India. Some recent changes.* INDIGO Policy.

Krishna, V., & Mishra, R. (2016). *Policy brief: India science and technology cooperation with EU and other select countries.* Vienna: Centre for Social Innovation (ZSI).

Kumar, N., & Puranam, P. (2012). *India inside. The emerging global challenge to the West.* Boston: Harvard Business Review.

Kumaraswamy, P. (2012). Looking west 2: Beyond the Gulf. In D. Scott (Ed.), *Handbook of India's international relations* (pp. 179–188). London: Routledge.

Kyodo News Service. (2008, September 25). Japanese official calls for space cooperation with China. *Kyodo News Service.* Retrieved from Space Daily: http://www.spacedaily.com/reports/Shenzhou_7_Astronauts_Brace_For_Space_Walk_999.html

Lahcen, A. (2013). *EUMETSAT-NOAA collaboration in meteorology from space. Review of a long-standing Trans-Atlantic partnership.* Vienna: European Space Policy Institute.

Laxman, S. (2009). *Moonshot India. Chandrayaan-1 the mission complete.* Mumbai: Navneet Publications.

Laxman, S. (2016a, February 15). *Plan to largely privatize PSLV operations by 2020: ISRO chief.* Retrieved September 20, 2016, from The Times of India: http://timesofindia.indiatimes.com/india/Plan-to-largely-privatize-PSLV-operations-by-2020-Isro-chief/articleshow/50990145.cms

Laxman, S. (2016b, February 15). *Plan to largely privatize PSLV operations by 2020: ISRO chief.* Retrieved November 10, 2016, from The Times of India: http://timesofindia.indiatimes.com/india/Plan-to-largely-privatize-PSLV-operations-by-2020-Isro-chief/articleshow/50990145.cms

Lee, J. (2010, June 4). *India's rise helps complement U.S. interests.* Retrieved January 7, 2017, from Hudson Institute: https://hudson.org/research/7051-india-rsquo-s-rise-helps-complement-u-s-interests

Leffler, C. (2017, January 17). *Speech by Christian Leffler on EU foreign policy & future of EU-India relations.* Retrieved from European External Action Service: https://eeas.europa.eu/sites/eeas/files/christian_lefflers_speech_at_think_tank_twinning_initiative.pdf

Lele, A. (2013a). *Asian space race: Rhetoric of reality?* New Delhi: Springer.

Lele, A. (2013b, September 9). *Commentary on GSAT-7: India's strategic satellite.* Retrieved November 8, 2016, from Space News: http://spacenews.com/37142gsat-7-indias-strategic-satellite/

Lele, A. (2014). *Mission Mars: India's quest for the red planet.* New Delhi: Springer.

Lele, A. (2015, September). Space collaboration between India and France – Towards a New Era. *Asie.Visions, 78.*

Lele, A. (2016). Power dynamics of India's space programme. *Astropolitics: The International Journal of Space Politics and Policy, 14*(2), 120–134.

Lele, A. (2017a). *Fifty years of the outer space treaty. tracing the journey.* New Delhi: Pentagon Press.

Lele, A. (2017b). India's policy for outer space. *Space Policy, 39–40,* 26–32.

Lele, A. (2017c). India's strategic space programme: From apprehensive beginner to ardent operator. In R. P. Rajagopalan (Ed.), *Space India 2.0 commerce, policy, security and governance perspectives* (pp. 179–192). New Delhi: Observer Research Foundation.

Listner, M. (2011, March 28). *India's ABM test: A validated ASAT capability or a Paper Tiger.* Retrieved March 30, 2017, from The Space Review: http://www.thespacereview.com/article/1807/1

Lok Sabha – Ministry of Space. (2016). *"Questions: Lok Sabha," Government of India.* Retrieved September 2, 2016, from Lok Sabha: http://164.100.47.192/Loksabha/Questions/QResult15.aspx?qref=35204&lsno=16.

Maass, R. (2017, March 14). *India test fires BrahMos extended range missile.* Retrieved from Space Daily: http://www.spacedaily.com/reports/India_test_fires_BrahMos_Extended_Range_missile_999.html

Madhumathi, D. (2016, June 4). ISRO now looks for satellite system vendors. *The Hindu.*

Mallisena. (1933). Syādvādamanjari, *19*, 75–77. *Dhruva, A.B*, 23–25.

Malone, D. M. (2011). *Does the elephant dance? Contemporary India foreign policy.* New Delhi: Oxford University Press.

Mars Daily. (2014, September 24). *Quoted from: "India wins Asia's Mars race as spacecraft enters orbit.* Retrieved September 30, 2016, from Mars Daily: http://www.marsdaily.com/reports/India_wins_Asias_Mars_race_as_spacecraft_enters_orbit_999.html

Matheswaran, M. (2015, May 9). *Hot on Mars, but short everywhere else.* Retrieved December 12, 2016, from Business Line: http://www.thehindubusinessline.com/opinion/hot-on-mars-but-short-everywhere-else/article7224259.ece

Mathieu, C. (2008). *Assessing Russia's space cooperation with China and India.* Vienna: European Space Policy Institute.

Mazzucato, M. (2015). The innovative state. *Foreign Affairs.*

McKinsey Global Institute. (2014). *India's technology opportunity: Transforming work, empowering people.* New Delhi: McKinsey.

MedIndia. (2017). Indian population clock. Retrieved December 21, 2017, from *MedIndia*: http://www.medindia.net/patients/calculators/pop_clock.asp

Mehrotra, S. (2015). *Realising the demographic dividend policies to achieve inclusive growth in India.* London: Harvard University Press.

Meredith, R. (2008). *The elephant and the dragon: The rise of India and China and what it means for all of us.* New York: Norton.

Microsoft Venture. (2016, November 10). *Enterprise readiness of Indian startup ecosystem.* Retrieved November 15, 2016, from Microsoft Accelerator: https://www.microsoftaccelerator.com/blog/entry/EnterpriseReadinessofIndianStartupEcosystemReport%7C6205

Miller, J. B. (2013, May 29). The Indian piece of Abe's security diamond. Retrieved May 30, 2017, from *The Diplomat*: https://thediplomat.com/2013/05/the-indian-piece-of-abes-security-diamond/

Ministry of Home Affairs – Government of India. (2011). *2011 census data.* Retrieved October 23, 2016, from Office of the Registrar General & Census Commissioner: http://www.censusindia.gov.in/2011-Common/CensusData2011.html

Ministry of Home Affairs – Government of India. (2016, April). *National Geospatial Policy [NGP 2016].* Retrieved December 9, 2016, from http://www.dst.gov.in/sites/default/files/Draft-NGP-Ver%201%20ammended_05May2016.pdf

Missile Technology Control Regime. (n.d.). Retrieved from http://mtcr.info/

Mohan, C. R. (2013). Changing global order: India's perspective. In A. J. Tellis & S. Mirski (Eds.), *Crux of Asia: China, India and the emerging global order.* Washington, DC: Carnegie Endowment for International Peace.

Mohan, R. (2017, February 10). *Delhi, Tokyo, Canberra.* Retrieved February 11, 2017, from The Indian Express: http://indianexpress.com/article/opinion/columns/tpp-donald-trumpmalcolm-turnbullphone-call-leak-islamic-state-canberra-delhi-tokyo-4516471/

Moltz, J. C. (2012). *Asia's space race. National motivations, regional rivalries, and international risks.* New York: Columbia University Press.

Mukherjee, A., Wymenga, P., Van Den Bosse, E., Goyal, T. M., & Goswami, R. (2013). *The long road towards an EU-India free trade agreement.* Brussels: European Parliament.

Murali, M. (2015, January 26). *Dhruva space, German startup Berlin space technologies ink MoU on satellites.* Retrieved June 8, 2017, from Economic Times: http://economictimes.indiatimes.com/small-biz/startups/dhruva-space-german-startup-berlin-space-technologies-ink-mou-on-satellites/articleshow/46015590.cms

Murthi. (1999). *India space programme.* New York: Springer.

Murthi, K. S. (2007). Legal environment for space activities. *Current Science, 93*(12), 1823–1827.

Murthi, S., & Rao, M. (2015). Future Indian (New) space – Contours of a national space policy that positions a new public-private regime. In *3rd Manfred Lachs international conference on NewSpace commercialisation and the Law.* Montreal: NIAS.

Murthi, S., & Rao, M. (2016). Privatising space missions: The critical route to boost India's space economy. *New Space Journal.*

Murthi, K. S., Sankar, U., & Madhusudhan, H. (2007, December). Organisational systems, commercialisation and cost-benefit analysis of Indian space programme. *Current Science, 93*(12), 1812.

Naidu, S. (2009). India engagement in Africa: Self-interest or mutual partnership? In R. Southhall & H. Melber (Eds.), *The new scramble for Africa: Imperialism, development in Africa.* Scottsville: University of KwaZulu Natal Press.

Nair, S. U. (2015). Initiatives on India's Human Spaceflight. In P. M. Rao (Ed.), *From fishing hamlet to red planet: India's space journey.* New Delhi: Harper Collins.

Narayana, K. (2015). The spaceport of ISRO. In P. M. Rao (Ed.), *From fishing hamlet to red planet: India's space journey.* New Delhi: Harper Collins.

Narayanamoorthy, N. (2015). PSLV: The workhorse of ISRO. In P. M. Rao (Ed.), *From fishing hamlet to red planet: India's space journey.* New Delhi: Harper Collins.

NASSCOM. (2016). *Start-up report – Momentous rise of the Indian start-up ecosystem.* Retrieved from NASSCOM: http://www.nasscom.in/knowledge-center/publications/start-report-momentous-rise-indian-start-ecosystem

National Aeronautics Space Administration. (2016). *NASA-ISRO SAR Mission (NISAR).* Retrieved February 8, 2017, from NASA-JPL: https://nisar.jpl.nasa.gov/

National Innovation Council – Government of India. (2011). *Decade of innovations: 2010–2020 roadmap.* Retrieved November 3, 2016, from National Innovation Council: http://innovation-councilarchive.nic.in/index.php?option=com_content&view=article&id=36:decade-of-innovation&catid=7:presentation&Itemid=8

National Intelligence Council. (2012). *Global trends 2030: Alternative worlds.* Washington, DC: National Intelligence Council.

National Intelligence Council. (2017). *Global trends. paradox of progress.* Washington, DC: National Intelligence Council.

Navalgund, R. (2015). Remote sensing applications. In P. M. Rao (Ed.), *From fishing hamlet to red planet: India's space journey.* New Delhi: Harper Collins.

Nayak, A. (2015). Potential fishing zones. Science to service. In P. M. Rao (Ed.), *From fishing hamlet to red planet: India's space journey.* New Delhi: Harper Collins.

NDTV. (2013, August 16). *Human space flight mission off ISRO priority list.* Retrieved September 28, 2016, from NDTV: http://www.ndtv.com/article/india/human-space-flight-mission-off-isro-priority-list-406551

Nye, J. (2004). *Soft power: The means to success in world politics.* New York: Public Affairs.

Observer Research Foundation. (2013, May 1). *Expert policy dialogue on India-France relations.* Retrieved September 10, 2016, from ORF: http://www.orfonline.org/research/experts-policy-dialogue-on-india-france-relations/

Observer Research Foundation. ORF KC 2017 (2017). Retrieved April 10, 2017, from: https://www.youtube.com/watch?v=VqxUwugJL9E&index=1&list=PLHu-cm5A_l80hS3DSfMd22exoCNFppODg

Observer Research Foundation. (2016). *2nd ORF Kalpana Chawla annual space policy dialogue.* New Delhi: Observer Research Foundation.

OECD. (2011). *The space economy at a Glance 2011.* Paris: OECD Publishing.

OECD. (2014a). *India – addressing economic and social challenges through innovation.* Paris: OECD Publishing.

OECD. (2014b). *India policy brief: Innovation.* Paris: OECD Publishing.

OECD. (2014c). *The space economy at a Glance 2014.* Paris: OECD Publishing.

OECD. (2016). *Space and innovation.* Paris: OECD Publishing.

OECD. (2017). *GDP long-term forecast.* Retrieved April 29, 2017, from OECD Data: https://data.oecd.org/gdp/gdp-long-term-forecast.htm

Office of the Register General and Census Commissioner. (2011). *Percentage of households to total households by amenities and assets.* Retrieved October 1, 2016, from Census of India: http://www.censusindia.gov.in/2011census/hlo/Houselisting-housing-PCA.html

Ollapally, D. M., & Rajagopalan, R. (2013). India: Foreign policy perspectives of an ambiguous power. In H. Nau & D. Ollapally (Eds.), *Worldviews of aspiring powers: Domestic foreign policy debates in China, India, Iran, Japan and Russia.* New Delhi: Oxford University Press.

Othman, M. (2017). National space policy and administration. In C. Johnson (Ed.), *Handbook for new actors in space.* Denver: Secure World Foundation.

Our World in Data. (2017). *Future world population growth.* Retrieved April 30, 2017, from Our World in Data: https://ourworldindata.org/future-world-population-growth/#un-population-projection-by-country-and-world-region-until-2100

Pan-African E-Network. (2016). *About the project.* Retrieved January 17, 2017, from Pan-African E-network Project: http://www.panafricanenetwork.com

Pandit, R. (2008, June 17). *Space command must to check China.* Retrieved March 28, 2017, from Times of India: http://timesofindia.indiatimes.com/india/Space-command-must-to-check-China/articleshow/3135817.cms

Paracha, S. (2013). Military dimensions of the Indian space program. *Astropolitics: The International Journal of Space Politics and Policy, 11*(3), 156–186.

Pardesi, M. S. (2005). *Deducing India's grand strategy of regional hegemony from historical and conceptual perspectives.* Singapore: Institute of Defense and Strategic Studies. Retrieved from Singapore: Institute of Defense and Strategic Studies: http://www.rsis.edu.sg/publications/WorkingPapers/WP76.pdf

Patairiya, M. K. (2013, November 22). Why India is going to Mars? *The New York Times.*

Paul, T. (2014, April). India's soft power in a globalizing world. *Current History, 113,* 157–162.

People Research on Indian Consumer Economy. (2014). *Snapshots – ICE 360° Survey.* Retrieved November 20, 2016, from People Research on Indian Consumer Economy: http://www.ice360.in/en/projects/data-and-publications/snapshots-ice360-survey-2014

Perumal, R. (2015). Evolution of the geosynchronous satellite launch vehicle. In P. M. Rao (Ed.), *From fishing hamlet to red planet: India's space journey.* New Delhi: Harper Collins.

Peter, N. (2008). Developments in space policies programmes and technologies throughout the world and Europe. In K.-U. Schrogl, C. Mathieu, & N. Peter (Eds.), *ESPI yearbook 2006/2007: A new Impetus for Europe* (p. 96). Vienna: Springer.

Peter, N. (2009). Developments in space policies programmes and technologies throughout the world and Europe. In Kai-Uwe, Schrogl, C. Mathieu, & N. Peter (Eds.), *ESPI yearbook on space policy 2007/2008: From policies to programmes* (p. 81). Vienna: Springer.

Pew Research Center. (2015, July 8). *Global population by income.* Retrieved October 10, 2016, from Per Research Center: http://www.pewglobal.org/interactives/global-population-by-income/

Pillai, C. M. (2016, April 25). *India as the pivotal power of the 21st century security order.* Retrieved January 12, 2017, from Center for International Maritime Security.

Pisharoty, P. (2015). Historical perspective of remote sensing. Some reminiscences. In P. M. Rao (Ed.), *From fishing hamlet to red planet: India's space journey.* New Delhi: Harper Collins.

Planning Commission – Government of India. (2014, May 31). *GDP at factor cost at 2004–05 Prices, Share to Total GDP and % Rate of Growth in GDP.* Retrieved September 20, 2016, from Planning Commission Database: Planning Commission. http://planningcommission.nic.in/data/datatable/0814/table_4.pdf

PM Jan Dhan Yojana. (2016). *Startup India scheme.* Retrieved June 8, 2017, from PM Jan Dhan Yojana: http://pmjandhanyojana.co.in/start-up-india-stand-up-india-scheme/

Population Institute. (2009). *2030: The "Perfect Storm" scenario.* Retrieved December 9, 2016, from Population Institute: https://www.populationinstitute.org/external/files/reports/The_Perfect_Storm_Scenario_for_2030.pdf

Prasad, M. (2015a). ISRO and international cooperation. In M. Rao (Ed.), *From fishing hamlet to red planet: India's space journey*. New Delhi: Harper Collins.

Prasad, N. (2015b, July 6). Is India turning a blind eye to space commerce? *The Space Review*. Retrieved December 12, 2016, from The Space Review: http://www.thespacereview.com/article/2782/1

Prasad, N. (2016a, March). Demystifying space business in India and issues for the development of a globally competitive private space industry. *Space Policy, 36*, 1–11.

Prasad, N. (2016b). Diversification of the Indian space programme in the past decade: Perspectives on implications and challenges. *Space Policy, 36*, 38–45.

Prasad, N. (2016c). Industry participation in India's space programme: Current trends & perspectives for the future. *Astropolitics: The International Journal of Space Politics and Policy, 14*(2–3), 234–255.

Prasad, N. (2016d, July 2016). ISRO/Antrix-Devas arbitration fiasco explained. *New Space India*.

Prasad, N. (2017a). Developing a space startup incubator to build a new space ecosystem in India. In N. Prasad & R. Rajagopalan (Eds.), *Space India 2.0. commerce, policy, security and governance perspectives*. New Delhi: Observer Research Foundation.

Prasad, N. (2017b). Space 2.0 India – Leapfrogging Indian space commerce. In N. Prasad & R. P. Rajagopalan (Eds.), *Space India 2.0. commerce, policy, security and governance perspectives*. New Delhi: Observer Research Foundation.

Prasad, N. (2017c). Traditional space and new space in industry in India: Current outlook and perspectives for the future. In N. Prasad & R. P. Rajagopalan (Eds.), *Space India 2.0 commerce, policy, security, and governance perspectives*. New Delhi: Observer Research Foundation.

Prasad, N., & Basu, P. (2015). *Space 2.0: Shaping India's leap into the final frontier*. New Delhi: Observer Research Foundation.

Prasad, N., & Rajagopalan, R. P. (2017). Creation of a defence space agency. A new chapter in exploring India's space security. In S. Singh & P. Das (Eds.), *Defence primer 2017. Today's capabilities, tomorrow's conflicts*. New Delhi: Observer Research Foundation.

PwC. (2013). *Easing India's capacity crunch. An assessment of demand and supply for television satellite transponders*. Hong Kong: Casbaa.

PwC. (2016). *Capacity crunch continues assessment of satellite transponders' capacity for the Indian broadcast and broadband market*. Hong Kong: Casbaa.

Radjou, N., & Prabhu, J. (2010). *Jugaad innovation*. New Delhi: Jossey-Bass.

Rajagopalan, R. P. (2011, October). Debate on space code of conduct: An Indian perspective. *ORF Occasional Paper 26*.

Rajagopalan, R. P. (2013, October 13). Synergies in space: The case for an Indian aerospace command. *ORF Issue Brief, 59*.

Rajagopalan, R. P. (2016). *India's space program. challenges, opportunities and strategic concerns*. Washington, DC: National Bureau of Asian Research.

Rajagopalan, R. P. (2017a). Need for an Indian military space policy. In R. P. Rajagopalan (Ed.), *Space India 2.0 commerce, policy, security and governance perspectives* (pp. 199–212). New Delhi: Observer Research Foundation.

Rajagopalan, R. P. (2017b, January). The need for India's space policy is real. *ORF Space Alert, 5*(1), 7–8.

Rajagopalan, R., & Sahni, V. (2008). India and the great powers: Strategic imperatives, normative necessities. *South Asian Survey, 15*(1), 5–32.

Rajgopalan, R., & Jayasimha, S. (2016, December 6). *Non-Geostationary satellite constellation for Digital India*. Retrieved from Observer Research Foundation: http://www.orfonline.org/expert-speaks/digital-india-non-geostationary-satellite-constellation/

Rajwi, T. (2016, March 29). *RLV-TD Likely to touch record mach 5 speeds: ISRO Officials*. Retrieved from The New Indian Express: http://www.newindianexpress.com/cities/thiruvananthapuram/RLV-TD-Likely-to-Touch-Record-Mach-5-Speeds-ISRO-Officials/2016/03/29/article3351385.ece

Rao, R. (2012, February 1). Why does India need an Aerospace Command? *IPCS*(3570).

Rao, P. (2015a). Applications of communication satellites. In P. M. Rao (Ed.), *From fishing hamlet to red planet: India's space journey*. New Delhi: Harper Collins.

Rao, U. (2015b). Origins of satellite technology in India: The story of Aryabhata. In P. M. Rao (Ed.), *From fishing hamlet to red planet: India's space journey*. New Delhi: Harper Collins.

Rao, M., & Murthi, S. (2012, September). *Perspectives for a National GI Policy – Including a National GI Policy Draft*. Retrieved November 5, 2016, from National Institute of Advanced Studies: http://ficci.in/sector/report/20034/Perspectives-for-National-GI-Policy-NIAS.pdf

Rao, M. K., Murthi, S., & Raj, B. (2015). *Indian Space – Towards a "National Eco-System" for Future Space Activities*. Jerusalem: 66th International Astronautical Federation.

Rao, M. K., Murthi, S. K., & RaJ, B. (2016). Indian space – towards a "National Ecosystem" for future space activities. *New Space, 4*(4), 228–236.

Reddy, V. S. (2017). Exploring space as an instrument in India's foreign policy and diplomacy. In R. P. Rajagopalan & P. Narayan (Eds.), *Space India 2.0 commerce, policy, security and governance perspectives* (pp. 165–176). New Delhi: Observer Research Foundation.

Reuters. (2015, June 13). *European space chief suggests making room for India, China on station*. Retrieved February 15, 2017, from Reuters: http://in.reuters.com/article/space-station-china-idINKBN0OT0J520150613

Reuters. (2016, January 25). *India to build satellite tracking station in Vietnam that offers eye on China*. Retrieved from Reuters: http://in.reuters.com/article/india-vietnam-satellite-china-idINKCN0V309W

Rompuy, H. V. (2010, September 14). *We have strategic partners, now we need a strategy*. Retrieved from European Council: https://www.consilium.europa.eu/uedocs/cms_data/docs/pressdata/en/ec/116494.pdf

Sachdeva, G. (2015). *Evaluation of the EU-India strategic partnership and the potential for its revitalisation*. European Union.

Sachdeva, G. S. (2016). Space doctrine of India. *Astropolitics: The International Journal of Space Politics and Policy, 14*(2–3), 104–119.

Samson, V. (2010, May 10). *India's Missile Defense/ASAT nexus*. Retrieved from The Space Review: http://www.thespacereview.com/article/1621/1

Samson, V. (2017). India-US: New dynamism in old partnership. In R. P. Rajagopalan & N. Prasad (Eds.), *Space India 2.0 commerce, policy, security and governance perspectives* (pp. 235–244). New Delhi: Observer Research Foundation.

Samson, V., & Bharath, G. (2011). Introduction. *India Review, 10*(4), 351–353.

Sankar, U. (2006). *The economics of India's space program. An exploratory analysis*. New Delhi: Oxford University Press.

Saran, S., Pejsova, E., Price, G., Gupta, K., & Wilkins, J.-J. (2016). *Prospects for EU-India security cooperation*. Observer Research Foundation.

Satak, N., Putty, M., & Bhat, P. (2017). *Exploring the potential of satellite connectivity for Digital India*. New Delhi.

Satellite Industry Association. (2016, September). *State of the satellite industry report*. Retrieved December 20, 2016, from SIA: http://www.sia.org/wp-content/uploads/2014/05/SIA_2014_SSIR.pdf

SatSure. (2017). *Solutions*. Retrieved May 20, 2017, from SatSure: http://www.satsure.in/#solutions

Schwab, K. (2016). *The global competitiveness report 2015–16*. Geneva: World Economic Forum.

Scott, D. (2012). *Handbook of India's international relations*. New York: Routledge.

Scott, H. (2016, December). India rising: The evolution of the Indian space enterprise. *New Space, 4*(4), 227–227.

Sen, A. (2006). *The argumentative Indian. Writings on Indian history, culture and identity*. London: Penguin Books.

SES. (2017). *Networks in Asia*. Retrieved March 1, 2017, from SES: https://www.ses.com/asia/asia

Sharma, G. (2011). *Space security: Indian perspective*. New Delhi: Vij Books.

Sheehan, M. (2007). *The international politics of space*. New York: Routledge.

Singh, H. (2009). *India's strategic culture: The impact of geography*. New Delhi: KW Publishers.

Smith, J. M. (2014, February 10). India and China: The end of cold peace? *The National Interest*.

Smith, M., & Xie, H. (2010). The EU and China: The logics of strategic partnership. *Journal of Contemporary European Research, 10*(4), 432–448.

Somasekhar, M. (2007, September 27). Italian space agency, ISRO to expand ties. *The Hindu Business Line*.

Sourbès-Verger, I. (2016, December 30). *EU-India cooperation on space and security* (IAI working papers 16|38).

Space Daily. (2007, January 28). *India to Set Up Aerospace Defence Command*. Retrieved April 3, 2016, from Space Daily: http://www.spacedaily.com/reports/India_To_Set_Up_Aerospace_Defence_Command_999.html

Space Foundation. (2016). *The space report 2016: The authoritative guide to global space activity*. Colorado Springs: Space Foundation.

Sputnik News. (2017, March 9). *BrahMos aerospace to be Indian DRDO's commercial wing abroad*. Retrieved from Space Daily: http://www.spacedaily.com/reports/BrahMos_Aerospace_to_Be_Indian_DRDOs_Commercial_Wing_Abroad_999.html

Staley, S. (2006, June). The rise and fall of Indian socialism. Why India embraced economic reforms. *Reason*.

Start-Up Europe. (2017). *Start-Up Europe India network*. Retrieved from Start-Up Europe India Network: https://startupeuropeindia.net

Statista. (2017). *Percentage of households owning a car in selected countries in 2014, by country*. Retrieved June 8, 2017, from The Statistics Portal: https://www.statista.com/statistics/516280/share-of-households-that-own-a-passenger-vehicle-by-country/

Stepan, A., Linz, J., & Jadav, Y. (2010). The rise of the state-nation. *Journal of Democracy, 21*(3), 50–68.

Suzuki, K. (2013). The contest for leadership in East Asia: Japanese and Chinese approaches to outer space. *Space Policy, 29*(2), 99–105.

Suzuki, K. (2017). An Asian space partnership with Japan? In R. R. Pillai & N. Prasad (Eds.), *Space India 2.0. commerce policy, security and governance perspectives*. New Delhi: Observer Research Foundation.

Swaminathan, M. (2015). Genesis and growth of remote sensing applications in Indian Agriculture. In P. M. Rao (Ed.), *From fishing hamlet to red planet: India's space journey*. New Delhi: Harper Collins.

Tanham, G. (1992). *Indian strategic thought. An interpretative essay*. Santa Monica: RAND Corporation.

Telecom Regulatory Authority of India. (2016, May 26). *Pre-consultation paper on infrastructure sharing in broadcasting TV distribution sector*. Retrieved November 15, 2016, from Telecom Regulatory Authority of India: http://www.trai.gov.in/Content/ConDis/20774_0.aspx

Television Post. (2013, December 4). *Dish TV to add transponder capacity, SES' new satellite launches successfully*. Retrieved March 1, 2017, from Television Post: http://www.television-post.com/dth/dish-tv-to-add-transponder-capacity-ses-new-satellite-launches-successfully/

Tellis, A. J. (2015). *Unity in difference. Overcoming the U.S. – India Divide*. Washington, DC: Carnegie Endowment for International Peace.

The Economic Times. (2016a, May 12). ISRO has set up committees for production of satellite launch vehicles with private sector. *The Economic Times, 2016*.

The Economic Times. (2016b, February 17). Make in India: ISRO to double number of missions to 12 per Year. *The Economic Times*.

The Economist. (2010, April 15). First break all the rules. *The Economist*.

The Hindu. (2012, March 14). Half of India's homes have cell phones, but not toilets. *The Hindu*.

The Hindu Business Line. (2014, November 28). *DTH satellite services: Dept of space lost out to Foreign Players*. Retrieved November 10, 2016, from The Hindu Business Line: http://www.thehindubusinessline.com/economy/dth-satellite-services-dept-of-space-lost-out-to-foreign-players/article6644047.ece

The Innovation Policy Platform. (2016). *India*. Retrieved October 20, 2016, from The Innovation Policy Platform: https://www.innovationpolicyplatform.org/content/india

The New Indian Express. (2016, February 6). Parliament panel favours budget hike for ISRO. *The New Indian Express*.

The World Bank. (2013). *India: Climate change impacts*. Retrieved June 8, 2017, from World Bank: http://www.worldbank.org/en/news/feature/2013/06/19/india-climate-change-impacts

The World Bank. (2016a). *Internet users*. Retrieved October 11, 2016, from The World Bank: http://data.worldbank.org/indicator/IT.NET.USER.P2?locations=IN

The World Bank. (2016b). *India*. Retrieved October 2, 2016, from The World Bank: https://data.worldbank.org/country/india

The World Bank. (2016c, October 1). *India – Country snapshot*. Retrieved November 3, 2016, from The World Bank: http://documents.worldbank.org/curated/en/341851476782719063/pdf/109249-WP-IndiaCountrySnapshots-highres-PUBLIC.pdf

The World Bank. (2016d). *Urban population – Data*. Retrieved November 25, 2016, from The World Bank: http://data.worldbank.org/indicator/SP.URB.TOTL?locations=IN

Thomas, R. G. (1986). India's nuclear and space programs: Defense or development? *World Politics, 38*(2), 315–342.

Times of India. (2004, November 8). *India, EU close to agreement on Galileo project*. Retrieved January 12, 2017, from Silicon India News: http://www.siliconindia.com/shownews/India_EU_close_to_agreement_on_Galileo_project-nid-25955-cid-3.html

Track Dish. (2015, August 10). *38 Indian channels started on Eutelsat 70B*. Retrieved March 1, 2017, from Track Dish: http://www.trackdish.com/38-indian-tv-channel-started-on-eutelsat-70b/

United Nations. (1982). *Report of the second United Nations conference on the exploration and peaceful uses of outer space. Vienna, 9-21 August 1982, UN, A/Conf.101/10*. United Nations.

United Nations. (2014). *World urbanisation prospects: The 2014 revision, highlights (ST/ESA/SER.A/352)*. New York: Department of Economic and Social Affairs, United Nations.

United Nations. (2015). *UNDP human development reports*. Retrieved November 2, 2016, from India Human Development Indicators: http://hdr.undp.org/en/countries/profiles/IND

United Nations. (2016). *World population prospects. The 2016 Revision*. New York: United Nations Department of Economic and Social Affairs.

United Nations. (2016, June 14). *"Open Universe" proposal, an initiative under the auspices of the committee on the peaceful uses of outer space for expanding availability of and accessibility to open source space science data*. Retrieved July 7, 2017, from UNOOSA: http://www.unoosa.org/res/oosadoc/data/documents/2016/aac_1052016crp/aac_1052016crp_6_0_html/AC105_2016_CRP06E.pdf

University Corporation for Atmospheric Research. (2017). *INDOEX: Indian Ocean experiment*. Retrieved March 31, 2017, from Earth Observing Laboratory Data: https://data.eol.ucar.edu/project?INDOEX

University of Central Florida. (2013). Global perspectives: India; *Population, Technology and Education*.

Unnithan, S. (2012, April 27). *India has all the building block for an anti-satellite capability*. Retrieved March 18, 2017, from India Today: http://indiatoday.intoday.in/story/agni-v-drdo-chief-dr-vijay-kumar-saraswat-interview/1/186248.html

Vasagam, R. (2015). APPLE in retrospect. In M. P. Rao (Ed.), *From fishing hamlet to red planet. India's space journey*. New Delhi: Harper Collins.

Vasani, H. (2016, June 14). *India's anti-satellite weapons*. Retrieved June 15, 2016, from The Diplomat: http://thediplomat.com/2016/06/indias-anti-satellite-weapons/

Verma, R. (2016, February 25). *"Bringing U.S.-India space cooperation to the edge of the universe" special address by U.S. Ambassador to India Richard Verma at the ORF Kalpana Chawla*. Retrieved from US Embassy and Consulates in India: https://in.usembassy.gov/bringing-u-s-india-space-cooperation-to-the-edge-of-the-universe-special-address-by-u-s-ambassador-to-india-richard-verma-at-the-orf-kalpana-chawl/

Vernile, A. (2017). *The rise of the private actor in the space sector*. Vienna: European Space Policy Institute.

Vicziany, M. (2015). EU-India security issues: fundamental incompatibilities. In P. Winand, M. Wicziany, & P. Datar (Eds.), *The European Union and India: Rhetoric or meaningful partnership*. Cheltenham: Edward Elgar Publishing.

Vikram Sarabhai Space Centre. (2017). *Crew Module Atmospheric Re-entry Experiment (CARE)*. Retrieved June 18, 2017, from Vikram Sarabhai Space Centre: http://www.vssc.gov.in/VSSC_V4/index.php/care

Weiss, M. (2014, October). France-India: A fraternal relation. *CNES Mag* (pp. 54–57).

World Economic Forum. (2016a). *The global competitiveness report 2015–2016*. Davos: WEF.

World Economic Forum. (2016b). *The global competitiveness report 2016–2017*. Davos: WEF.

Wülbers, S. A. (2011). *The paradox of EU-India relations: Missed opportunities in politics, economics, development, cooperation, and culture*. Plymouth: Lexington Books.

Zinnov. (2015). *The Indian promise*. Retrieved October 12, 2016, from Zinnov.com: http://zinnov.com/the-indian-promise/